Principles
of
Pyrometallurgy

Principles
of
Pyrometallurgy

C. B. ALCOCK

University of Toronto, Canada

1976

ACADEMIC PRESS

LONDON · NEW YORK · SAN FRANCISCO

A Subsidiary of Harcourt Brace Jovanovich, Publishers

ACADEMIC PRESS INC. (LONDON) LTD.
24/28 Oval Road,
London NW1

United States Edition published by
ACADEMIC PRESS INC.
111 Fifth Avenue
New York, New York 10003

Library of Congress Catalog Card Number: 73 19000
ISBN: 0 12 048950 3

*Printed in Gt. Britain by Page Bros (Norwich) Ltd,
Norwich and London.*

PREFACE

It is now 25 years since the Faraday Society's Discussion entitled "The Physical Chemistry of Process Metallurgy" took place. In this period, the experimental and theoretical efforts of a great number of high temperature scientists have produced the information which forms the basis for the modern teaching of this subject at the University undergraduate level. The present time is one in which the industrial scene in metal-making seems set for change under the pressure of social and economic factors as well as the intrinsic problems of mineral concentrate and energy production. In this book I have therefore tried to distil the essential physico-chemical aspects of pyrometallurgy in order to provide future students with the information to enable them to see why present-day processes work, and to learn how to marshal information in the design of projected new processes. It should be apparent from the book that design in pyrometallurgy requires more than an acquaintance with the broad sweep of physico-chemical principles in high temperature systems. As a branch of inorganic chemistry, the complexity of the reactions which are shown by metals presents specific facets in the chemistry of each one. It is only by a combination of general principles with specific details that process metallurgists can separate the metal from the dross, and this task can only get more difficult. The structure of the book was therefore dictated by what is current industrial practice, but the treatment of the problems in any particular process has been chosen to exemplify for future use as well as to quantify the familiar.

Any undertaking such as this requires either a great deal of effort and self-discipline by the author, or a compensating measure of help from those around him. I must gratefully acknowledge my ability to choose the latter option through discussion and text readings which my colleagues at this University have always made freely available. I wish to thank especially Drs. J. M. Toguri, A. McLean. S. Zador and K. T. Jacob amongst these.

Not readily finding the intrinsic will to finish this task given the ubiquity of excuses that University life provides in good measure, I have needed non-technical assistance as well in arriving at this point, and the generously expended skill of my secretary, Miss I. G. Warnock, whose patience had to be very enduring, is also acknowledged with my thanks. My indebtedness to my wife, Valerie, who has been a constant source of encouragement and support during the production of the text, cannot be simply expressed, but was of the greatest value to me.

May 1976   C. B. Alcock

CONTENTS

Part II. Metal Extraction Reactions Producing Liquid or Gaseous Metals

Part III. Metal Refining Processes

INTRODUCTION

The quantitative description of metal-making processes has advanced to the state where most of the processes which are now in industrial use can be described within a logical physico-chemical framework. The pace of development in this field has largely been determined by the rate of improvement of our experimental capabilities in high temperature chemistry; the theoretical contribution in the building of our present knowledge has been small in providing the fundamental basis for *a priori* calculation. This is not surprising, since the materials with which the metallurgist works are usually complex, and the precision of the information which is required to describe a process is frequently too high to be evaluated through theoretical prediction. However, the practical situation can now be assessed from the substantial results of experimental studies which cover almost every situation to be found on the present industrial scene.

The role of the physico-chemical study of pyrometallurgy has often been consigned to a secondary position of interest by those engaged directly in the metal manufacturing processes. This has probably come about for two reasons. The first and most obvious reason is that economic factors as well as physical chemistry play a very important part in industrial decision-making. Those who direct the production aspects of industry seldom have equally developed skill in the physico-chemical aspects as well as in economics. As a result, the decision-making tends to be under financial direction, and the decision makers draw their scientific advice in a digested form. The scientists are therefore not usually involved directly in the choice of a process from a number of options. The second reason is that high temperature chemists have been fully occupied up till now in the business of understanding the processes already in use and their contributions to knowledge seem always to be *post hoc*. At present, it is true to say that their efforts have been more of value in teaching the neophyte laboratory workers than in predicting potentially new processes. To some extent, this state of affairs has been brought about by empirical progress which has built up a formidable amount of knowledge over decades by the use of works trials.

These aspects of industrial development together with the financial constraints of process innovation probably account for the fact that the physical chemist has had no really outstanding impact on metal-making to date. Nevertheless, he has now provided experimental tools for the appraisal of new processes, and has developed the 'tools of thought" which can be transferred from the analysis of an established process to prognosis when new methods are being sought.

The precision of measurement which has been achieved in laboratory studies is now probably higher than the level of applicability of the results in modern industry. The distribution of temperature in an operational furnace is very much greater than the limits of control which can be achieved in an experimental tube furnace, and it is probably unreal to expect that the industrial apparatus will ever be brought into these same limits except in the case of small scale operation with very high cost materials. The homogeneity of chemical composition which must be ensured in a laboratory study of thermodynamic properties, or the simple geometry of laboratory kinetic studies will not be reproduced on the large scale in the foreseeable future. The applications of scientific data in industry do not demand a precision of knowledge to the typical $\pm 5\%$ accuracy which is readily achieved under laboratory conditions, and it is therefore not very realistic to pursue the small scale study to a higher degree of precision if the effort is being made with a large scale process as the final objective. However, the experimentalist frequently achieves a wider understanding of natural laws as a by-product of the pursuit of accuracy, and totally new directions of progress can arise out of the painstaking study of a process which is already sufficiently well characterized for the immediate needs of industry. It is probably for this reason that the advances of modern high temperature chemical research appear to be directed largely towards goals which are not related to the pyrometallurgical industry, although this has been a mainspring of inspiration in the past. It seems probable, though, that the increasing needs for control devices will provide a new and fruitful line of contact between the laboratory workers and the industry. Thus, the very practical value of basic research in solid state electrochemistry which has been carried out during the past fifty years has only been realized within the last decade, when the continuous measurement of the oxygen contents of liquid metals appeared to be feasible in the industrial furnace for the first time.

In this book, it is intended to present analyses of a number of typical industrial processes in the terms of the physical chemist. No attempt has been made to make the description exhaustive of all of the high temperature metal extraction operations at present in use, but it is hoped that a sufficient number of examples are presented to help the process metallurgist appreciate the state of quantitative knowledge in the field, and to be able to design new processes with the aid of physico-chemical data.

The approximate nature of some of the data which are used and the rough way in which some of the precise data are sometimes applied in the text may seem ill-considered to the laboratory workers, but they must remember that in the context of productive industry it is only an approximate treatment which is worthwhile more often than not.

The experimental techniques which have been developed in the study of extractive metallurgy will be included in the description of the processes to be given here. This is to show what experimental resources can now be brought to bear on a high temperature problem, and the inclusion of this material is also intended to suggest a number of possible techniques to the interested development research worker who has a particular problem to solve. The references at the end of each section should indicate further reading in this matter.

The book is divided into three main sections which represent the distinguishable operations of the industry. The first section concerns the processes in which solid–gas reactions are used principally in the preparation of materials for the metal-extracting reactions; there is also one important reaction where the direct reduction to a solid metal is made. In the second section, the crude metal-extracting reactions are considered. The account brings together the metals which are won by one particular technique such as the blast furnace or the Kroll process. The refining methods which have been used on typical metals on the industrial scale are discussed in the third section.

The treatment is intended for those who have already received the basic courses in classical thermodynamics which nearly all students of Metallurgy, Materials Science and Chemical Engineering must assimilate nowadays before passing on to courses in Extractive Metallurgy. For the interested graduate, a brief refresher in any of the standard textbooks of Physical Chemistry is recommended if he is not facile in thermodynamic analysis. In order to keep the size of this book down to reasonable proportions, and hence a reasonable price to the student, references are given at the end of each section to other works and original literature sources which are normally available to the student of Extractive Metallurgy. A parallel study of these reference materials will augment the value of this book very considerably, but it is the author's hope that the main ideas which are germane to the analysis of extraction processes are to be found in this work.

Part I

Solid State Processes

B

1

THE RAW MATERIALS OF PYROMETALLURGY

INTRODUCTION

The minerals from which metals are made are usually either sulphides or oxides. The processes which are applied to these materials in the winning of metals therefore revolve around the high temperature chemistry of metallic sulphides and oxides. The physical appearance of the starting material for the extraction process is most frequently that of a fine powder, since separation processes must be carried out on the raw material from the mine in order to remove the so-called gangue, the useless solids which frequently form the major part of the mine products. The crushing, grinding and flotation processes of mineral concentration need not concern us here since these are usually carried out at low temperatures and do not involve a significant chemical change of the raw material which goes on to the pyrometallurgical operations.

Table I shows some of the most important mineral sources of metals, and it soon becomes clear that, in general, changes in composition and the forms of chemical combination are needed before metal extraction can be carried out. This is because there are only very few processes which can be based on the direct reduction of sulphides for reasons which will be gone into later, and oxides are usually the most readily dealt with in metal extraction.

An important class of reactions which is usually involved in the preliminary treatment of minerals, before the metal emerges, is that between solids and gases. As examples, galena, sphalerite and molybdenite are all roasted in air to give lead, zinc and molybdenum oxides which can be reduced in one further stage to the metal. The naturally occurring minerals of copper

TABLE 1.

Oxide Minerals	Sulphide Minerals
Magnesite $MgCO_3$	Chalcopyrite $CuFeS_2$
Dolomite $(Mg, Ca)CO_3$	Bornite Cu_3FeS_3
Haematite Fe_2O_3	Cinnabar HgS
Siderite $FeCO_3$	Sphalerite ZnS
Magnetite Fe_3O_4	Galena PbS
Chromite $FeO.Cr_2O_3$	Pyrite FeS_2
Cassiterite SnO_2	Pentlandite $(Ni, Fe)_9S_8$
Ilmenite $FeO.TiO_2$	Stibnite Sb_2S_3
Rutile TiO_2	Molybdenite MoS_2
Zircon $ZrSiO_4$	
Baddeleyite ZrO_2	
Columbite–tantalite $(Fe, Mn)O. (Nb, Ta)_2O_5$	
Scheelite $CaWO_4$	
Wolframite $(Fe, Mn)WO_4$	
Pitchblende UO_2	

all have a very high sulphur dissociation pressure at the temperatures at which liquid copper (m.p. 1083°C) can be made, and these ores are roasted before the extraction of the metal to reduce the sulphur content. Because of the chemistry of the copper–oxygen–sulphur ternary system, it is possible in this particular case to produce the pure metal directly from the roasted product without chemical combination of copper with oxygen. This special reaction will be dealt with in Part II.

There are a number of elements which are best obtained from the metallic chlorides rather than the naturally occurring oxides. These are the newer metals which have only become available during the present century on anything approaching a useful scale of production, e.g. titanium, zirconium, niobium, uranium and a number of others.

The need for some conversion before primary metal extraction may be conveniently carried out brings the subject of roasting to the forefront of our consideration. The science of roasting involves the physical chemistry of gas–solid reactions in connection with the oxidation of sulphides and the chlorination of oxides.

Another type of reaction, the direct production of metallic powder by reduction of solid powders with hydrogen or carbon monoxide, can be discussed in the same general terms as a roasting reaction, and so the physico-chemical backgrounds of these reactions will be included in this first part.

In all reactions which involve solids at high temperatures, the change of contact area with a gas phase which accompanies the reduction of the surface area of the solid, the process of sintering, is of great importance to the feasibility and speed of high temperature operations as also is the transfer of heat between the solid and gaseous phases. These topics are also discussed in Part I.

The gathering of information for the understanding of metalmaking processes usually proceeds along a well-defined path. The thermodynamic information makes it possible to determine the *extent* to which a reaction will eventually proceed under a given temperature and pressure. An understanding of the *rate* at which the reaction can occur requires a study of the kinetics, again, as a function of temperature and pressure. Once the rate has been established, it is next important to understand the way in which the reaction proceeds, or the *mechanism*, in order that the physical conditions which are applied to the process can be optimized in the industrial context. This approach to understanding will now be exposed for solid-state reactions in the subsequent sections of this Part.

2

THE THERMODYNAMIC ASPECTS OF THE ROASTING PROCESSES

THE OXIDATION OF SULPHIDES

The roasting of sulphide ores in the solid state can have one of two major objectives. The first of these is the conversion of material to the oxide form as a preliminary to a metal extraction reaction, e.g. lead sulphide roasting. The alternative objective is the formation of readily water-soluble sulphates which can be employed in subsequent hydrometallurgical processes. The operating conditions under which either of these two objectives can be attained are determined by consideration of the thermodynamics of the roasting reactions, and the duration of the roast depends on the kinetics of the solid–gas reactions.

In both oxidizing and sulphating roasting, the chemical composition of the gas phase is of major importance, and the gas–solid equilibria for the appropriate metal–sulphur–oxygen systems must be known before accurate predictions can be made concerning the range of operating conditions which may be used for the production of the oxide or the sulphate.

The simplest way of representing all of the thermodynamic information for a given system is by means of a number of isothermal diagrams which show the ranges of gas compositions over which each phase can exist singly, or in equilibrium with another phase. These are called stability diagrams. According to the phase rule, at a fixed temperature and fixed total pressure of the gaseous phase, a maximum number of three condensed phases can co-exist in the three component system, metal–sulphur–oxygen. It has been found to be most convenient to show the stability diagram for a given metal–sulphur–oxygen system, at a given temperature, with the logarithms of

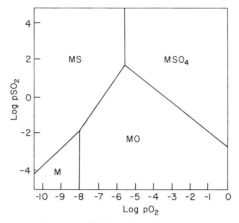

Fig. 1. A stability diagram for a metal(M)–sulphur–oxygen system at a fixed temperature.

p_{SO_2} and p_{O_2} as the ordinates (Fig. 1). In such a diagram, the boundary between the regions in which a metal sulphate and oxide exist separately will be a straight line the slope of which can simply be calculated. Consider the reaction for the formation of a simple sulphate

$$MO + SO_2 + \tfrac{1}{2}O_2 \rightarrow MSO_4$$

The equilibrium constant K_{MSO_4} is given by

$$K_{MSO_4} = \frac{1}{p_{SO_2} \cdot p_{O_2}^{\frac{1}{2}}}$$

when MO and MSO_4 are pure phases whose thermodynamic activities are one and hence the boundary between the oxide and the sulphate on the stability diagram is determined by the equation

$$-\log p_{SO_2} = \log K_{MSO_4} + \tfrac{1}{2}\log p_{O_2}$$

The boundary has a slope in the diagram of minus one-half. This is so even in the apparently complex case where ferric sulphate is formed from Fe_2O_3

$$Fe_2O_3 + 3SO_2 + \tfrac{3}{2}O_2 \rightarrow Fe_2(SO_4)_3$$

$$K_{Fe_2(SO_4)_3} = \frac{1}{p_{SO_2}^3 \cdot p_{O_2}^{\frac{3}{2}}}$$

$$-\log p_{SO_2} = \tfrac{1}{3}\log K_{Fe_2(SO_4)_3} + \tfrac{1}{2}\log p_{O_2}$$

Other two-phase boundaries are also simply obtained in the following manner. The line separating the pure metal and metal oxide regions represents a constant p_{O_2}, the dissociation pressure of the M–MO mixture. This line is therefore parallel to the log p_{SO_2} ordinate. The boundary between the metal and metal sulphide regions where p_{S_2} is constant, has a slope of unity since for the reaction

$$\tfrac{1}{2}S_2 + O_2 \rightarrow SO_2 \qquad K_{SO_2} = \frac{p_{SO_2}}{p_{S_2}^{\frac{1}{2}} \cdot p_{O_2}}$$

and therefore log $p_{SO_2} = \log K + \tfrac{1}{2}\log p_{S_2} + \log p_{O_2}$.

Finally, the boundary between metal sulphide and metal oxide has the slope three halves since the reaction may be written

$$MS + \tfrac{3}{2}O_2 \rightarrow MO + SO_2$$

and hence

$$\log p_{SO_2} = \log K + \tfrac{3}{2}\log p_{O_2}.$$

The boundary between metal sulphate and the metal sulphide corresponds to a constant oxygen pressure according to the equation

$$MS + 2O_2 \rightarrow MSO_4$$

A more complicated diagram is obtained when a metal can form a number of sulphides and oxides, and some metals also form quite stable basic sulphates, e.g. $CuO . CuSO_4$. These features can all be accommodated in the isothermal diagrams, and the directions of the boundaries across each diagram can be deduced by consideration of the appropriate equilibrium constants for the reactions in which SO_2 and oxygen both appear. Obviously,

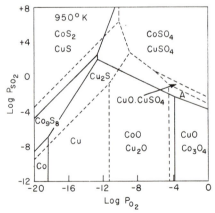

Fig. 2. Superimposed stability diagrams for Cu–S–O (broken lines) and Co–S–O (solid lines) at 950° K. Within the area A the water soluble $CoSO_4$ and the insoluble phase CuO are stable, and hence differential sulphation may be achieved. (After Ingraham, loc. cit.)

the boundary between a lower and upper oxide, e.g. Cu_2O–CuO or that between a lower and upper sulphide, e.g. Cu_2S–CuS will have slopes corresponding to constant p_{O_2} in the former case and constant p_{S_2} in the latter. The lines are thus parallel to the Cu–Cu_2O and Cu–Cu_2S boundaries respectively. A comparison of the two stability diagrams at 950 K (Fig. 2) of the Cu–S–O and Co–S–O systems is shown.

In most cases of practical importance, the data which are required in drawing the diagrams are available in two-term free energy equations of the form

$$\Delta G^\circ = A + BT$$

which are used in the construction of an Ellingham free energy diagram. These equations are usually based on the measurement of the relevant gas–solid equilibria and there should be, in principle, a third term, at least, in the free energy equation to reflect the finite value of the heat capacity change. The equation should then read

$$\Delta G^\circ = A' + B'T + C'T \ln T.$$

It is rare indeed, however, that the experimental results are sufficiently accurate to merit such treatment, or, alternatively, that the equation covers such a wide temperature range that the effects of such a term are of practical significance.

THE MEASUREMENT OF THE FREE ENERGIES OF FORMATION OF SIMPLE OXIDES AND SULPHIDES

The free energy of formation of a simple metal oxide from metal and oxygen, can, in principle, be obtained from calorimetric measurements of the heat of the reaction and the entropies of each component since

$$\Delta G^\circ = \Delta H^\circ - T\Delta S^\circ.$$

The measurement of the entropy content of a substance entails the determination of the heat capacity at constant pressure and the heats of transformation over the whole temperature range from the absolute zero up to the required temperature. Because of this, the direct determination of these two components of the standard free energy change by calorimetry has not been of major importance in establishing metallurgical thermodynamics except where no other method was available.

The alternative method for obtaining these values is the measurement of the equilibrium constant for the reaction since

$$\Delta G^\circ = -RT \ln K; \quad \partial \ln K / \partial 1/T = \frac{-\Delta H^\circ}{R}$$

For the simple dissociation of an oxide

$$MO \rightarrow M + \tfrac{1}{2}O_2$$

$$K_{MO} = \frac{a_M \cdot p_{\frac{1}{2}O_2}}{a_{MO}}$$

and the measurement of the oxygen pressure above a mixture of the pure metal and the pure oxide ($a_M = a_{MO} = 1$) would suffice. Unfortunately, it is only the most unstable oxides which have high enough dissociation pressures to be measurable by this simple technique, e.g. PdO; the oxides of metallurgical interest such as FeO, PbO, etc., are too stable.

The free energies of formation of the more stable oxides, for example, FeO, can be measured by combining the formation reaction with that for water vapour from hydrogen and oxygen, or for CO_2 from CO and oxygen thus

$$Fe + \tfrac{1}{2}O_2 \rightarrow FeO \qquad \Delta G^\circ_{\,I}$$

$$\frac{H_2 + \tfrac{1}{2}O_2 \rightarrow H_2O}{Fe + H_2O \rightarrow FeO + H_2} \qquad \frac{\Delta G^\circ_{\,II}}{\Delta G^\circ} = \Delta G^\circ_{\,I} - \Delta G^\circ_{\,II}$$

$$K = \frac{p_{H_2}}{p_{H_2O}} \cdot \frac{a_{FeO}}{a_{Fe}}$$

and similarly

$$Fe + CO_2 \rightarrow FeO + CO, \qquad K = \frac{p_{CO}}{p_{CO_2}} \cdot \frac{a_{FeO}}{a_{Fe}}$$

The measurement of these equilibrium constants cannot be made with a static gas phase since the gaseous mixture will be subject to thermal segregation in the temperature gradient of the furnace. This results in the gas at the low temperature end of the apparatus having a higher partial pressure of the gas with the larger molecular weight, e.g. H_2O in H_2/H_2O, than exists in the high temperature, equilibrium, mixture over the solid phases. The most successful technique is to pump the gas around a closed system which includes the furnace reaction tube, at a speed which is sufficient to overcome thermal segregation. Typically in H_2/H_2O mixtures, this can be achieved if the gas has a flow rate of greater than 100 cm^2 min^{-1} through a 2 cm bore furnace tube. The gaseous phase can then be analysed by sampling the contents of the room temperature section of the system by conventional chemical methods. The experimental limit of this technique is usually found to be that the gas ratio cannot be greater than 10^4:1, either H_2/H_2O or H_2O/H_2, because the analysis of the gas for the minor component is then too difficult to be made precisely. This range, however, is sufficient to enable the gathering of information for the oxides of most of the important base metals, such as iron, nickel, copper, tin and lead.

The corresponding results for simple metal sulphides can be similarly obtained from studies of the equilibria

$$MS + H_2 \rightarrow M + H_2S \qquad K = \frac{p_{H_2S}}{p_{H_2}} \cdot \frac{a_M}{a_{MS}}$$

Again, the limitations of this technique are set by the problems of gas analysis. The upper limit in the case of H_2S/H_2 mixtures is about $1:1$ at high temperatures where problems involving sulphur deposition in the apparatus become difficult. The lower limit of this ratio, using standard gas analytical methods, is about $1:10^4$.

A recently introduced electrochemical method can be used for the measurement of the thermodynamics of metal oxides. This involves the use of a solid electrolyte which conducts electricity by the migration of oxygen ions only. The solid solutions of CaO or Y_2O_3 in ZrO_2 or ThO_2 have the fluorite (CaF_2) structure and, hence, the anion sub-lattice contains a number of vacancies (V_O^{2-}). The vacancy concentration is related to the concentration of CaO or Y_2O_3, being equal to the molar concentration of CaO or half the concentration of Y_2O_3 ($YO_{1.5}$).

$$CaO \rightarrow \{CaO\}_{\substack{ThO_2 \\ ZrO_2}} + V_{O^{2-}}$$

$$YO_{1.5} \rightarrow \{YO_{1.5}\}_{\substack{ThO_2 \\ ZrO_2}} + \tfrac{1}{2}V_{O^{2-}}$$

A solid-state electrochemical cell can be set up in which the electrolyte is connected to a gas electrode of fixed oxygen pressure on one side and to a pellet made by compressing a mixture of the powders of the metal and its oxide on the other side. Electrical contacts to the electrodes are usually made with platinum wires which are led out of the furnace and to a high-impedance voltmeter.

$$M/MO \,|\, Electrolyte \,|\, Gas \text{ with fixed } p_{O_2}$$

The container for the gas is also made of electrolyte material, and the M/MO pellet can be held in an atmosphere of argon or in vacuum. The free energy change of the reaction

$$M + \tfrac{1}{2}O_2 \rightarrow MO$$

is given by

$$\Delta G = \tfrac{1}{2} RT \ln \frac{p_{O_2}}{p'_{O_2}} = 2\,FE$$

where p'_{O_2} is the pressure of oxygen in the gas electrode, and p_{O_2} is the dissociation pressure of the metal–metal oxide mixture. Once a few metal–metal oxide systems had been studied accurately, they could then serve as reference electrodes in further cells, e.g. Pt electrode Fe/FeO|CaO–ZrO$_2$ electrolyte|Ni/NiO Pt electrode. This electrochemical method can be used from one atmosphere of oxygen dissociation pressure down to about 10^{-16} atm at 1000 K using the CaO–ZrO$_2$ electrolyte. The corresponding range for the thoria-based electrolyte appears, at present, to be 10^{-6}–10^{-30} atm of oxygen dissociation pressure at this temperature.

In a few favourable cases, Sodi and Elliott (1968) have been able to establish thermodynamic quantities for sulphides using these electrolytes by measuring the EMF's of cells in which one electrode consists of the metal–metal sulphide held in an atmosphere of a fixed partial pressure of SO$_2$. The oxygen pressure of this electrode is determined by the SO$_2$ and the S$_2$ dissociation pressure of the metal–metal sulphide system through the equilibrium constant.

$$K_{SO_2} = \frac{p_{SO_2}}{p^{\frac{1}{2}}_{S_2} \cdot p_{O_2}}$$

THE FREE ENERGY OF SULPHATE FORMATION

The measurement of sulphate equilibria is a little more complicated because of the complex nature of the gas phase which is involved. A mixture of SO$_2$ and O$_2$ must contain a partial pressure of SO$_3$ under equilibrium conditions; for a given mixture of SO$_2$ and oxygen which is passed into a furnace at the room temperature end, there will be a change of composition along the temperature gradient as this equilibrium is established. However, thermodynamic data for the reaction

$$SO_2 + \tfrac{1}{2}O_2 \rightarrow SO_3$$

are available, and the calculation of the partial pressure of each species in a high temperature system, though tedious, can be easily made.

The total dissociation pressures of the sulphates of a number of base metals are large enough for direct measurement, and data are available from measurements which have been made under static gas conditions. Here again the possibility of thermal segregation must be considered, and alternative methods involving flowing gases have been employed. These are not recirculation methods because of the corrosive nature of SO$_3$, and usually once-through passage of the gas is used. The formation or dissociation of the sulphate has been followed by thermogravimetry or by differential thermal analysis. Since the formation of a sulphate from oxide and gas is usually exothermic, a signal can be generated in a differential thermocouple, one leg of which is immersed in the solid, and the other samples the temperature of the surrounding gas and furnace walls.

The signal is observed whilst the furnace temperature is gradually raised or lowered, the ingoing composition of the $SO_2 + O_2$ gas mixture to the furnace being kept constant throughout.

The results which have been obtained for the base-metal sulphates indicate that no serious errors arise from thermal segregation of the gases in the direct, static, dissociation pressure measurements.

THE BORN–HABER CYCLE AND THE OXIDATION OF SULPHIDES

The experimental results which can now be gathered together for the simple oxides and sulphides show that in most instances the assumption that these are ionic solids is a reasonably close approximation to the truth.

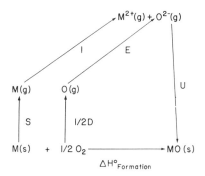

Fig. 3. The Born–Haber cycle for the energy of formation of a metal oxide

$$\Delta H^{\circ}_{\text{formation}} = S + \tfrac{1}{2}D + I + E - U.$$

In this case, it becomes a relatively simple matter to bring out the trends in bonding by the use of the Born–Haber cycle (Fig. 3). The heat of formation of a metal oxide, for example, can be obtained by the summation of the heat changes for a number of elementary processes. These are as follows, starting from solid metal and diatomic oxygen gas:

(i) Evaporation of the metal to the monatomic vapour;   S

(ii) Ionization of the metallic atoms to the normal valency state $(Z_M e+)$;   I

(iii) Dissociation of O_2 molecules to atoms;   D

(iv) Addition of electrons to the O atoms to form O^{2-} ions (Z_Oe-);
    E

(v) Condensation of all the ions to form the lattice of the compound. This term is called the lattice energy and is mainly composed of two contributions

$$E_{lattice} = \frac{-AZ_MZ_Oe^2}{r_{M-O}} + \frac{B}{r_{M^n-O}}$$

The first term represents the coulombic attraction interaction between the ions. A is the Madelung constant expressed in terms of the interionic distance r_{M-O} for the given crystal structure, and the second term represents the repulsive interaction between the ion-cores. The exponent, n, typically has values around 9.

If we now consider the relative stabilities of the oxide and the corresponding sulphide of a metal, the heat change for the exchange reaction

$$MS + \tfrac{1}{2}O_2 \rightarrow MO + \tfrac{1}{2}S_2$$

largely determines the magnitude of this quantity since the entropy change of the exchange reaction is practically zero. In this connection, the Born–Haber analysis suggests that in the value of the heat of the exchange reaction two separate effects can be distinguished. In the cycles for sulphide and oxide formation, the different terms relating to the anion, the energies of dissociation and ionization, remain the same no matter what the metal. The energy change for the process in the gaseous phase

$$S^{2-} + \tfrac{1}{2}O_2 \rightarrow O^{2-} + \tfrac{1}{2}S_2$$

can be obtained from the following information:

$\tfrac{1}{2}S_2 \rightarrow S$	$\Delta E = 41$ kcal
$S + 2e^- \rightarrow S^{2-}$	$\Delta E = 100$
$\tfrac{1}{2}S_2 + 2e^- \rightarrow S^{2-}$	$\Delta E = 141$
$\tfrac{1}{2}O_2 \rightarrow O$	$\Delta E = 59$
$O + 2e^- \rightarrow O^{2-}$	$\Delta E = 172$
$\tfrac{1}{2}O_2 + 2e^- \rightarrow O^{2-}$	$\Delta E = 231$

The value for this anionic electron exchange energy for the reaction as written above is therefore $+90$ kcal, and this number must be used for *all* exchange reactions involving sulphides and oxides.

The other important term to be considered is the lattice energy term. The ionic radius of the S^{2-} ion (1·84 Å) is larger than that of the O^{2-} ion (1·40 Å)

TABLE II.

Metal	Cation radius Å	$-\Delta H^\circ$ sulphide	$-\Delta H^\circ$ oxide kcal mol^{-1}
Na	0·95	93	104
Cs	1·69	81	76
Mg	0·65	83	143
Ba	1·35	106	133
Cu (cuprous)	0·96	20	40
Fe (ferrous)	0·75	23	64

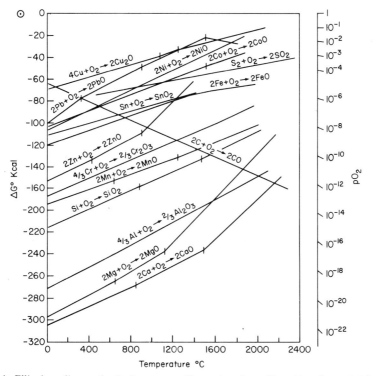

Fig. 4a. Ellingham diagram for the free energy of formation of metallic oxides (after F. D. Richardson and J. H. E. Jeffes, *J. Iron Steel Inst.* **160**, 261 1948). The oxygen dissociation pressure of a given M–MO system at a given temperature is obtained by joining ⊙ on the top left hand to the appropriate point on the M–MO free-energy line, and extrapolating to the scale on the right hand ordinate for p_{O_2}.

and hence there will be an increase of the Madelung attraction energy term in the lattice energy when oxide ion replaces sulphide ion. When the cation has a small radius compared with these ions, the effect of the replacement will be more marked than when the cation radius is approximately the same as the anionic radii. This effect can be clearly seen from Table II. In the case where this effect is most marked, that of the small cationic radius, the difference in lattice energy between the oxide and the sulphide is about -150 kcal mol^{-1}, and this term outweighs the anionic electron exchange term ($+90$ kcal) in making the oxide more stable than the sulphide. It can be concluded by considering the sum of these two effects, i.e. of anionic exchange and the lattice energy term, that the oxide is more stable than the sulphide the smaller the ionic radius of the cation when the bonding is predominantly ionic in nature. The only qualitative exception which can be seen in Table II is in the relative stabilities of caesium oxide and sulphide. In this example,

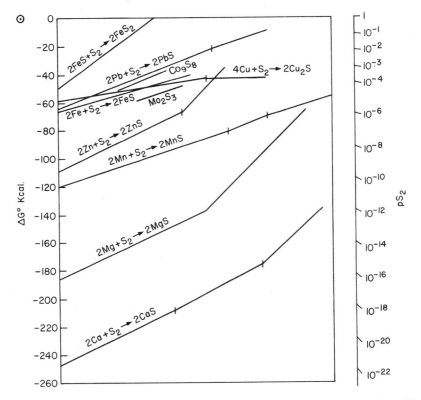

Fig. 4b. Ellingham diagram for metal sulphides (Richardson and Jeffes, *J. Iron Steel Inst.*, **171**, 165, 1952). This diagram can be read as for the corresponding oxide diagram by joining ⊙ to M–MS and extrapolating to the p_{S_2} scale.

TABLE III. Two-term free energy equations for the oxides and sulphides of metals which occur in sulphide minerals

COBALT	$\Delta G°$ cal mol^{-1}
$2Co + O_2 \rightarrow 2CoO$	$-111,800 + 33·8T$
$6CoO + O_2 \rightarrow 2Co_3O_4$	$-87,600 + 70·8T$
$\frac{9}{4}Co + S_2 \rightarrow \frac{1}{4}Co_9S_8$	$-79,240 + 39·81T$
$\frac{1}{2}Co_9S_8 + S_2 \rightarrow \frac{3}{2}Co_3S_4$	$-51,640 + 34·03T$
$Co_3S_4 + S_2 \rightarrow 3CoS_2$	$-50,160 + 50·4T$

Note: There is a eutectic in the cobalt–sulphur system at 1153 K.

COPPER (up to 1357 K, the melting point of the metal)	
$4Cu + O_2 \rightarrow 2Cu_2O$	$-80,160 + 34·23T$
$2Cu_2O + O_2 \rightarrow 4CuO$	$-68,140 + 49·62T$
$4Cu + S_2 \rightarrow 2Cu_2S$	$-62,840 + 14·70T$
$2Cu_2S + S_2 \rightarrow 4CuS$	$-45,200 + 54·0T$

IRON	
$2Fe + O_2 \rightarrow 2FeO$	$-124,100 + 29·90T$
$6FeO + O_2 \rightarrow 2Fe_3O_4$	$-149,240 + 59·80T$
$4Fe_3O_4 + O_2 \rightarrow 6Fe_2O_3$	$-119,240 + 67·24T$
$2Fe + S_2 \rightarrow 2FeS$	$-71,820 + 25·12T$
$2FeS + S_2 \rightarrow 2FeS_2$	$-86,700 + 90·0T$

MOLYBDENUM	
$Mo + O_2 \rightarrow MoO_2$	$-131,500 + 33·95T$
$2MoO_2 + O_2 \rightarrow 2MoO_3$	$-77,400 + 39·0T$
$\frac{4}{3}Mo + S_2 \rightarrow \frac{2}{3}Mo_2S_3$	$-85,700 + 36·41T$

NICKEL	
$2Ni + O_2 \rightarrow 2NiO$	$-111,900 + 40·58T$
$3Ni + S_2 \rightarrow Ni_3S_2$	$-79,240 + 39·01T$
$2Ni_3S_2 + S_2 \rightarrow 6NiS$	$-51,400 + 25·21T$

Note: There is a eutectic in the nickel–sulphur system at 918 K.

LEAD (liquid metal)	
$2Pb + O_2 \rightarrow 2PbO$	$-106,600 + 51·4T$
$2Pb + S_2 \rightarrow 2PbS$	$-75,160 + 38·25T$

ZINC (liquid metal up to the boiling point, 1180 K)	
$2Zn + O_2 \rightarrow 2ZnO$	$-168,900 + 51·10T$
$2Zn + S_2 \rightarrow 2ZnS$	$-123,500 + 48·54T$

the size of the cation is very large, and the lattice energy contribution fails to outweigh the electron exchange term. With this exception, it is generally found that the oxidation of sulphides is an exothermic process. Table III brings together the results for the corresponding simple sulphides and oxides of those metals which occur in sulphide minerals for the temperature range in which sulphide roasting to oxides is normally carried out (600–1500 K). These equations are also represented in Ellingham diagrams (Fig. 4a and b) where the limits of the experimental methods which were described earlier are indicated.

THE IRON–SULPHUR–OXYGEN SYSTEM AND CHEMICAL POTENTIAL DIAGRAMS

One very important aspect of the thermodynamics of these phases which is not shown in the simple stability or the Ellingham diagrams, is the effect of the metal/sulphur or metal/oxygen ratios in metal sulphides or oxides on the chemical potentials of the elements. This aspect is of interest in connection with the mechanisms of sulphide oxidation, as will be shown below. The system for which the most complete information has been obtained is the iron–sulphur–oxygen system, and these results will now be used as an example.

The sulphide FeS has a wide range of non-stoichiometry from a composition close to $FeS_{1.00}$ on the metal-rich side. On the sulphur-rich boundary, the phase is in equilibrium with pyrite FeS_2. Across the range of metal/sulphur ratios the sulphur pressure, p_{S_2}, increases from about 10^{-9} atm to approximately 1 atm. Thus, when the iron content of a pyrrhotite, $FeS_{1.00}$, is decreased by about 10 atom percent, the sulphur pressure increases by about a factor of 10^9.

It follows from the Gibbs–Duhem relationship

$$X_{Fe} \, d \ln a_{Fe} + X_S \, d \ln a_S = 0$$

where X is the atom fraction and a the thermodynamic activity of each species, that the iron activity decreases markedly across the stoichiometry range of $FeS_{1+\delta}$ from unity on the iron-rich side to approximately 10^{-4} on the sulphur-rich side of composition.

It is important to realize that although the activity of sulphur and that of iron have undergone such a marked change across the range of composition of $FeS_{1+\delta}$, the activity of FeS, which can be calculated from the equation

$$K_{FeS} = \frac{a_{FeS}}{a_{Fe} \cdot p_{S_2}^{\frac{1}{2}}}$$

remains sensibly constant, only decreasing to 0·7 at the sulphur-rich side. Here the standard state for FeS is normally chosen to be the metal-rich, practically stoichiometric sulphide.

The stability diagram shows that if iron is removed from $FeS_{1.00}$, which is held in an atmosphere of fixed oxygen pressure, the SO_2 pressure will increase in going to the FeS–FeS_2 boundary. Thus, if p_{O_2} is kept constant

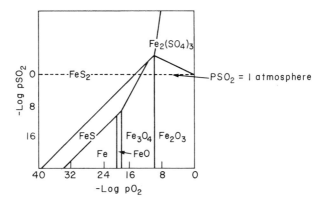

Fig. 5. Stability diagram for the Fe–S–O system at 1000° K

at 10^{-24} atm (Fig. 5), the SO_2 pressure is about 10^{-15} atm on the metal-rich edge, and about 10^{-10} atm on the sulphur-rich edge. This must follow since

$$K_{SO_2} = \frac{p_{SO_2}}{p_{S_2}^{\frac{1}{2}} \cdot p_{O_2}}$$

What cannot be discovered readily from the diagram is the SO_2 pressure which occurs during the oxidation of FeS, and at what point the SO_2 pressure at the sulphide–oxide interface reaches one atmosphere. The piece of information which is needed to complete the picture can be obtained from a consideration of the iron and oxygen chemical potentials in the oxide phase which is formed immediately adjacent to the sulphide. We may assume that the iron activity is the same at the surface of the sulphide as in the neighbouring face of the oxide, and this may therefore be taken as approximately 10^{-4}, following the previous discussion (FeS–FeS_2 equilibrium).

From the tabulated thermodynamic data, the equilibrium constant for the formation of FeO is found to have the value 10^{10} at 1000 K.

$$K_{FeO} = \frac{a_{FeO}}{a_{Fe} \cdot p_{O_2}^{\frac{1}{2}}}$$

and hence the oxygen pressure is about 10^{-20} atm over the oxide in equilibrium with pure iron. The oxide of this composition might be formed at the very beginning of the oxidation of the stoichiometric sulphide, but the oxygen pressure rises to the value 10^{-12} atm when the activity of iron is decreased to the value of 10^{-4} during the later stages of the oxidation. It should be remembered that this is the value of a_{Fe} on the sulphur-rich side of $FeS_{1+\delta}$. It can be seen on the stability diagram that at this sulphur-rich boundary, the SO_2 pressure of the system $FeS_{1.10}-p_{O_2} = 10^{-12}$ atm would have the value of approximately 10^3 atm. Such a pressure would clearly rupture the oxide scale on the sulphide.

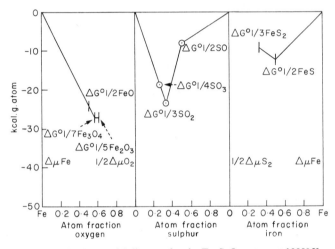

Fig. 6. Chemical potential diagram for the Fe–S–O system at 1000° K.

There are breaks in the stability diagram in the boundary between FeS and Fe, FeO and Fe_3O_4. The metastable equilibria can be obtained by extrapolating the straight lines of one boundary, say the FeS–FeO boundary, into the next phase field, i.e. below the FeS–Fe_3O_4 boundary. Since the oxygen pressure in the FeS–Fe_3O_4 system is lower than that in the extrapolated FeS–FeO boundary, it follows that, under conditions of local equilibrium at the sulphide-oxide interface, Fe_3O_4 would be the phase to be formed next to $FeS_{1+\delta}$ during oxidation rather than FeO.

It will be clear from the discussion that the stability diagram is not really suitable for the consideration of oxidation mechanisms, and we will now consider another representation, the integral free energy of mixing diagram, which is more readily applied. The diagrams for the Fe–S and Fe–O systems are shown in Fig. 6, and are drawn for 1 g atom of material throughout at 1000 K. This means that the value for Fe_3O_4 corresponds to one-seventh of the free energy of formation of one mole of that substance, and that for FeS_2

corresponds to one-third of the value for one mole and so on. The tangent to the free energy of mixing curve has intercepts at each ordinate corresponding to the chemical potential of the species shown on the ordinate. The condition of coexistence of a sulphide and oxide of iron at thermodynamic equilibrium is that the iron chemical potential, $RT \ln a_{Fe}$, is the same for both phases. It can be seen immediately from the diagrams that Fe_3O_4 is the phase which can coexist with sulphur-rich FeS, since the tangent to the Fe–S diagram near the two-phase equilibrium line $FeS_{1.1}$–FeS_2 yields an iron potential which, if transposed to the Fe–O diagram, is the origin of a tangent to Fe_3O_4 at a lower oxygen potential, and hence the more stable system, than one which touches FeO.

THE DIFFERENTIAL SULPHATION OF METAL OXIDES

In order to make a comparison between metal sulphates to determine the conditions when one metal may be retained as oxide and the other as sulphate with a binary mixture of the oxides as starting material, the graphical representation of the data which is more useful to apply than the stability diagram is the Ellingham free energy diagram for the formation of the sulphates from the oxides SO_2 and oxygen. The data for some of the most common elements are shown in Fig. 7.

The standard free energies of formation of the sulphates of a number of common metals such as copper, nickel and iron differ from one another mainly by the difference in the standard heats of formation. The Ellingham diagram for sulphate formation consists of a number of almost parallel straight lines which represent the changes for the reaction

$$2MO + 2SO_2 + O_2 \rightarrow 2MSO_4$$

in the general case. The slopes of these lines correspond approximately to the entropies of formation of the sulphates according to this reaction equation, and the numerical values are very similar for the formation of all of the metal sulphates.

The field of temperature and gas compositions in which the formation of metallic sulphates can be formed can readily be obtained from the diagram, and it is obvious that a sulphate with the greater, more negative, free energy of formation is more stable than one with a smaller, less negative free energy of formation. For kinetic reasons, it is not worth considering sulphation as a practical process below 600 K, and above 1200 K the gas must usually be too rich in SO_2 for practical operations. During such a roast, the gas phase will contain a partial pressure of SO_3 which is determined by the temperature, and the SO_2 and oxygen partial pressures. Thus, another diagram can be drawn for the formation of sulphates from metal oxide and SO_3 gas.

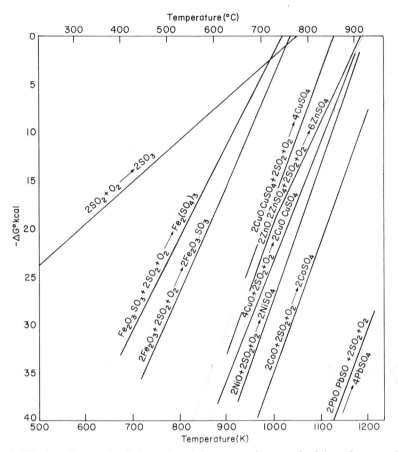

Fig. 7. Ellingham diagram for the formation of the common base metal sulphates from metallic oxide, sulphur dioxide and oxygen.

Providing we are only concerned with the formation of the pure sulphate from the pure oxide, then the standard free energy change is equal to the chemical potential of SO_3.

$$MO + SO_3 \rightarrow MSO_4$$

$$\Delta G^\circ = RT \ln p_{SO_3} \text{ when } a_{MO} = a_{MSO_4} = 1$$

Over a temperature range of a few hundred degrees, the heat and entropy changes for a sulphation reaction will remain sensibly constant, and hence

$$\log p_{SO_3} = \frac{+\Delta H^\circ}{4 \cdot 575 T} - \frac{\Delta S^\circ}{4 \cdot 575} = + \frac{A}{T} - B.$$

Figure 8, which shows the diagram for the common metal sulphates in the practical temperature range 700–1200 K, also shows the range of SO_3 pressures which can be obtained by combinations of SO_2 and oxygen from partial pressures of 10^{-3} to 1 atm for each component over this temperature range. The free energy equation

$$SO_2 + \tfrac{1}{2}O_2 \rightarrow SO_3 \quad \Delta G^\circ = -22,600 + 21{\cdot}36T \text{ cal mol}^{-1}$$

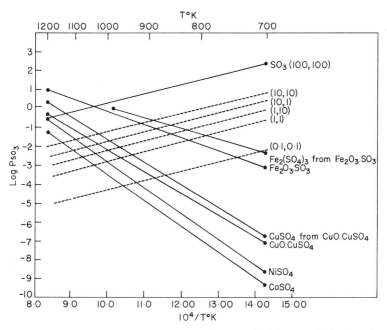

Fig. 8. SO_3 dissociation pressures (atmos.) of some base metal sulphates and SO_2/O_2 mixtures as a function of temperature. The bracketed figures indicate the percentages of one atmosphere of SO_2 and O_2 partial pressures for each line.

was used in making this diagram, and obviously

$$\log p_{SO_3} = \frac{-\Delta G^\circ}{4{\cdot}575T} + \log p_{SO_2} + \tfrac{1}{2} \log p_{O_2}$$

It can be seen that SO_2 and oxygen are only minor constituents of the gas at one atmosphere total pressure at the low temperature range of operation.

It may be concluded from this diagram that the separation of cobalt, nickel and copper from one another and from iron should be possible if roasting can be made to attain thermodynamic equilibrium. It may also be noted that the basic sulphates are usually insoluble in water and are thus not satisfactory as products for hydrometallurgical leaching processes.

THE COMPOSITION OF THE GAS PHASE IN SULPHIDE ROASTING

One final aspect of the thermodynamics of sulphide roasting which should be considered, is the complexity of the gaseous phase which is encountered during oxidation or sulphation roasting. The gaseous chemical species which can be formed by reaction between sulphur and oxygen include sulphur monoxide as well as SO_2 and SO_3. Under oxidizing conditions, there are therefore at least five gaseous species whose partial pressures must be

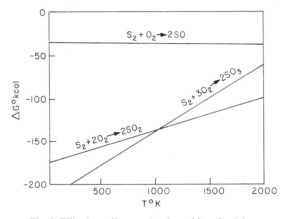

Fig. 9. Ellingham diagram for the oxides of sulphur.

evaluated for a given roasting condition in order to determine what the final product will be at thermodynamic equilibrium, viz. S_2, O_2, SO, SO_2 and SO_3. The representation of the thermodynamic data by means of the Ellingham diagram for these gases is shown in Fig. 9, but it can readily be seen that the rapid assessment of the gas composition in any sulphide roasting process by the use of data presented in this form can only be done with a considerable degree of manipulation. Similarly, the stability diagram is only of limited value for the total analysis of the gas phase. The chemical potential diagram is, again, a representation of the data which is simple to apply as can be judged from Fig. 10. Here the data for sulphur–oxygen system are represented for the three compounds SO, SO_2 and SO_3 for one gram atom of each species. The broken and full lines connect the values for 500 and 1000 K respectively. Since the ordinates for the diagram are the sulphur and oxygen potentials per gram atom, $\frac{1}{2}S_2$ and $\frac{1}{2}O_2$ respectively, a line joining any selected values of $\Delta\mu_{S_2}$ and $\Delta\mu_{O_2}$ will represent a gas mixture of partial pressures of S_2 and O_2 derived from these chemical potentials

$$\Delta\mu_{S_2} = RT \ln p_{S_2}; \ \Delta\mu_{O_2} = RT \ln p_{O_2}$$

The partial pressure of SO, SO_2 and SO_3 for these values of $\Delta\mu_{S_2}$ and $\Delta\mu_{O_2}$ can be found according to the simple geometrical properties of the diagram (Fig. 11).

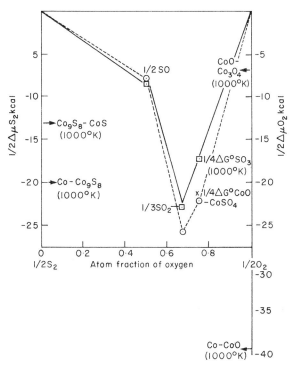

Fig. 10. Chemical potential diagram for the Co–S–O system at 1000° K with values for the S–O system at 500° K, ⊙ and 1000° K, ☐.

As an example of the application of the data to a specific system, the data for the cobalt–sulphur–oxygen system are also shown in fig. 10. Here the sulphur potentials of Co–Co_9S_8 and Co_9S_8–CoS are shown at 1000 K on the sulphur potential ordinate and the oxygen potentials for Co–CoO and CoO–Co_3O_4 at 1000 K are shown on the oxygen potential ordinate. The free energy of formation of $CoSO_4$ from CoO and SO_3 mol at 1000 K is shown at the oxygen atom fraction 0·75 corresponding to the oxygen/sulphur ratio in SO_3 equal to 3:1.

Applying the construction shown above to obtain a "feel" of the gas composition, it can be seen that SO is an unimportant species in the roasting of cobalt sulphides unless the sulphur potential is very much greater than that

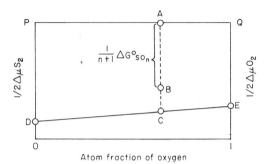

Atom fraction of oxygen

Fig. 11. The statement that the line joining any oxygen potential $\frac{1}{2} RT$ lin p_{O_2}, written as $\frac{1}{2} \Delta\mu_{O_2}$, with any sulphur potential $\frac{1}{2} \Delta\mu_{S_2}$ passes the point representing $1/n + 1 \, \Delta G^\circ_{SO_n}$ at a distance $1/n + 1 \, RT$ ln p_{SO_n} where $\Delta G^\circ_{SO_n}$ is the standard free energy of formation of SO_n per mole, is proved as follows, referring to Fig. 11.

$$PD = \tfrac{1}{2} \Delta\mu_{S_2}$$
$$QE = \tfrac{1}{2} \Delta\mu_{O_2}$$
$$AB = 1/n + 1 \, \Delta G^\circ_{SO_n}$$
$$AC = 1/n + 1 \times \tfrac{1}{2} \Delta\mu_{S_2} + n/n + 1 \times \tfrac{1}{2} \Delta\lambda_{O_2}$$
$$1/n + 1 \, \Delta G^\circ_{SO_n} = -1/n + 1 \, RT \ln K_{SO_n}$$
$$= -1/n + 1 \, RT \ln p_{SO_n} + 1/n + 1 \, RT \ln p_{\frac{1}{2}S_2} + n/n + 1 \, RT \ln p_{\frac{1}{2}O_2}$$
$$= -1/n + \Delta\mu_{SO_n} + 1/n + 1 \tfrac{1}{2}\Delta\mu_{S_2} + n/n + 1 \times \tfrac{1}{2} \Delta\mu_{O_2}$$

Hence $BC = AC - AB = 1/n + 1 \, RT \ln p_{SO_n}$.

of the Co_9S_8–CoS system. Furthermore, when CoO coexists with Co_9S_8 the principal gaseous species is SO_2 which can reach a pressure of one atmosphere when the oxygen/metal ratio is slightly greater than the value in equilibrium with cobalt metal.

This follows because any line originating in the band of chemical potentials between Co–Co_9S_8 and Co_9S_8–CoS on the left-hand ordinate will pass through the point for $1/3 \, SO_2$ and extrapolate to a value of $1/2\Delta\mu_{O_2}$ on the right-hand ordinate *above* the Co–CoO position (-40 kcal). Furthermore, any condition under which CoO will sulphate will exclude the co-existence of a sulphide of cobalt. Any line originating on the $1/2\Delta\mu_{O_2}$ ordinate and passing through $1/4\Delta G^\circ CoO$–$CoSO_4$ must intersect the $1/2\Delta\mu_{S_2}$ ordinate much below the Co–Co_9S_8 intercept. Again, SO will be an insignificant component of the gaseous phase, and it can be seen that as P_{O_2} is decreased in the gas phase, $1/2\Delta\mu_{O_2}$ becomes more negative, and the gas phase will become mainly a mixture of SO_3 with increasing partial pressures of SO_2.

REFERENCES

Alcock, C. B. (1971). *Canadian Metallurgical Quarterly* **10**, 287. Chemical potential diagrams.

Burgmann, W., Urbain, G. and Frohberg, M. G. (1968). *Révue de Métallurgie* **65**, 567. Thermodynamics of $FeS_{1+\delta}$ and FeS_2.

Ingraham, T. R. (1967). *In* "Applications of Fundamental Thermodynamics to Metallurgical Processes" (Ed. G. R. Fitterer). Gordon and Breach. Stability diagrams for Cu–S–O, Ni–S–O, Co–S–O and Fe–S–O.

Kellogg, H. H. and Basu, S. K. (1960). *Trans. AIME* **218**, 70. Stability diagrams and Pb–S–O system.

Kubaschewski, O., Evans, E. Ll. and Alcock, C. B. (1967). *In* "Metallurgical Thermochemistry". Pergamon Press. Survey of high temperature techniques for thermodynamic measurements.

Niwa, K., Wada, T. and Shiraishi, Y. (1957). *J. Metals* **9**, 269. The mechanism of the oxidation of ferrous sulphide.

Rosenqvist, T. (1954). *J. Iron and Steel Inst.* **176**, 37. Thermodynamics of systems in Fe, Co, Ni + S.

Sodi, J. and Elliott, J. F. (1968). *Trans. AIME* **242**, 2143. The free energy of formation of ReS_2 by solid electrolyte EMF studies.

3

SOLID STATE REACTIONS INVOLVING HALOGENS

THE CHLORINATION OF REFRACTORY METAL OXIDES

A number of the common elements such as iron, lead and tin are produced from their oxides by carbon reduction. Some of those metals which form oxides of intermediate stability, such as manganese, are made by reduction of the oxides by aluminium. In the case of the refractory metals, titanium, zirconium, etc., the metal is usually made by the Kroll process, in which a chloride or fluoride of the metal is prepared from the oxide mineral, and this is reduced by magnesium or calcium to yield the metal.

The principal reason for this change in method to a halide route is to be found from the properties of the refractory metal–oxygen systems. As an example of the difficulty, zirconium has, besides the dioxide ZrO_2, two ranges of solid solution of oxygen in the metal, usually designated in the high temperature range the β (dilute) and α (concentrated) solutions. The phase diagram shows the extents of these phase fields (Fig. 12). The oxygen potentials of the dilute solid solution phases are very low, and the oxygen content of the metal is only reduced to a level which will not adversely affect the physical properties of the metal by calcium reduction.

A similar situation occurs in the case of titanium, but now there are a number of lower oxides, together with TiO_2 and a dilute solid solution of oxygen in the metal in the phase diagram of the titanium–oxygen system (Fig. 13). Here again, the oxygen in the dilute solution is very firmly bound, and reduction with a very strong reducing agent such as calcium is necessary.

The metal vanadium dissolves a considerable amount of oxygen and forms many oxides. The similarity with titanium is quite marked in that there are

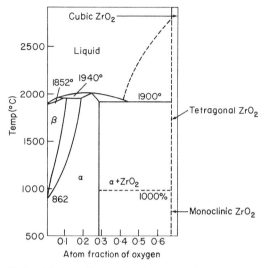

Fig. 12. Partial phase diagram for the system zirconium–oxygen.

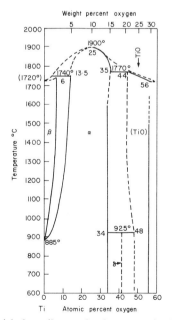

Fig. 13. Partial phase diagram for the system titanium–oxygen.

many intermediate oxides, the so-called Magnéli phases which have the general formula V_nO_{2n-1} between VO_2 and VO. It is thought that an ordering of the oxygen atoms in the terminal solid solution occurs at the composition corresponding to a lower oxide V_4O at low temperatures.

Niobium forms only a dilute terminal solid solution with oxygen, and the oxides NbO, NbO_2 and Nb_2O_5.

In all of these metal–oxygen systems, the oxygen potentials of the dilute solid solutions are so low that the only possible reducing agent which would yield substantially oxygen-free metal, e.g. 0·1 wt.%, would be calcium, whereas the chlorides can be reduced to yield a satisfactory product with the more readily available magnesium. Another reason for the change to halide metallurgy is that the oxides of the refractory metals as well as calcium oxide have very high melting points. The reduction of a refractory metal oxide by liquid calcium would result in the formation of solid products only. The kinetics of such a reduction would be slow by comparison with other possible processes which involve the reduction of the volatile refractory metal chloride by liquid calcium or magnesium to yield the low melting $CaCl_2$ (m.p. 1045 K) or $MgCl_2$ (m.p. 987 K).

Kubaschewski and Dench (1955–56) have studied the equilibria between the dilute solutions of oxygen in the refractory metals and alkaline earth metals in contact with their oxides. Thus, a sample of titanium would be encapsulated in an iron crucible with calcium/calcium oxide and held at constant temperature until equilibrium was reached. The Group IIA metal and its oxide would then be removed by leaching the product with dilute hydrochloric acid, and the titanium sample analysed for its oxygen content. In this way, the oxygen potential as a function of the oxygen content of the metal can be established at a number of single points of alkaline earth metal/metal oxide oxygen potential, e.g. Mg/MgO, Ca/CaO, Ba/BaO, etc. (Fig. 14).

Komarek and Silver (1962) increased the scope of this technique by using a large iron crucible so that the refractory metal could be held at a fixed temperature at one end of the container, and the temperature of the alkaline earth metal could be controlled separately at the other end. When calcium was used in these experiments, the calcium pressure could be fixed around the metal sample, the corresponding oxygen pressure would therefore also change in a controlled manner. Thus, in equilibrium with CaO

$$Ca(g) + \tfrac{1}{2}O_2 \rightarrow CaO \qquad K_{CaO} = \frac{1}{p_{Ca} \cdot p_{O_2}^{\frac{1}{2}}}$$

In this way, a larger number of results corresponding to a wide range of oxygen contents in the refractory metals were established by these workers, and the thermodynamics of the dilute solid metal–oxygen systems are now quite well defined.

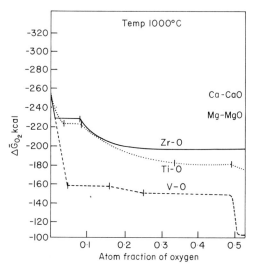

Fig. 14. Oxygen potentials in some refractory metal–oxygen systems at 1000° C. (After Kubaschewski and Dench, loc. cit.)

CHLORINATION EQUILIBRIA

The thermodynamic data for oxide chlorination show that the equilibrium constant for reactions such as

$$MO_2 + 2Cl_2 \rightarrow MCl_4 + O_2 \qquad K = \frac{a_{MCl_4}}{a_{MO_2}} \cdot \frac{p_{O_2}}{p_{Cl_2}^2}$$

are usually much less than, or at best approximately equal to, unity. It follows that the chlorine utlization efficiency would be very low if such a direct chlorination were attempted. The efficiency can be raised considerably if a reducing agent is present which would lower the oxygen partial pressure, and yet not react significantly with chlorine. Carbon is very effective in this role, consequently the oxide to be chlorinated is first pelletized with carbon. The reaction which is employed is then

$$MO_2 + C + 2Cl_2 \rightarrow MCl_4 + CO_2$$

or

$$MO_2 + 2C + 2Cl_2 \rightarrow MCl_4 + 2CO$$

The effect which the addition of carbon will have can be easily assessed by inserting the oxygen pressure of the C–CO$_2$ (1 atm) equilibrium for p_{O_2} in the equilibrium constant above for chlorination temperature below 900 K and the C–CO (1 atm) equilibrium above 1000 K.

There are important thermodynamic considerations which must be brought to bear in the choice of an unreactive lining material for the chlorination reactor. Silica can be used successfully in most cases, but in the chlorination of beryllia to form gaseous $BeCl_2$, graphite must be used. The corrosive action of $BeCl_2$ on silica can be predicted thermodynamically from the following argument:

For the chlorination of cristobalite, to form gaseous $SiCl_4$

$$SiO_2 + 2Cl_2 \rightarrow SiCl_4 + O_2; \quad \Delta G° = 62,000 - 10·5T,$$

and for beryllia, to form gaseous $BeCl_2$

$$2BeO + 2Cl_2 \rightarrow 2BeCl_2 + O_2; \quad \Delta G° = 123,200 - 47·8T$$

On combining these two equations, we obtain the one for the attack of silica by $BeCl_2$ vapour:

$$SiO_2 + 2BeCl_2 \rightarrow SiCl_4 + 2BeO; \quad \Delta G° = -61,200 + 37·3T$$

which has a negative standard free energy change in the temperature range of industrial operation.

On the other hand, for titania, and gaseous $TiCl_4$

$$TiO_2 + 2Cl_2 \rightarrow TiCl_4 + O_2; \quad \Delta G° = 38,500 - 13·5T$$

and for the attack of silica by $TiCl_4$ vapour,

$$SiO_2 + TiCl_4 \rightarrow SiCl_4 + TiO_2; \quad \Delta G° = 23,500 + 2·8T$$

which is positive in the same temperature range.

It may thus be concluded that the attack of a silica lining in a TiO_2 chlorination plant will be virtually absent except as a result of the direct access of chlorine to the lining, whereas in the BeO operation, a silica lining cannot be employed. In this case, a carbon lining has been used in an industrial reactor.

SEPARATION OF HALIDES BY VOLATILIZATION

The relative vapour pressures of metal halides are determined mainly by their heats of vaporization, since the entropies of vaporization are fairly constant. For a given metal, it is generally true that the higher the valency the greater the volatility. Thus, ferric chloride is more volatile than ferrous chloride, niobium pentachloride than $NbCl_3$ and so on. The elementary process which is considered when relative volatilities of compounds are discussed is that of volatilization of the pure solid to the corresponding equilibrium vapour. However, it must be realized that all the gaseous chlorides of a metal are present to a greater or less extent above a system containing the metal at a given activity and a partial pressure of chlorine. The partial pressure of each metal chloride species is a function of the chlorine potential of the gas phase. Thus, for a metal which forms three gaseous chlorides,

$$M + Cl_2 \rightarrow MCl_2(g); \qquad p_{MCl_2} = K_{MCl_2} \cdot a_M \cdot p_{Cl_2}$$

$$M + \tfrac{3}{2}Cl_2 \rightarrow MCl_3(g); \qquad p_{MCl_3} = K_{MCl_3} \cdot a_M \cdot p_{Cl_2}^{\frac{3}{2}}$$

$$M + 2Cl_2 \rightarrow MCl_4(g); \qquad p_{MCl_4} = K_{MCl_4} \cdot a_M \cdot p_{Cl_2}^{2}$$

and the relative proportions of each vapour species can be altered by a change in the chlorine potential and the metal activity as well as the temperature.

This point can be demonstrated very well on the chemical potential diagram for a metal which forms, as an example, two gaseous chlorides MCl and MCl_4. The points on the diagram A and B represent the free energy of formation of one half of a mole of MCl and one fifth of a mole of MCl_4. For a given activity of metal, M, and a given chlorine pressure, p_{Cl_2}, the straight line joining the chemical potentials corresponding to these activities on the diagram can be used to calculate the partial pressure of MCl and MCl_4 which co-exist with these two potentials.

In the general case, a line which is drawn across a free energy of mixing diagram for a given metal activity and given chlorine pressure will miss the point on the diagram corresponding to $1/(n + m)$ times the free energy of formation of the compound, M_nCl_m, by an amount equal to $1/(n + m)$ $\Delta\mu_{M_nCl_m}$ (see Fig. 15). It follows also that the relative partial pressures of the vapour chlorides will depend on the chlorine potential even when equilibrium

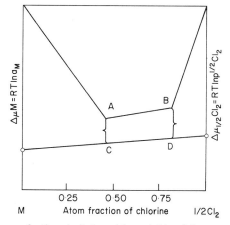

Fig. 15. Schematic diagram for the calculation of the activities of the compounds MCl and MCl_4 for the given chemical potentials of metal and chlorine.

C

is with a compound, such as the oxide or sulphide rather than a metallic phase

$$MO + Cl_2 \rightarrow MCl_2 + \tfrac{1}{2}O_2 \qquad K = \frac{p_{MCl_2} \cdot p_{O_2}^{\frac{1}{2}}}{p_{Cl_2}}$$

$$MO + 2HCl \rightarrow MCl_2 + H_2O \quad K = \frac{p_{MCl_2} \cdot p_{H_2O}}{p_{HCl}^2}$$

In the latter case, the gas phase will contain small partial pressure of hydrogen and the p_{H_2}/p_{H_2O} ratio will be fixed by the presence of the oxide MO, and the p_{H_2} combined with p_{HCl}^2 fixes p_{Cl_2}. The metal activity, which together with the chlorine pressure fixes the partial pressures of the metal chlorides, is of course determined by the presence of the metal oxide at a fixed activity, and the p_{H_2}/p_{H_2O} ratio in the gaseous phase.

It can now be seen that the use of volatile halides can be exploited as a means for separating metals provided adequate chlorine potential control can be exercised. For example, in the chlorination of zirconium ores to form $ZrCl_4$ the iron which is present is usually converted to ferric chloride which passes out of the chlorinator along with the volatile $ZrCl_4$. The products of

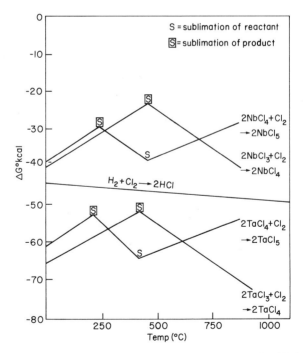

Fig. 16. Ellingham diagram for the chlorides of niobium and tantalum. The chlorine potential of a 1:1 H_2/HCl mixture is located on the free energy line for $H_2 + Cl_2 \rightarrow 2HCl$.

chlorination are then heated in hydrogen which reduces the volatile $FeCl_3$ to the much less volatile $FeCl_2$, but does not alter the chemical composition of the zirconium halide. In practice, the impure $ZrCl_4$ is first heated to 200°C under a hydrogen atmosphere when $SiCl_4$ is removed, and then heated to 400°C when $ZrCl_4$ sublimes on to cooling coils in the upper part of the assembly. This relatively crude method of separation is successful because of the much greater chemical stability of $ZrCl_4$, and therefore resistance to reduction, when compared with both $FeCl_3$ and $FeCl_2$.

The elements niobium and tantalum form a number of chlorides, viz. $NbCl_2$, $NbCl_3$, $NbCl_4$ and $NbCl_5$, etc. The thermodynamics of these compounds run according to expectation in that the pentachlorides are quite volatile but the trichlorides are much less so. It can be seen from the Ellingham diagram for some of the higher chlorides that the volatile tantalum pentachloride can be separated from niobium trichloride in an atmosphere of H_2/HCl of approximately 1:1 or a chlorine pressure of about 10^{-10} atm at 600°C (see Fig. 16).

THE MEASUREMENT OF VAPOUR PRESSURE AT HIGH TEMPERATURES

The range of pressure over which it is desirable to have information about the behaviour of metallic chlorides is from about 10^{-10} atm up to 1 atm pressure. There are a number of experimental methods which can be used to determine the vapour pressure, but no single technique covers the whole span of this range. The classical methods for the measurement of the boiling point have been used in high temperature studies. Fischer and Rahlfs (1932) determined the boiling points of aluminium halides by measuring the weight loss of samples under a controlled inert gas pressure. When the gas pressure is made slightly less than the vapour pressure of the substance which is under investigation, the weight loss of the substance accompanying vaporization increases considerably. These workers also determined the temperature of condensation of the vapour above the boiling liquid which was held under a fixed pressure of an inert gas.

Fischer, Gewehr and Wingchen (1939) measured the vapour pressures of halides by placing a sample of each of the compounds under a molybdenum bell which was floated on liquid tin. The gas pressure which was needed to prevent the upward movement of the bell at a given temperature was then equal to the vapour pressure of the halide.

Finally, a relatively high vapour pressure may be measured by isolating a sample of the halide with an inert liquid in the closed limb of a U-tube. The pressure of an opposing gas which is needed to keep the manometric fluid at constant level is then measured at the desired sample temperature. Biltz and Meyer (1928) used a eutectic salt mixture as the manometric fluid in a quartz apparatus for measuring the vapour pressures of some amalgams. For measurements on halides, it would be necessary to use a low-melting alloy or metal which is inert to the halides.

These methods can be used from one atmosphere pressure down to a few millimeters Hg pressure, and depend for their successful application on the availability of satisfactory container materials which are leak tight and transparent at the temperature of study. Quartz has been extensively used in the measurement of high vapour pressures, but transparent refractory materials, e.g. MgO, are now becoming available in the form of container tubes, and these could be employed in the higher temperature range, i.e. above 1000°C.

For the intermediate range of pressures, around 0·1–1 mmHg, there are not a large number of methods available for use, but a direct measurement of the vapour density can be made optically if a transparent container is available or by radioactive measurements where a suitable isotope can be obtained. In both cases, the principle of the technique is to present a column of the vapour which is in thermal equilibrium with the condensed solid or liquid, either for the measurement of absorption of a light beam, or for the measurement of the radioactive emission from the radioactive species which is contained in the vapour molecule. Both of these methods need some calibration, so the vapour pressure must also be established by some other technique at a suitable pressure, e.g. by manometry at higher pressures than one millimetre.

Low pressure techniques, which can be used between 10^{-10} and 10^{-4} atm, have received a good deal of attention over the past two decades. The gas transportation method is very similar to the technique which is used for the determination of H_2/H_2O equilibria with metal oxides. The gas, which must be chemically inert, is passed over a sample of the material of which the vapour pressure is to be found, and either the weight loss of the sample, or the weight of material which is transported by the gas must be determined. It is important to bring the gas to equilibrium with the condensed phase sample in one pass only, since it is impossible to recycle the gas unless the whole apparatus is heated. Provided that the vapour pressure is not too high, and hence the upper limit of 10^{-4} atm on the use of the method, saturation can be achieved by ensuring that the gas bubbles through the sample if it is liquid, or passes through a tube which is almost filled by a solid sample, in the even temperature zone of the heater. The vapour pressure is calculated from the equation

$$p = \frac{W}{M} \frac{22 \cdot 4}{V} \text{ atm}$$

where W is the weight transported; M is the molecular weight of the vapour specie; V is the volume in litres of the transporting gas which is passed, at STP (1 atm pressure and 0°C).

It is usually found that the apparent vapour pressure which is obtained from the measurements is too high at very low carrier flow rate due to diffusion effects, both thermal and concentration. At high flow rates, the apparent pressure is too low due to undersaturation of the carrier gas. There is usually a range of flow rates over which the apparent vapour pressure is constant and this value is the correct one.

The other important method for the determination of vapour pressures in the range 10^{-10}–10^{-4} atm is the Knudsen effusion technique. In this method, the sample is held in a crucible which is heated in a vacuum furnace to a fixed temperature. The crucible is gas-tight except for a small orifice through which the vapour of the condensed phase can escape. Providing that the diameter of the orifice is less than about one tenth of the mean free path of the vapour inside the crucible, the vapour molecules effuse into the vacuum furnace as a collision-free beam. Ideally, the orifice should be knife-edged so that there can be no reflection after collision of molecules crossing the orifice area. The weight loss of the sample can then be used to calculate the vapour pressure of the sample by means of the Knudsen equation:

$$ W = pAt \sqrt{\left(\frac{M}{2\pi RT} \right)}, $$

where A is the area of the orifice; t is the duration of the experiment; R is the gas constant.

Calculations have been made of the effects of the use of a channel rather than a knife-edge orifice. The so-called Clausing factor by which the right-hand side of the Knudsen equation must be multiplied is a function of the radius to length ratio of the channel, and values have been tabulated by Schulz and Searcy (1962).

In order to ensure that equilibrium was achieved within the crucible, it is customary in studies employing the Knudsen effusion technique to show that the results do not vary over a significant range of orifice areas.

In all vapour pressure measurements, it is important to show that the results of the measurements are not affected by a change in the container material, since many vapour species can be formed by side-reactions which might not normally be expected. Thus, Grieveson and Alcock (1960) found that the use of beryllia as a container for the measurement of the vapour pressure of elementary silicon failed because of the reaction

$$ BeO(s) + Si(l) \rightarrow Be(g) + SiO(g) $$

The pressures which were obtained in BeO boats were three times the correct value which was obtained in a molybdenum disilicide container.

The solution of both the transportation and Knudsen equations requires a knowledge of the molecular weight of the vapour species. This is not always easy to ascertain because a number of vapours co-exist in significant amounts in equilibrium with some inorganic solids. The dimers and trimers of the simple alkali halides are known to be significant components of the vapours above these compounds, and $(CuCl)_3$ is the principal species above cuprous chloride at high temperatures. The most valuable technique for elucidating the complexity of the vapour phase in any system is the combination of the Knudsen technique with the mass spectrometer. The collision-free beam which effuses from the Knudsen crucible can be directed toward the ionization source of the spectrometer. Here the molecules and atoms are ionized by bombardment with electrons of a known energy. The ions which are produced are accelerated into the spectrometer and analysed into mass components either by the separation produced by a magnetic field, or by the time of flight of the ions through the spectrometer.

Unfortunately, the ion current which is produced by bombarding a given flux of vapour molecules depends upon the ionization cross-section of the species, and this is rarely known to better than $\pm 50\%$. The spectrometer is valuable mostly as a means of appraising the complexity of vaporization processes rather than for the accurate measurement of vapour pressure. Ratios of the ion currents, for example of monomers and dimers, can be used to advantage in the determination of the thermodynamics of solutions containing halides. If this ratio is known for the pure phase and for the component ACl in a binary mixture of two chlorides, ACl + BCl, the activity of ACl in the mixture can be derived since

$$a_{ACl} = kp_{ACl} = k'\sqrt{(p_{A_2Cl_2})}$$

and

$$\frac{p_{ACl}}{p_{A_2Cl_2}} = \frac{\sigma_{A_2Cl_2}}{\sigma_{ACl}} \cdot \frac{I^+_{ACl}}{I^+_{A_2Cl_2}},$$

σ is the cross section for ionization of each vapour species and I^+ is the ion current which is produced at the spectrometer current detector.

CHEMICAL TRANSPORT REACTIONS

A use of halogens in extractive metallurgy and materials science which is of growing interest is the transport of material from one solid to another by means of gaseous halides. The solid which is transported may be an element or a compound. The initial solid can be a very impure material from which a particular element is transported to a pure sample, or a polycrystalline starting material which is to be re-deposited in single crystal form. These are two examples out of many possibilities such as alloy formation, cladding or shaping operations, vapour phase doping, etc., which are presented through developments of chemical transport technique.

A common procedure in the growth of single crystal specimens of inorganic compounds is exemplified by the growth of NbO_2 samples by Schafer and Huesker (1962). A small polycrystalline sample of the dioxide is sealed together with a small amount of iodine in a quartz tube which is placed in a furnace having a gradient in temperature. The oxide reacts with iodine to form a volatile metal iodide, and this vapour species migrates, together with oxygen molecules, to the hot end of the quartz container. Here the iodide decomposes yielding NbO_2 by reaction with oxygen. Providing the correct conditions of temperature gradient are chosen so that deposition is slow, the product of decomposition contains sizeable single crystals of NbO_2.

In this example, one element of the two combined in the starting material, niobium, forms a gaseous halide; whilst the other element, oxygen, is naturally gaseous.

Compounds such as the metal silicides can also be trasported via the vapour phase when both combined elements form gaseous halides, e.g.

$$TaSi_2(s) + 6I_2(g) \rightarrow TaI_4(g) + 2SiI_4(g)$$

Clearly, the conditions for co-transport of tantalum and silicon require that these elements have about the same affinity for iodine.

In the transport of SiC, it is preferable to use HCl gas as the transporting agent rather than elementary chlorine. This is because silicon has a much greater affinity for chlorine than does carbon, and carbon must be transported as methane

$$SiC(s) + 4HCl(g) \rightarrow SiCl_4(g) + CH_4(g)$$

These examples added to such obvious applications as the van Arkel refining of zirconium through the volatile iodide, indicate the wide applicability of chemical transport reactions.

The science of vapour transport is based upon our knowledge of the thermodynamics of gaseous halides, and in choosing the appropriate conditions of operation for a transport reaction consideration must be given to these data before kinetic factors may be taken into account. Clearly, the operator would need criteria for the choice of the temperature and temperature gradient in which to make the reaction, and this brings the volatility of the halides into the problem.

A collection of data for the gaseous chlorides of the elements shows a general trend which is of importance in this aspect of chemical transport.

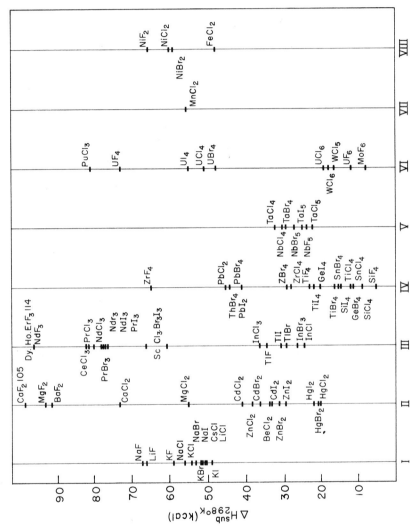

Fig. 17. Heats of sublimation of metal halides arranged in Periodic Table groups.

From Fig. 17, it can be seen that the polyvalent elements have their lowest heats of vaporization, and hence greater volatility in the highest valency state. In comparing amongst the groups, Group IIIA chlorides are the most difficult to vaporize, whereas in the B group elements, the volatilities are considerably higher than in the corresponding A groups.

In any projected chemical transport application, consideration must first be given to the relative volatilities of the valency states of a given element, and when more than one element is to be vaporized the stabilities of the various possible combinations of valency state must be considered to arrive at the most probable combination. Thus, one could envision the chemical transport of gallium arsenide involving GaCl or $GaCl_3$, together with elementary arsenic or $AsCl_3$ or $AsCl_5$. Consideration of the respective bond energies shows that GaCl and As_4 are the appropriate species in this case.

BOND ENERGIES AS A FUNCTION OF METAL VALENCY

A survey of the experimental data for the heats of formation of gaseous metal chlorides reveals that the bond energies between metal and chlorine atoms decrease as the metal valency increases. In order to see this clearly, the heats of formation of the chlorides from the metal and chlorine atomic gaseous species must be compared for a given metal. The normal heat of formation which is used in metallurgical thermodynamics is the heat change when the solid or liquid metal reacts with diatomic chlorine gas. The heats of vaporization of the metal and dissociation of the gas must be added to this heat change in order to calculate the bond energies.

TABLE IV. Bond energies of some gaseous species

$$M(g) + nCl(g) \rightarrow MCl_n(g) \quad cal.$$

AlCl	118	$SiCl_2$	205	PCl_3	233	UCl_4	433	$FeCl_2$	188
$AlCl_3$	305	$SiCl_4$	370	PCl_5	312	UCl_6	540	$FeCl_3$	246
GaCl	117	$SnCl_2$	188	$SbCl_3$	228				
$GaCl_3$	263	$SnCl_4$	309	$SbCl_5$	301				

Some typical values of experimental results are shown in Table IV, where it can be seen that the formation of another bond reduces the bond energy per atom as the valency of the metallic species increases.

An approximately 10% decrease in bond energy is to be expected when another bond is formed between chlorine and a metal atom. Because of the large values of most metal–chlorine bond energies, it is feasible to devise processes in which a solid metal can be vaporized by reaction with its own higher chloride. This is the basis of the process which was proposed by Gross *et al.* (1948) for the production of aluminium. The reaction between aluminium and $AlCl_3$ leads to the formation of $AlCl$. The increase in bond energy between aluminium and chlorine here compensates for the energy of vaporization of aluminium metal. Thus, the formation of the lower halide can be said to have made the metal effectively more easily volatilized. The entropy change of the reaction

$$2Al(s) + AlCl_3 \rightarrow 3AlCl$$

was found by Gross *et al.* to be 61·2 entropy units, and the heat change was $+93$ kcal and, hence, the equilibrium shifts to the right with increasing temperature. It follows that aluminium metal will be deposited by the reacted gas as the temperature is lowered. For the chemical transport of pure metals, there is a change in valency of the metal gaseous species involved, as in the reaction above, and the pressure in the system is maintained at about one atmosphere. Reactions in which a compound is formed which is subsequently completely decomposed are more frequently carried out at reduced pressure. An example of this is in the refining of zirconium.

Alcock and Jeffes (1967) have examined the thermodynamic conditions for a chemical transport process to function efficiently. Here, the criterion for efficiency which was adopted was that the maximum amount of material should be transported across a given temperature gradient in a given volume of the transporting gas mixture. The vaporization and subsequent deposition of a metal by volatile halide formation requires both favourable circumstances for the formation of a transporting species, such as $AlCl$, and favourable circumstances for the decomposition of the transporting gas at another temperature. If we consider the general transporting reaction

$$M(s) + n/2Cl_2 \rightarrow MCl_n(g); \quad K = \frac{p_{MCl_n}}{a_M \cdot p_{Cl_2}^{n/2}}$$

two points emerge quite simply. The first of these is that the greatest change in p_{MCl_n} occurs around values of K equal to unity when there are the same number of gaseous molecules on both sides of the transporting equation. When there is a decrease in the number of gas molecules in the transport reaction ($n > 2$), then the best value of K decreases as the total pressure increases, and it is greater than unity as the pressure increases above one atmosphere.

The value of K is determined by the free energy change, and thus the operating total pressure for a given reaction which is chosen will determine the temperature at which the reaction should be carried out. Alternatively, if the lower limit of the temperature is selected for a given operation for

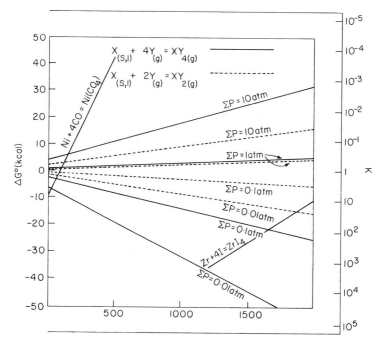

Fig. 18. Most favourable values of the equilibrium constant, K, as a function of total pressure and temperature for the reactions $X(s, l) + 4Y(g) \rightarrow XY_4(g)$ and $X(s, l) + 2Y(g) \rightarrow XY_2(g)$.

kinetic reasons, then this choice will determine the optimum total pressure which should be used.

It follows that a reaction such as the Van Arkel refining reaction, which is strongly exothermic, should be carried out at reduced pressure because the temperature of operation must not be too high.

$$Zr + 4I(g) \rightarrow ZrI_4(g) \qquad \Delta G° = -35 \, \text{kcal at } 1500 \, \text{K}$$

Conversely, a reaction which is only feebly exothermic should be carried out at high pressure, e.g.

$$Ni + 4CO \rightarrow Ni(CO_4), \qquad \Delta G° = 0 \, \text{at } 360 \, \text{K}$$

The second point to emerge from this study is that the balance between the heat change of the reaction and the entropy change multiplied by the temperature which determines the free energy change at the given temperature and, hence, the equilibrium constant, restricts the choice for the best point of operation. Clearly, a reaction having a large entropy change could also have a large heat change and be useful in some temperature and pressure range as a transporting reaction. A reaction having zero heat change will have the same value of the equilibrium constant at all temperatures and, therefore, this reaction is useless as a transporting reaction. Finally, a reaction having a large heat change, but a small entropy change, will not be very useful in transporting via the gas phase. This is because K will be either very large (ΔH negative) or very small (endothermic reaction) and, hence, the values of K which are achieved as the temperature of the gases is altered will always be very large or very small, and the change of K will not be sufficient to transport significant amounts of material.

THE CHOICE OF HALOGEN AS A REAGENT

Apart from the relatively cheap availability of chlorine for the halogenation of metal oxides, this element presents a number of advantages over the other halogens for metal extraction processes. It is found, in surveying the thermodynamics of metal halides, that the gaseous chlorides of a metal are usually about $25\,kcal\,mol^{-1}$ halogen more stable than the corresponding bromides, and about $50\,kcal\,mol^{-1}$ halogen more stable than the corresponding iodides. It follows that in a reaction involving the halogenation of a metal oxide, the reaction with chlorine will produce the highest partial pressure of halide.

Since the fluorides are more stable than the chlorides, it might be expected that fluorine would find a large industrial application. The drawbacks in the use of this halogen are firstly the cost and difficulty of working with fluorides, and secondly the fact that the heats of vaporization of fluorides are usually substantially larger than those of the other corresponding halides. The volatilities of fluorides are much smaller than those of the corresponding chlorides, bromides and iodides, as a result of this fact. It follows that separation of the fluorides from the furnace charge in the halogenator is only achieved at a higher temperature than with the other halides. However, as will be shown later, in the production of some refractory metals, the thermal advantages of using fluorides in the metal extraction reaction outweighs the difficulties of fluorination of the oxide minerals.

Finally, it can be stated that bromides and iodides have an advantage over chlorides when the vapour phase refining of metals and many chemical transport reactions are to be undertaken. This is because it is best to use a reaction at a temperature where the corresponding standard free energy change is nearly zero in order to work the process under conditions of high efficiency. Gaseous bromides and iodides are usually less stable than the corresponding chlorides by about 23 and 54 kcal per mol Br_2 and I_2 respectively. Because of the high heats of formation of many of the metal chlorides it would be necessary to work at a high temperature in order to bring the standard free energy change of the transporting reaction to zero. By employing a bromide or better still the iodide, this temperature is lowered drastically.

This is because of the entropy change for a given transport reaction does not change significantly when comparing chloride, bromide and iodide, and hence the relative heat changes determine the sequence of free energy changes. Since these are lowest in the iodides, the free energy change becomes zero at a lower temperature than in the corresponding reaction for bromide and chloride. It may sometimes happen that the ideal situation for the use of iodide gaseous molecules is too low and that the kinetics of the reaction will be too slow. It is then that the bromide reaction which will reach the highest efficiency at a higher temperature than this may be employed.

REFERENCES

Ackermann, R. J. and Rauh, R. G. (1962). *J. Chem. Phys.* **36**, 448. Time of flight–mass spectrometry.

Alcock, C. B. and Hooper, G. W. (1961). *Phys. Chem. of Proc. Met.*, Part 1, p. 325. Interscience, N.Y. Transpiration method for vapour pressure measurement.

Alcock, C. B. and Jeffes, J. H. E. (1967). *Trans. Inst. Min. Met.*, **76**, C246. Thermodynamics of vapour transport reactions.

Biltz, W. and Meyer, F. (1928). *Zeit. Anorg. Allg. Chem.* **176**, 23. Manometric measurement of vapour pressures.

Chupka, W. A. and Inghram, M. G. (1955). *J. Phys. Chem.* **59**, 100. Magnetic resolution mass spectrometer.

Fischer, W., Gewehr, R. and Wingchen, H. (1939). *Z. Anorg. Allg. Chem.* **242**, 161. Vapour pressure "floating bell" technique for moderately high pressures.

Fischer, W. and Rahlfs, O. (1932). *Zeit. Anorg. Allg. Chem.* **205**, 1. Vapour pressure of aluminium halides.

Grieveson, P. and Alcock, C. B. (1960). "Special Ceramics", p. 183. Heywood and Co., London. Vapour pressure of silicon.

Gross, P., Hayman, C., Kent, P. J. C. and Levi, D. D. (1948). *Disc. Far. Soc.* No. 4, Aluminium chloride disproportionation reaction.

Jeffes, J. H. E. and Alcock, C. B. (1968). *J. Mat. Sci.* **3**, 635. Thermodynamics of vapour transport reactions of intermetallic compounds.

Komarek, K. L. and Silver, M. (1962). *Proc. I.A.E.A. Symposium,* "Thermodynamics of Nuclear Materials", Vienna, p. 749. Thermodynamics of Zr–O, Ti–O and Hf–O alloys.

Kubaschewski, O. and Dench, W. A. (1955–56). *J. Inst. Met.* **84**, 440. Dissociation pressures in Zr–O system at 1000 K.

Schafer, H. and Huesker, M. (1962). *Z. Anorg. Allg. Chem.* **317**, 321. Vapour transport of NbO_2.

Schafer, H. and Kahlenberg, F. (1958). *Zeit. Anorg. Allg. Chem.* **294**, 242, (1960). **305**, 178, 291, 327. Thermodynamics of niobium chlorides.

Schafer, H. (1964). "Chemical Transport Reactions", Academic Press, London and New York. A general account of vapour phase transport.

Schulz, D. A. and Searcy, A. W. (1962). *J. Chem. Phys.* **36**, 3099. Clausing factor for the effusion technique.

4

THE KINETICS OF SOLID STATE REACTIONS—MOBILITY IN SOLIDS AND GASES

TRANSPORT PHENOMENA

The thermodynamic data for the separate systems which are involved in a solid-state reaction are valuable in giving qualitative guidance to the most probable reaction path. Providing that local equilibrium is always achieved at the interfaces, the compositions of the phases at the bounding surface can be obtained from a consideration of thermodynamic data. This information gives no clue about the speed with which the reaction will proceed, and further data concerning the transport properties are needed for the completion of the general kinetic picture. The specific rate of any given reaction involves the particular mechanism and is a composite product of the thermodynamic and the transport properties. The progress of a reaction depends on the rate of arrival of reactant particles at the interface and the transport of the products away from the interface. The number of particles reaching unit area of interface in unit time is called the flux, and in problems such as the flux of current carriers in electrical conduction or the flux of heat in thermal conduction, this quantity is related to a specific conductivity multiplied by a driving force. The respective equations for electrical and thermal fluxes are:

$$z_e J = -\sigma \frac{dV}{dx}$$

$$\text{and } J = -K \frac{dT}{dx}$$

Driving force

$\dfrac{dV}{dx}$ is the electrical potential gradient

$\dfrac{dT}{dx}$ is the temperature gradient

σ and K are the specific electrical and thermal conductivities and z_e is the charge of the current carrier.

Depending on the choice of units for the conductivity, the flux may be given directly as the number of particles, ions, electrons or phonons, or as the number of moles, amperes, calories, and so on, which cross unit area in unit time.

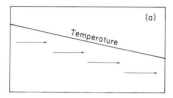 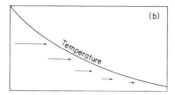

Fig. 19a. Steady state thermal flux and temperature distribution. The length of the arrows indicates the magnitude of the flux in each section.

Fig. 19b. Unsteady state flux and temperature distribution. The length of the arrows indicates the magnitude of the flux in each section.

When there exists a steady state in a system, as for example when heat flows from the hot to the cold face of a furnace wall and the temperature difference across the system is independent of time, the differential expression above can then be expressed in a simple integrated form

$$Q = -K.A.\frac{(T_1 - T_2)}{X}$$

where Q is the heat transferred from T_1 to T_2 through a wall of area A and thickness X (Fig. 19a).

When the system is in an unsteady state, the temperature distribution through the wall, in this example, changes with time. This would be the situation when the furnace is being heated up at the beginning of a campaign. The flux of heat energy arriving at a given depth in the furnace wall from the interior of the furnace will be greater than that leaving this depth to dissipate in the surroundings. There will therefore be an accumulation of energy at this point which will raise the temperature (Fig. 19b). The change in temperature at a given depth x below the inner face of the furnace wall is only obtained in the form of a second-order differential equation which must be solved for the given physical conditions of the system, the so-called boundary conditions. The equation is now following the

$$\frac{\partial T}{\partial t} = K\frac{\partial^2 T}{\partial x^2}$$

The boundary conditions for this particular example would be the temperature of the wall throughout its depth at the beginning of the heat-up and the temperature at the inner face during the whole period of time. The particular values of these temperatures would then determine the form of the solution of the equation for the given situation. The coefficient of thermal conductivity may be treated as a constant for the system, since the chemical composition of the material of construction of the furnace wall is the same throughout.

The driving forces dV/dx and dT/dx are related to the work which is done in transporting electrical charge or heat, and the chemical reaction flux is driven in an analogous way by the chemical potential gradient $d\Delta\mu_i/dx$. The flux in the chemically reacting system is given by the Einstein equation

$$J_i = -\frac{B_i c_i}{N}\frac{d\Delta\mu_i}{dx}$$

where c_i is the concentration of the ith species of activity a_i, and B_i is the mobility coefficient, N is Avogadro's number.

If we substitute

$$\frac{d\Delta\mu_i}{dx} = RT\frac{d\ln a_i}{dx} = RT\left[\frac{d\ln\gamma_i}{dx} + \frac{d\ln c_i}{dx}\right]$$

from the thermodynamic equation relating the activity with the activity coefficient and the concentration

$$a_i = \gamma_i c_i$$

then the equation for the flux which is given above becomes

$$J_i = -\frac{B_i RT}{N}\cdot\frac{dc_i}{dx}\left[1 + \frac{d\ln\gamma_i}{d\ln c_i}\right]$$

Finally, if γ_i is independent of c_i and we insert the diffusion coefficient which is defined as follows

$$D_i = \frac{B_i RT}{N}$$

we have the simple expression for the flow of atoms through a system in which there is a concentration gradient

$$J_i = -D_i\frac{dc_i}{dx}\qquad [\text{Fick's first law}]$$

and there exists a corresponding unsteady state equation

$$\frac{\partial c_i}{\partial t} = D_i\frac{\partial^2 c_i}{\partial x^2}\qquad [\text{Fick's second law}]$$

It will be noted, however, that it is only rarely that D_i can be used as a constant of the system, and it will usually vary with composition because the activity coefficient, γ_i, will vary. This complication makes the solution of chemical transport problems more intricate than the corresponding heat transfer analogues, since the thermal conductivity coefficient remains reasonably constant for a given system.

DIFFUSION IN PURE METALS

The self-diffusion coefficients of metals which control the movement of atoms within a pure metal vary over a very wide range of values at any given temperature, but the value for most metals approaches $10^{-8}\,\mathrm{cm^2\,s^{-1}}$ near the melting point. The coefficient for any given metal shows the Arrhenius temperature dependence and may be written as a function of temperature according to the equation

$$D = D_0 \exp -\Delta H^*/RT$$

where D_0 is a constant for each element, and ΔH^* is called, by analogy with the Arrhenius equation for gas reaction kinetics, the "activation energy". Since the quantity which is involved relates to a condensed phase in this instance, it is more correct to call this term an "activation enthalpy". The pre-exponential term has a fairly constant value for most metals of approximately unity, and the activation energies follow the same trend as the heats of vaporization (Table V). The latter observation gives a hint as to the nature of the most important mechanism of diffusion in metals, which is vacancy migration. It is now believed that the process of self-diffusion in metals mainly occurs by the exchange of sites between atoms and neighbouring vacancies in the lattice. The number of such vacancies at a given temperature should clearly be determined by the free energy of vacancy formation. The activation enthalpy for self-diffusion is the sum of the energy to form a vacancy and the energy to move the vacancy.

It has been found that the heat of formation of vacancies is approximately half the enthalpy of activation for diffusion, and hence it may be concluded that the heat of formation of vacancies is roughly equal to the enthalpy of vacancy movement. This is demonstrated by measurements of the electrical resistance of wire samples which are heated to a high temperature and then quenched to room temperature. At the high temperature, the equilibrium concentration of vacancies is established and this concentration is retained on quenching (Fig. 20). The electrical resistance includes the contribution from scattering of conduction electrons by the vacancies as well as by ion-core and impurity scattering. If the experiment is repeated at a number of high

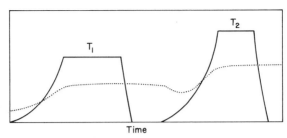

Fig. 20. The variation of vacancy concentration in a metal sample which is alternately heated and quenched ($T_2 > T_1$).

——— = Temperature.
- - - - - - = Vacancy concentration.

temperature treatments, then the effects of temperature on the vacancy contribution can be isolated since the other two terms will be constant providing that the temperature at which the resistance is measured is always the same. The energy to form vacancies is then found from the temperature coefficient of this contribution

$$R_{vacancy} = R_0 \exp -\frac{\Delta H_f^*}{RT}$$

where R_0 is a constant of the system.

Typical values of the energy to form vacancies are for silver, 23 kcal and for aluminium, 18 kcal mol^{-1}. These values should be compared with the values for the activation enthalpy for diffusion which are given in Table V. It can also be seen from this Table V that the activation enthalpy for self-diffusion which is related to the energy to break metal–metal bonds and form

TABLE V. Data for diffusion coefficients in elementary metals

Element	D_0	ΔH^* (kcal mol^{-1})	$\Delta H^\circ_{sublimation}$ (kcal mol^{-1})
Sodium	0·24	10·5	26·4
Silver	0·40	44·0	66·8
Magnesium	1·0	32·0	33·6
Aluminium	1·7	34·0	76·2
Lead	0·28	24·2	47·0
α-Zirconium	$5·9 \times 10^{-2}$	52·0	141·6
Vanadium	0·36	73·7	121·9
Tantalum	0·12	98·7	189·8
Molybdenum	0·1	92·2	142·0
Tungsten	0·7	121·0	200
γ-Iron	0·22	64·0	95·2

a vacant site is related semi-quantitatively to the energy of sublimation of the metal, in which process all of the metal atom bonds are broken.

The measurements of self-diffusion coefficients in metals are usually carried out by the sectioning technique. A thin layer of a radioactive isotope of the metal is deposited on one face of a right cylindrical sample and the diffusion anneal is carried out at constant temperature for a fixed time. After quenching, the rod is cut into a number of thin sections at right angles to the axis, starting at the end on which the isotope was deposited, and the content of the radioisotope in each section is determined by counting techniques.

The diffusion process in which a thin layer of radioactive material is deposited on the surface of a sample and then the distribution of the radioactive species through the metal sample is analyzed after diffusion, obeys Fick's second law with the following boundary conditions:

$$D\frac{\partial^2 c}{\partial X^2} = \frac{\partial c}{\partial t} \qquad \begin{array}{lll} c = c_0 & x = 0 & t = 0 \\ c = 0 & x > 0 & t = 0 \end{array}$$

D can be regarded as a constant of the system in this experiment since there is no change of chemical composition involved in the exchange of radioactive and stable isotopes between the sample and the deposited layer. The solution of this equation with these boundary conditions is

$$c = \frac{c_0}{\sqrt{\pi Dt}} \exp -\frac{x^2}{4Dt}$$

It follows that a plot of the logarithm of the concentration of the radioactive isotope in each section against the square of the mean distance of the section below the original surface transfer ($x = 0$) should be linear with slope $-1/4Dt$. Since t, the duration of the experiment, is known, D may be calculated.

THE MEASUREMENT OF DIFFUSION COEFFICIENTS IN SIMPLE COMPOUNDS

In order to study the diffusion of the cation or anion in a solid compound, it is preferable to prepare the compound in dense form, preferably as a single crystal specimen. This is because the diffusion process may either occur through the bulk of the solid involving a simple atomic movement, or along grain boundaries. The atom movement within the volume of the solid is usually the result of the change of position of an ion to a neighbouring vacant site on the same crystallographic sublattice. The diffusion coefficient depends on the frequency of such exchanges, and therefore the value of the coefficient increases as the number of vacant sites increases. Since the vacancy concentration is a function of the metal/non-metal ratio, care must be exercised in producing the compound in a well-defined state both from the physical point of view and with regard to the stoichiometric ratio of metal to non-metal.

The usual technique employed makes use of a radioactive tracer of the component, for which the diffusion coefficient is required. In the case of $FeS_{1+\delta}$ for example, there is an isotope for iron Fe^{55} which has a half-life of 2·7 years and emits X-radiation. The radiation is the characteristic K_α radiation of manganese, which is emitted by the iron nucleus after the capture of an orbital electron,

$$Fe^{55} + e^- \rightarrow Mn^{55} + K_\alpha$$

Because of the brittleness of inorganic compounds, the sectioning technique which is employed for metals cannot normally be used. An alternative procedure is to polish off sections of material from the right cylindrical sample and to collect the material which is removed for counting section by section, or to measure the residual activity after each section has been polished away.

In the study of $FeS_{1+\delta}$, a thin layer of the radioactive sulphide containing the Fe^{55} isotope is deposited on a polished face of the diffusion sample and the diffusion couple so formed is heated at a fixed temperature and in a controlled atmosphere of H_2/H_2S. In this way the metal/sulphur ratio is fixed in the sample during the diffusion process.

After the anneal at constant temperature for a known time, the sample is analyzed for the distribution of radioactive iron by measuring the residual activity in the solid after each thin layer is removed.

The intensity of radiation emitted from the surface after a number of layers have been removed may only be used as a quantity in determining the diffusion coefficient when the absorption law for the radiation in the sample material is known accurately. In the case of X-radiation, the exponential absorption law

$$I = I_0 \exp -\mu x$$

where μ is the absorption coefficient, is obeyed accurately, hence the total intensity of radiation which is emitted from the surface after the removal of a number of layers down to x' below the original surface, is given by

$$I = \int_{x=x'}^{x=\infty} c_x \exp -\mu(x - x') \, dx = \int_{x=x'}^{x=\infty} \frac{c_0}{(\pi D t)^{\frac{1}{2}}} \exp -\frac{x^2 + \mu(x - x')}{4Dt} \, dx$$

With other radiation, such as β-rays, the absorption law is not strictly the exponential law, and the evaluation of the integral is therefore more difficult.

The study of sulphur diffusion in $FeS_{1+\delta}$ has been made with the use of the pure β-emitting isotope S^{35} in the polish-residual count technique. A modified exponential law was used, in the sense that the range of the β-particles was determined empirically. This is in effect a cut-off of the exponential absorption law at a distance below the exposed surface corresponding to the maximum penetration of the FeS sample by the β-rays of maximum energy which are emitted by the radioactive sulphur atoms.

TABLE VI. Self-diffusion coefficients in typical oxides at 1000°C

Compound	Crystal structure	D_{cation} (cm^2 s^{-1})	D_{anion} (cm^2 s^{-1})
BeO	NaCl	10^{-12}	10^{-16}
MgO	NaCl	10^{-15}	10^{-17}
Al$_2$O$_3$	Hex.	10^{-17}	10^{-17}
Cr$_2$O$_3$	Hex.	10^{-15}	10^{-16}
ThO$_2$	CaF$_2$	10^{-17}	10^{-11}
UO$_2$	CaF$_2$	10^{-19}	10^{-9}

Another method for the measurement of sulphur diffusion is to count the decrease in the radioactivity of a fixed volume of the H_2S/H_2 mixture which is in chemical equilibrium with the sulphide, but which initially contains a known amount of radiosulphur. The decrease in activity of the gas comes about as a result of the exchange of sulphur atoms between the solid FeS sample and the gaseous phase. The radioactive atoms which are introduced into the solid in this manner then diffuse away from the surface and into the interior of the solid by the normal anionic diffusion process. The solution of Fick's second order equation for the unsteady state in this particular experiment is then

$$\frac{A_t}{A_0} = \exp Z^2 (1 - \operatorname{erf} Z)$$

where $Z = \sqrt{Dt}/lm$, A_t and A_0 are the radioactivity of the gas at times $t = t$ and $t = 0$ respectively, l is the thickness of the sulphide sample, m is the number of moles of sulphur in the gas divided by the number of moles of sulphur in the solid and $\operatorname{erf} Z$ is the error function

$$\operatorname{erf} Z = \frac{2}{\sqrt{\pi}} \int_0^Z e^{-t^2} . dt$$

This technique is also used extensively in the study of oxygen diffusion in oxides. The isotope which is used is O^{18} which is not radioactive, but the O^{16}/O^{18} ratio in the gas phase can be determined by mass spectrometry.

Values at a temperature of 1273 K are given for a number of important oxides for both the cation and anion in Table VI. It can be seen that one ion usually migrates much more rapidly than the other, and to understand this, the relationship between the diffusion coefficients and the structures and compositions of simple compounds must be understood.

The enthalpy of activation for diffusion in oxides and sulphides is related to the heat of formation and movement of the appropriate vacancies in an exactly analogous fashion to the behaviour in metals which has been considered earlier. It is rather more simple to separate the effects of formation and movement in the case of the compounds than in metallic systems. This is because the concentration of vacant sites can be altered quite readily by the addition of aliovalent ions. It was mentioned earlier that the introduction of CaO or Y_2O_3 into solid solution in ZrO_2 or ThO_2 increases the concentration of vacant anion sites. The change in the activation enthalpy for diffusion of the anions can be readily observed in these solids, and it can be shown that the enthalpy is approximately halved. This is an indication that, as in the case of metallic systems, the activation enthalpy is about one half due to ion movement, and one half due to vacancy formation.

In the solid solutions of CaO in ZrO_2 or ThO_2 the migration of the anions is very much more rapid than that of the cations and on application of an electric potential across a sample of any of these solid solutions, the electric current which passes is practically entirely due to anion migration. The electrical conductivity, is related to the diffusion coefficient D, through the Nernst–Einstein relationship

$$\frac{\sigma}{D} = \frac{nz^2e^2}{kT}$$

Here n is the number of ions cm^{-3} of charge ze approximately 10^{22} and D is expressed in $cm^2 s^{-1}$. k is Boltzmann's constant which has the value $1.38 \times 10^{-16} \, erg \, deg^{-1}$. The value of σ is usually quoted in $ohm^{-1} cm^{-1}$, but this must be converted to electrostatic units by multiplying by 300×10^9 (the velocity of light in $cm \, s^{-1}$) to convert e.m.u. units of current to e.s.u. and practical volts to e.s.u. Then the ratio σ/D can be shown to have the value of approximately 10^5 at temperatures around 1000 K for a divalent species such as O^{2-}.

It follows directly from this equation that when the electrical conductivity of a solid results from the migration of one of the ionic species, that is the transport number of one species is practically unity, then measurements of the conductivity may be used to obtain values for the self-diffusion coefficient of that ion. The transport number t_{cation} of the cation in a solid compound is defined as the ratio of the partial electrical conductivity due to the movement of the cation, divided by the total electrical conductivity. For most non-stoichiometric solids, t_{ion} is very small when compared with $t_{electrons}$ or $t_{positive}$ holes.

In the case of KCl, t_{K^+} is practically unity, and the comparison of the activation enthalpy for the electrical conductivity of pure KCl, with that for the solid containing a dilute solution of $SrCl_2$ shows a difference of a factor of two (Fig. 21).

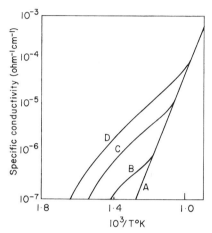

Fig. 21. Specific conductivity of KCl samples containing different levels of added $SrCl_2$.

A = pure KCl.
B = 5×10^{-5} mole fraction of $SrCl_2$.
C = 10^{-5} mole fraction of $SrCl_2$.
D = 10^{-4} mole fraction of $SrCl_2$.

(After Kelting and Witt, *Zeit für Physik*, **176**, 697 (1949).)

The addition of the aliovalent Sr^{2+} on the cation sub-lattice of KCl is accompanied by the formation of a corresponding number of vacant cation sites on this lattice. Since there is now no measurable thermal effect on the conductivity corresponding to an increase in the number of vacancies as the temperature is increased, only the activation enthalpy to move vacancies is measured in the solid solution. The experiments therefore demonstrate again the fact that the activation enthalpy to move the vacancies is about the same as that to form them.

DIFFUSION COEFFICIENTS AND POINT DEFECTS IN COMPOUNDS

The compound $FeO_{1+\delta}$ cannot be prepared with the value of $\delta = 0$ there always being an excess of oxygen in the compound. Since the oxide can be regarded as ionic, it follows that there must be some trivalent Fe^{3+} ions in the compound to retain electroneutrality. The value of δ can be changed by alteration of the oxygen pressure above a sample, and it has been found that, for a certain range of composition, with δ very small, the variation of the Fe^{3+} ion concentration with oxygen pressure can be expressed by the simple functional relationship

$$cFe^{3+} \propto p_{O_2}^{\frac{1}{6}}$$

We may formulate the addition of oxygen according to the stoichiometric equation

$$\tfrac{1}{2}O_2 + 2Fe^{2+} \rightarrow O^{--} + 2Fe^{3+} + V_{Fe^{2+}}$$

where $V_{Fe^{2+}}$ represents the vacant site on the cation sub-lattice which must be formed for each oxygen atom which is added. It is then appropriate to write the mass action equation

$$K = \frac{c_{Fe^3+}^2 \, c V_{Fe^{2+}}}{p_{O_2}^{\frac{1}{2}}}$$

since $c_{Fe^{2+}}$ and $c_{O^{2-}}$ remain essentially constant. The concentration of vacancies is half that of the ferric ions and hence on substitution we obtain.

$$\tfrac{1}{2}c_{Fe^{3+}}^3 = Kp_{O_2}^{\frac{1}{2}}$$

which reproduces the relationship given above. It is thus possible in a number of instances to calculate the effects of changing chemical potential of one component, i.e. oxygen or sulphur, on the concentration of vacancies in metallic oxides and sulphides by a simple mass action approach. The increase in the concentration of vacancies in the cation sub-lattice which occurs when extra oxygen atoms are added makes the mobility of the ferrous ion on that sublattice increase, and hence the self-diffusion coefficient also increases correspondingly. Himmel, Birchenall and Mehl (1953) showed experimentally that the diffusion coefficient of iron increases directly as a function of $p_{O_2}^{\frac{1}{6}}$ as anticipated from this treatment.

It is not necessary for a compound to depart from stoichiometry in order to contain point defects such as vacant sites on the cation sub-lattice. All compounds contain intrinsic defects even at the precisely stoichiometric ratio. The Schottky defects, in which an equal number of vacant sites are present on both cation and anion sub-lattices, may be present at a given temperature in such a large concentration that the effects of small departures from stoichiometry are masked. Thus, in $MnO_{1+\delta}$ it is thought that the concentration of defects (Mn^{3+} ions) is so large that when there are only small departures from stoichiometry, the concentrations of Mn^{3+} ions which arises from these departures is negligibly small. The dependence of the non-stoichiometry is then as $p_{\frac{1}{4}O_2}$ in this region. When the departure from non-stoichiometry is large and the concentration of defects arise from non-stoichiometry, the "extrinsic" defect is large compared with the intrinsic concentration, the substance follows the $p_{O_2}^{\frac{1}{4}}$ law.

Another example of the effects of non-stoichiometry on diffusion is to be found in $UO_{2+\delta}$. Here the defect consists of interstitial oxygen ions of high mobility, and it is found that the diffusion coefficient of oxygen increases with increasing values of δ. The oxygen dependence is more complicated in this case than in FeO and the following is an explanation of the observed effects. The mass action equation can be written thus

$$\tfrac{1}{2}O_2 + 2U^{4+} \rightarrow O_i^{2-} + 2U^5 \qquad K = \frac{c_{U^5+}^2 \, c_{Oi}^{2-}}{p_{O_2}^{\frac{1}{4}}}$$

If it is now assumed that the U^{5+} ions and the interstitial O^{2-} ions are "associated", i.e. not free to migrate in the lattice independently of one another, then we write for the equilibrium constant which involves the defect complex $U^{5+}-O^{2-}-U^{5+}$

$$K = c\frac{[U^{5+} - O_i^{2-} - U^{5+}]}{p_{O_2}^{\frac{1}{4}}}$$

and δ varies as $p_{O_2}^{\frac{1}{4}}$. This idea of associated defects has found some value in systematizing results for non-stoichiometric behaviour which do not follow the simple mass action rule.

Generally speaking, it is found that diffusion properties of the components of simple inorganic compounds are sensitive to small departures from stoichiometry, but the precise manner in which the dependence will relate to the non-stoichiometry is still a matter in need of further study. No general rules, such as the direct application of the mass action law, seem to be sufficient, and in many cases association of defects must be invoked to bring the measured effects into line with a simple chemical equation formulation of mass action.

The further understanding of the nature of non-stoichiometric compounds is more likely to be found by direct crystallographic studies than by formulations of mass action equations which are made to fit the experimental results for a given system. For example, the solid $FeO_{1+\delta}$ and $UO_{2+\delta}$ both present a very complex behaviour when δ takes on large values and the "explanation" of this behaviour in terms of the mass action law becomes less plausible the greater the value of δ. X-ray and neutron diffraction studies on FeO and UO_2 have presented structural information which inclines towards the idea of "micro-domains" in non-stoichiometric oxides. The defects in FeO have an associated structure of vacant cation sites and interstitial Fe^{3+} ions which are related to the structure of Fe_3O_4. The interstitial oxygen ions which occur in pairs in UO_2 present a similarity to the structure of U_4O_9. In both cases it appears that small regions within the compounds having a structural resemblance to the neighbouring oxide phases are to be found in much the same way as the Guinier–Preston zones occur in copper-aluminium alloys. There is therefore only one phase present, and there can be no incoherent interface between the micro-domain and the parent crystal.

When some oxides are reduced to yield non-stoichiometric phases, another phenomenon can occur. It appears that the lower valency compound structure instead of occurring in micro-domains within the parent oxide, forms a planar configuration of the appropriate lower oxide crystal structure throughout the crystal. These planes appear crystallographically as shear planes within the parent structure. As an example, the shear plane structure within reduced samples of rutile (TiO_2) has a structure with the cation coordination number which occurs in Ti_2O_3, that is to say 12 rather than six as in the rutile structure.

SHEAR STRUCTURE OXIDE SYSTEMS

The trioxides of molybdenum and tungsten, MoO_3 and WO_3 may be reduced to yield a number of intermediate compounds before the dioxides MoO_2 and WO_2 are formed. In the whole series of oxides, the basic unit structure of the systems consists of a metal ion coordinated octahedrally by six oxygen ions. When each octahedron touches its neighbours only at the corners, then each oxygen ion is shared by two metal ions, and the compound formula is MO_3. If oxygen ions are now removed by a chemical reduction process then isolated point defects are formed where the ions are removed. The situation in which these point defects are randomly distributed throughout the structure corresponds to the non-stoichiometric oxide case which we have previously described $MO_{3-\delta}$.

It has been found, however, that such a random distribution only occurs for a very small value of δ and that with increasing departure from stoichiometry a new phenomenon arises which can give rise to a large number of stable intermediate oxides between MO_3 and MO_2. If a number of vacancies can be imagined to line up on one row of oxygen ions, then the vacant sites can be annihilated by the translation of the neighbouring rows of octahedra which incorporate the defective row. Such a translation, which involves the movement of groups of metal atoms together, is called a crystallographic shear, and the shear plane is a new structural unit in the compound (Fig. 22). Since the vacant oxygen ion sites are removed as a result of the shear, a new stoichiometric compound can be formed, and the stoichiometry is determined by the spacing between the shear planes.

A number of compounds in the series W_xO_{3x-1} and W_xO_{3x-2} were

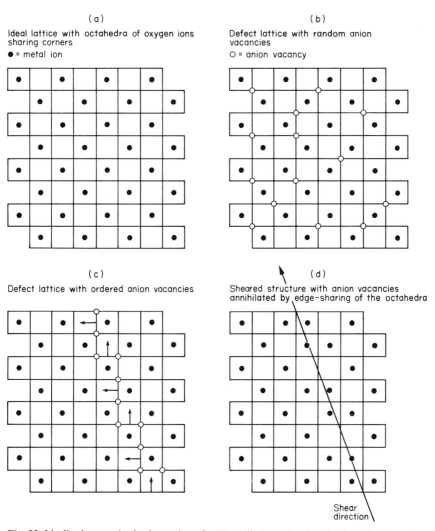

Fig. 22. Idealized stages in the formation of a Magnéli shear structure in the reduction of an oxide having oxygen octahedra which initially share only corners.

found by Magnéli (1953) in the original discovery of this phenomenon and many more examples have been found. Molybdenum also forms the Magnéli shear plane compounds and other examples are found in the groups of compounds which are formed by the reduction of TiO_2 to give Ti_nO_{2n-1} and VO_2 to give V_nO_{2n-1} series of oxides. The transport properties of ions in the shear plane structures have not been sufficiently elucidated to date for a worthwhile comment to be made.

ELECTRON TRANSPORT IN OXIDES

It will be noted that the departures from non-stoichiometry, result in a mixture of cationic valencies, e.g. Fe^{2+} and Fe^{3+} in FeO, U^{4+} and U^{5+} in UO_2, and so on. The introduction of higher valency cations in this manner reduces the number of electrons which are associated with the cation sub-lattice. There are now "positive holes" in what is in principle a full valency electron band in the stoichiometric compound. By their nature, the mobilities of positive holes in oxides are very much higher than those of the ions, about 10^3 times the mobility of the fastest moving ions, and therefore the electrical properties of non-stoichiometric oxides are influenced very markedly by the number of positive holes which they contain. From the mass action equations which have been previously formulated, it will be easily concluded that the magnitude of the positive hole contribution will increase with increasing oxygen pressure when the departure from stoichiometry is accompanied by the formation of higher valency cations. Conversely in some oxides, such as TiO_2 in which the departures from stoichiometry involve the reduction of Ti^{4+} ions to Ti^{3+}, the conductivity, now due to almost free electrons, increases with decreasing oxygen pressure. The two types of behaviour are characterized as "p-type" semiconduction when positive holes are introduced by cation oxidation and "n-type" semi-conduction when cation reduction occurs.

Since ionic mobilities never reach those of these electronic defects, the relative contributions of ions and electrons to the electrical conductivity of an oxide are very sensitive to the number of the latter. The total electrical conductivity can be written as the simple sum of the contributions from each species thus

$$\sigma_{\text{total}} = \sum_i \cdot \sigma_i = \sum_i n_i z_i e B_i$$

where the summation is for all ions and electron defects. It follows from this expression that even when one of the ionic species has a high mobility, the electronic defect will play the dominant role even when the concentration ratio of ions to electrons is about $10^3:1$ because of the greater mobility of the electron defects.

In solids such as ZrO_2 or ThO_2 containing aliovalent ions and vacancies on the anion sub-lattice (referred to earlier as solid oxide electrolytes) non-stoichiometry is not readily introduced by the reduction or oxidation of the cationic species. The absence of mobile electron defects in these solids there-fore allows electrolytic and not semiconducting behaviour. At oxygen pressures below 10^{-16} atm and temperatures around 1000°C, the calcia–zirconia electrolytes begin to show n-type semiconduction due to reduction of the Zr^{4+} ions.

Finally, a number of poorly conducting solids such as BeO and MgO show semi-conducting properties as a result of the thermal excitation of electrons from the full valence band to the empty conduction band. Because the mobilities of electrons in the conduction band are higher than those of the positive holes in the valence band, the solid behaves like an n-type solid, although the numbers of electrons and positive holes which arise from this source are equal.

TRANSPORT PROPERTIES OF GASES

The flow of atoms or molecules in gases which results from composition gradients or which can be detected through the movement of radioactive species in self-diffusion, results from a fundamentally different mechanism from that in condensed systems. The random motion of the molecules is characterized for a given species at a given pressure and temperature by the average distance which the molecules move between collisions with one another. The mean free path, λ, is defined, according to the simple kinetic theory, in terms of the properties of the molecules thus

$$\lambda = \frac{1}{2^{\frac{1}{2}}\pi n d^2}$$

where d is the molecular diameter and n is expressed in molecules cm^{-3}. The mean velocity of the molecules is given by the equation

$$\bar{c} = \left(\frac{8RT}{\pi M}\right)^{\frac{1}{2}}$$

where M is the molecular weight, and when this is combined suitably with the mean free path, the transport properties can be adequately described.

The *viscosity* comes about by momentum transfer across the volume of the gas when there is relative bulk motion between successive layers of gas, and the coefficient, v, is given by the equation

$$v = \tfrac{1}{3}\rho\bar{c}\lambda = \frac{m\bar{c}}{3\sqrt{2}\pi d^2} = 1.81 \times 10^{-5}\frac{(MT)^{1/2}}{d^2} \text{ poise}$$

where ρ is the density of the gas, and d is in Angstrom units (10^{-8} cm).

The viscosity increases approximately as $T^{\frac{1}{2}}$, and there is, of course, no vestige of the activation energy which characterizes the transport properties of condensed phases.

The *thermal conductivity* is obtained in terms of λ and $\bar{c}$ through the equation

$$k = \tfrac{1}{3}\rho\,c_v\,.\,\bar{c}\lambda = v\,.\,c_v.$$

where c_v is the specific heat at constant volume. The *self diffusion coefficient* is expressed as a function of λ and $\bar{c}$ through the equation

$$D = \tfrac{1}{3}\lambda . \bar{c}$$

All of the transport properties can be derived from the simple kinetic theory of gases, and their inter-relationship through λ and $\bar{c}$ leads one to expect that they are all characterized by a relatively small temperature coefficient. The simple theory suggests that this should be a dependence on $T^{\frac{1}{2}}$, but because of intermolecular forces, the experimental results usually indicates a larger temperature dependence even up to $T^{\frac{3}{2}}$ for the case of molecular inter-diffusion. As has been already pointed out, the Arrhenius equation which involves an enthalpy of activation is not appropriate to gases, because no "activated state" is involved in the transport processes. If, however, the temperature dependence of these processes is fitted to such an expression as an algebraic approximation, then an "activation enthalpy" of a few kilocalories is observed. It will thus be found that when the kinetics of a gas–solid reaction depends upon the transport properties of the gas phase, the apparent activation enthalpy will be a few kilocalories only.

Some typical results for common gases which are of importance in the context of sulphide oxidation are given in the following table. The special

TABLE VII. Thermophysical properties of common gases

Gas	Temp. °C	Viscosity (micropoise)	Sp. heat (cal g^{-1} deg C^{-1})	Thermal conductivity (cal cm^{-1} s^{-1} deg C^{-1} $\times 10^5$)
H$_2$	0	84	3·46	39·8
	800	210	3·52	107·0
Ar	0	209	0·124	3·80
	800	550	0·124	10·50
N$_2$	0	170	0·250	5·65
	800	415	0·272	15·50
O$_2$	0	189	0·225	5·76
	800	500	0·255	18·0
CO	0	166	0·250	5·49
	800	450	0·273	16·1
CO$_2$	0	139	0·226	3·43
	800	436	0·245	17·2
SO$_2$*	0	116	0·162	2·04
	500	310	0·188	8·11
H$_2$O*	100	125	0·460	5·71
	400	241	0·484	13·2

* Note the smaller range of temperature for SO$_2$ and H$_2$O. This was due to lack of high temperature viscosity data.

position of hydrogen which results from the small mass and size of the H_2 molecule should be particularly noted.

RECOMMENDED READING

Shewmon, P. G. (1963). "Diffusion in Solids". McGraw Hill, New York. Theory and experiment for diffusion in metallic systems.

Jost, W. (1960). "Diffusion", Academic Press, London and New York. Transport phenomena in solids and gases.

Greenwood, N. N. (1968). "Ionic Crystals, Lattice Defects and Non-stoichiometry". Butterworth.

Moore, W. J. (1967). "Seven Solid States". Benjamin Inc. A general account of solid state chemistry.

REFERENCES

Magnéli, A. (1953). *Acta Cryst.* **6**, 495. Shear structures in oxide systems.

5

REACTION MECHANISMS OF SOLID STATE PROCESSES

A complete knowledge of the thermodynamic and the transport properties of both the condensed phases and of the gas should make it possible to determine *a priori* the kinetics of a diffusion-controlled local-equilibrium reaction. The combination of these properties in flux calculations is, however, only a part of the overall picture of solid–gas reactions. In most circumstances, the detailed manner in which a reaction is allowed to proceed is a matter of choice for the operator. The way in which the gas phase moves through a reactor, the degree of fineness to which the solid is divided and its initial manner of presentation to the gaseous phase are variables which can be manipulated to suit a given set of circumstances. The extent to which the solid agglomerates during reaction, thus reducing the interfacial contact between the phases, or the extent to which particles are broken up by thermal or mechanical shock, thus increasing the interfacial area, are extremely difficult to predict and frequently, to control in a real industrial situation.

The scientific contribution to the analysis of metal-making reactions is confined, in the solid state, largely to the elucidation of model systems in terms of measurable parameters. The results of these studies can then indicate the more profitable directions in which the process engineer can turn his attention for optimization. This elucidation of model systems serves to show the ways in which reactions can occur and where departures from the elementary diffusion-controlled iocal-equilibrium pattern might be expected. This assessment of the interplay of the numerous contributions to the reaction mechanism often has a highly specific result which may not necessarily apply to many other systems. It is the objective of the laboratory worker to study enough specific systems so that a discernible pattern can be extracted which will serve as a guide to the engineer. In this section we shall bring together most of the aspects of isothermal reaction mechanisms at high temperatures which can be applied in solid–gas systems. It will be clearly understood that the results which are described should be used as a guide in analysing industrial operations, diverse as these are, and not as a quantitative description of tonnage operations.

THE SINTERING OF METALLIC POWDERS—SURFACE ENERGY AS A DRIVING FORCE

During the reduction of metallic oxides, the sintering of both the oxides and the metallic products plays an important role in determining the kinetics of the process. Sintering also plays an important part in the consolidation of the high-melting refractory metals, and is used as a method of fabrication.

TABLE VIII. Some typical values of surface free energy

Substance	Surface energy (erg cm^{-2})
Fe	1900
Cu	1600
Sn	1200
Ag	1100
Zn	105
MgO	1100
Al_2O_3	900

(Note: $1 \text{ erg cm}^{-2} \equiv 2\cdot39 \times 10^{-4} \text{ cal m}^{-2}$)

The process occurs at high temperatures and comes about because the total free energy content of a solid is reduced when the surface energy contribution is minimized. This follows simply from the fact that, when the surface area of a solid body is increased by, for example, fracture into a number of smaller pieces, the bonds originally joining the atoms which form the new surface layers, are broken. Clearly, work must be done on the system in order to create these new surfaces. This work is equal to the surface free energy which is created. We will not enquire into the methods for the measurement of surface free energies of solids, but the following table of results demonstrates the magnitudes of this quantity for some metals and oxides at room temperature (Table VIII).

The surface free energy has a heat change and an entropy change component, but there are very few measurements which show the respective

D

values of these for a given system. Suffice it to say that the entropy term is small (about 0.5 erg cm^{-2} deg^{-1}) and the use of a temperature-independent value for the surface free energy is a fair approximation.

The driving force for the sintering process may be shown to be extremely small, when compared with that of chemical reactions by the following calculation. If 1 g of copper metal spheres of the uniform particle diameter 100 microns (10^{-2} cm) sinter to form a solid, approximately half of the initial surface area remains at grain boundaries and free surface. The surface free energy change in this process is then approximately

$$\Delta G = \frac{n\pi d^2}{2} \cdot \frac{1600}{4.19 \times 10^7} \text{ cal g}^{-1}$$

where n is the number of particles. Since the density of copper is 8.9 g cm^{-3}, the value of n is 2.15×10^5 particles, and this yields the value

$$\Delta G = 1.28 \times 10^{-3} \text{ cal g}^{-1}$$

The quantitative study of the sintering of pure metals was given considerable impetus by the classical model studies of Kuczynski (1949). He measured the initial rate of growth of the neck which formed between two metal samples of well-defined geometry, as a result of the sintering process. If a piece of silver wire is wrapped around a cylinder of silver, neck growth occurs at the points of contact between the wire and the cylinder. The kinetics of the initial growth can be analysed, after Kuczynski, in terms of the diffusion of atoms from the volume of the wire and cylinder to the neck by volume diffusion. The neck has a circular section after a short period of growth, and the process of sintering reduces the radius of curvature of the outer surface of the neck (Fig. 23). The vapour pressure of a metal above a curved surface is given by the Gibbs–Thomson equation

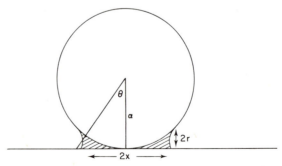

Fig. 23. An idealized representation of a model system for the study of sintering in metals. A wire of radius a forms a neck of sintered material with a flat surface of the same metal over a length $2x$ and having an external radius of curvature $2r$.

$$\ln \frac{p}{p^0} = \frac{2\gamma V}{rRT}$$

where γ is the surface tension, V is the molar volume, r is the radius of curvature and p^0 is the vapour pressure of the flat face. This equation may be proved in the following way:

If a mass of dm g is transferred from a flat surface of vapour pressure p^0 to a sphere of radius r and vapour pressure p of a given substance then the free energy of transfer ΔG_I is

$$RT \ln \frac{p}{p^0} \cdot \frac{dm}{M},$$

where M is the molecular weight of the substance.

The increase in surface area of the sphere dA on the addition of dm is given by

$$dA = \frac{d}{dr} 4\pi r^2 = 8\pi r \, dr,$$

also, since the weight of the sphere m is related to the volume by the equation

$$m = \tfrac{4}{3}\pi r^3 \rho \qquad (\rho \text{ is the density})$$

it follows that

$$dm = 4\pi r^2 \rho \, dr$$

and hence

$$dA = \frac{2dm}{r\rho}.$$

The free energy change when the surface is expanded by dA is

$$\Delta G_{II} = \gamma dA = \frac{2\gamma dm}{r\rho}.$$

But

$$\Delta G_I = \Delta G_{II}.$$

Hence

$$RT \ln \frac{p}{p_0} = \frac{2\gamma M}{r\rho}.$$

It follows that the ratio of the vapour pressure above two surfaces of radius of curvature r and a respectively is given by the equation

$$\ln \frac{p_r}{p_a} = \frac{2\gamma V}{RT} \left\{ \frac{1}{r} - \frac{1}{a} \right\},$$

but if the surface of radius r is *concave* and that of radius a is *convex*, then the equation above becomes

$$\ln \frac{p_r}{p_a} = \frac{2\gamma V}{RT} \left\{ \frac{1}{r} + \frac{1}{a} \right\}.$$

In Fig. 23 it can be seen that r is the radius of the concave surface of the neck which has grown by sintering between the wire of radius a and the flat surface. If θ is the angle between a radius to the periphery of the neck and a radius which is perpendicular to the flat surface

$$r = a(1 - \cos \theta) = 2a \sin^2 \frac{\theta}{2}.$$

If θ is small

$$2a \sin^2 \frac{\theta}{2} = 2a \frac{x^2}{4a^2} = \frac{x^2}{2a}.$$

Hence

$$r = \frac{x^2}{2a}.$$

The surface area of the neck which is exposed to the atmosphere is given by the equation

$$A = 2\pi x . 2 \left(\frac{x^2}{2a} \right)$$

and the volume of the neck by

$$V = \frac{\pi x^4}{2a}$$

in good approximation. If the atomic volume of the material being sintered is δ^3, then the flux of atoms, J, from the wire into the surface of the neck is given by

$$J = \frac{1}{\delta^3} \cdot \frac{dV}{dt} \cdot \frac{1}{A}. \qquad \text{particles cm}^{-2}.$$

This can then be applied, using the relationships given above and Fick's first law of diffusion to calculate the rate of neck growth due to the flux of vacancies away from the neck and into the bulk, according to Kuczynski, as follows:

The vacancy concentration at the concave neck is higher than that at the centre of the neck according to the Gibbs–Thomson equation thus

$$\ln \frac{c_r}{c_x} = \frac{2\gamma V}{RT} \left\{ \frac{1}{r} - \frac{1}{x} \right\}$$

If $c_r = c_x + \Delta c$ then

$$\ln \frac{c_r}{c_x} \simeq \frac{\Delta c}{c_x}$$

Since $r \ll x$, and the concentration gradient for vacancies is approximately given by $\Delta c/r$, then

$$J = \frac{1}{\delta^3} \cdot \frac{dV}{dt} \cdot \frac{1}{A} = D' \frac{\Delta c}{r} = \frac{1}{\delta^3} \cdot \frac{dV}{dx} \cdot \frac{dx}{dt} \cdot \frac{1}{A}$$

$$= \frac{1}{\delta^3} \cdot \frac{2\pi x^3}{a} \cdot \frac{a}{2\pi x^3} \cdot \frac{dx}{dt} = D'\Delta c \cdot \frac{2a}{x^2} = \frac{1}{\delta^3} \cdot \frac{dx}{dt}$$

where D' is the diffusion coefficient of vacancies. This latter is related to the volume self-diffusion coefficient of the material by the equation

$$D'c = D_v.$$

Hence the flux equation can be written, after substitution for Δc from the Gibbs–Thomson equation

$$D_v \frac{2\gamma N \delta^3}{RT} \left(\frac{2a}{x^2}\right)^2 = \frac{dx}{dt},$$

and separating the variables,

$$\frac{D_v \gamma \delta^3}{kT} 8a^2 \, dt = x^4 \, dx.$$

Hence

$$\frac{x^5}{a^2} = 40 \frac{D_v \gamma \delta^3}{kT} t.$$

It can also be shown that the transfer of matter from the wire to the neck by vaporization and condensation would follow a different rate equation. The equation for the difference in vapour pressure between a concave surface of radius r and a convex surface of radius a which was given above can be put in the form

$$\ln \frac{p + \Delta p}{p} = \frac{2\gamma V}{RT} \cdot \frac{1}{r} \qquad (r \ll a),$$

where p is the vapour pressure of the convex surface. Now, since Δp is small compared with p

$$\ln \frac{p + \Delta p}{p} \cong \frac{\Delta p}{p}.$$

This difference in pressures, Δp, is responsible for mass transfer from the convex to the concave surface; hence, the rate of mass transfer, G, is given by

$$G = k\Delta p = \frac{k'}{r} \qquad \text{at constant temperature (k and k' are proportionality constants)}$$

$$= k'' \frac{dV}{dt} \qquad \text{(V is again the volume of the neck)}$$

From the earlier geometric considerations $r = x^2/2a$ and on substitution and integration we find

$$\frac{x^3}{a} = \left(\frac{9\pi}{2MRT}\right)^{\frac{1}{2}} \cdot \frac{V^2\gamma P}{RT} \cdot t$$

where P is the vapour pressure of the substance.

In most instances of the sintering of metals, the vaporization–condensation mechanism is unimportant, and the volume diffusion process is the predominant one. The temperature coefficient of sintering by this mechanism will mainly be determined by the energy of activation for the diffusion process. Typical values for metals lie in the range 30–60 kcal and so it is to be expected that increasing the temperature will markedly raise the possibilities of metal sintering.

THE SINTERING OF METAL OXIDES—POWDER COMPACTS

The volume diffusion control which predominates in the kinetics of metal sintering, also plays a very important role in the sintering of oxide powders. The main feature which is new to this area of sintering is that, of course, two species, both cation and the anion, must move in order to transport molecules of the oxide. It would seem at first sight to be obvious that the diffusion of the slower moving species would determine the rate of transport of material.

The experimental study of oxide sintering cannot be carried out with the particular arrangement of wire wrapped around a cylinder which was used by Kuczynski (1949) for metallic systems. Instead, a number of studies have been made of the shrinkage of right cylinders made by pressing the powder of the material. The mathematics of the shrinkage of cylinders is clearly much more complicated than that in the Kuczynski technique, but Johnson and Cutler (1963) established a shrinkage equation for sintering via volume diffusion using the model system of an assembly of spheres of uniform diameter. They concluded that for this situation the equation

$$\frac{\Delta l}{l_0} = \frac{31}{\pi^2} \left(\frac{\gamma VD}{RTa^3}\right)^{0.46} t^{0.46}$$

describes the results satisfactorily. The terms in this equation are shown with the same symbols as in the previous results of Kuczynski, and $\Delta l/l_0$ is the fractional shrinkage along the axis of the cylinder. The particle diameter, a, which is used in the equation must be the average particle size, since it is always the case that real powder samples have a particle size distribution which depends on the method of preparation. The Johnson–Cutler (1963) equation, since it involves the diffusion coefficient, D, can also be used over a range of temperature and it follows that the most significant term in the temperature variation will involve the activation energy for diffusion.

The sintering of UO_2 is a process of technical importance, since it is applied in the manufacture of oxide fuel rods for nuclear reactors. Uranium dioxide, as has already been discussed, can accommodate a considerable excess of oxygen above the stoichiometric value, the oxygen ions occupying interstitial sites in the fluorite lattice. The self diffusion coefficients of both the cation and anion have been obtained from a number of studies, and although detailed quantitative agreement has not so far been achieved, the broad pattern of behaviour is now clear. At all values of stoichiometry, the diffusion coefficient of the oxide ion is very much larger, approximately 10^5 times, than that for the uranium ion. Both diffusion coefficients increase markedly when the oxygen to uranium ratio is more than two, the greatest change occurring between $UO_{2.00}$ and $UO_{2.10}$. The activation energy for sintering should be the same as that for the self-diffusion coefficient of uranium which is about $85–100\,\text{kcal mol}^{-1}$ (Fig. 24).

These expectations which are based upon the Kuczynski and Johnson and Cutler models for sintering, are largely borne out in practice, although, once again, considerable experimental difficulties cloud the picture of the precise relationship between the sintering and the diffusion studies. Lay and Carter (1969) report that the activation energy for sintering is to be found in the band of values 107 ± 11 kcal and that the diffusion coefficients which can be calculated from these results are in reasonably good agreement with the directly measured data from diffusion studies for the nearly stoichiometric oxide. For the oxygen-rich phase, the derived diffusion coefficient seems to be considerably larger than that from direct experiment, and there is a possibility that another mechanism than volume diffusion comes into play. It might be pointed out here that the volatility of the solid is enhanced by departure from non-stoichiometry by the formation of the vapour species UO_3. Therefore, it is possible that vaporization–condensation plays a part in the sintering of oxygen-rich UO_2.

Other studies which have been made of sintering in oxides support the general view that volume diffusion is the rate controlling step, but the precise agreement between the self-diffusion coefficients and activation energies for the slower-moving species with the corresponding values which are obtained from the sintering kinetics has not been satisfactorily achieved.

Zinc oxide, which vaporizes by dissociation relatively readily is one clear exception to this general statement. The kinetics of sintering of this substance clearly shows the control by evaporation and condensation above 1050°C.

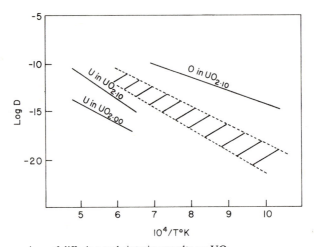

Fig. 24. A comparison of diffusion and sintering results on $UO_{2+\delta}$.

——————— = Direct results from diffusion studies.

///// = Range of results for diffusion coefficient calculated from sintering studies on powder specimens.

Below this temperature, and down to 600°C, volume diffusion controlled the process. In this experiment spheres of ZnO were pressed out of powder and used in the method of Kuczynski. The dependence of the neck diameter on x^3 as a function of time in the upper temperature range proved the evaporation–condensation model.

THE OXIDATION OF $FeS_{1+\delta}$—THE PARABOLIC TO LINEAR RATE LAW TRANSITION

The results for the self-diffusion of iron in $FeS_{1+\delta}$ show that this coefficient is orders of magnitude greater than that of sulphur and, at a given temperature, does not alter by as much as a factor of ten across the whole composition range. This is probably an example of a large intrinsic defect concentration masking the effects of compositional change.

It is thus to be expected that the oxidation of ferrous sulphide will proceed by the migration of iron ions and electrons out of the sulphide phase and into the oxide phase, leaving the sulphur-rich sulphide.

Niwa *et al.* (1957) showed that this is in fact the case during the early stages of oxidation at temperatures between 500 and 600°C, the oxide which is formed being Fe_3O_4. However, as has already been shown, this change in the sulphide composition raises the sulphur pressure at the sulphide-oxide

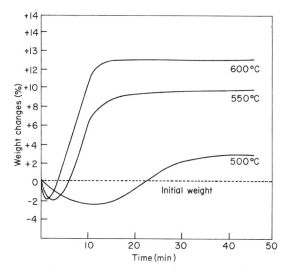

Fig. 25. Weight change of FeS samples as a function of time during roasting experiments. (After Niwa *et al.*)

interface until a partial pressure of SO_2 greater than one atmosphere can be generated. The oxide skin then ruptures, and the weight gain as a function of time changes from the parabolic relationship of a solid-state diffusion-controlled process to the linear gas-transport controlled one (Fig. 25).

THE PARABOLIC LAW OF DIFFUSION-CONTROLLED REACTIONS

In the experimental study of reactions between solids and gases, there are two important mechanisms that merit consideration here. The first of these occurs when the gas–solid interface is well defined and the area which is available for the reaction is constant during the whole course of the process.

If the product grows on a fixed interfacial area, and the reaction advances by the diffusion of ions or of ions and electrons through the product layer, then the rate of increase of the thickness, x, of this layer is given by the so-called parabolic law

$$\frac{dx}{dt} = \frac{k}{x} \quad \text{hence} \quad x^2 = 2kt$$

In a reaction which is diffusion controlled and which obeys the parabolic rate law, the mechanism which is proposed must not involve the net transport of electricity. Therefore, the migration of cations in one direction through the reaction product must be accompanied, equivalent for equivalent, by the migration of the neutralizing electrons for these ions. In the oxidation of a metal, the mechanism can be shown schematically thus

Metal	Oxide	gas
	$\rightarrow$	
	$M^{z+} + ze$	p_{O_2}

Interface reaction $M^{z+} + z/4\,O_2 + ze = M^{z+} + z/2\,O^{2-} = MO_{z/2}$

Alternatively, the migration of O^{2-} ions through the product layer must be balanced by the counter-migration of electrons.

As an example of the parabolic law for the diffusion-controlled reaction, we will consider the oxidation of nickel to NiO. This is the only oxide of nickel which can be formed at high temperatures and at oxygen pressures less than or equal to one atmosphere. The defect structure of the oxide is of vacant sites on the cation sub-lattice which are accompanied by the formation of Ni^{3+} positive holes.

The growth of the oxide layer when the metal is being oxidized is sustained by the transport of Ni^{2+} ions and electrons across the oxide layer from the metal–oxide interface to the oxide–gas interface. Expressing the fluxes of ions and electrons in equivalents $cm^{-2}\,s^{-1}$ the Einstein equation takes the forms

$$J_{\text{ions}} = \frac{-\sigma_{\text{ion}}}{z_{\text{ion}}^2 F^2} \frac{\partial \Delta\mu_{\text{ion}}}{\partial x} - \frac{\sigma_{\text{ion}}}{F} \frac{\partial V}{\partial x}$$

and

$$J_{\text{electron}} = \frac{-\sigma_{\text{electron}}}{F^2} \frac{\partial \Delta\mu_e}{\partial x} + \frac{\sigma_{\text{electron}}}{F} \frac{\partial V}{\partial x},$$

where $\partial\Delta\mu_i/\partial x$ and $\partial\Delta\mu_e/\partial x$ are the chemical potential gradients across the oxide layer for Ni^{2+} and electrons, and σ_{ion}, $\sigma_{\text{electrons}}$ are the partial electrical conductivities of NiO arising from ion and electron migration respectively.

$\partial V / \partial x$ is the electrical potential gradient which occurs across the oxide during the growth process.

From the requirement for electroneutrality, it follows that

$$J_{\text{ions}} = J_{\text{electrons}} \text{ in equivalents}$$

Hence

$$J_{\text{ions}} = \left(\frac{1}{\sigma_{\text{ions}}} + \frac{1}{\sigma_{\text{electrons}}} \right) \cdot \frac{1}{z_i F^2} \left(\frac{\partial \Delta \mu_i}{\partial x} + z_i \frac{\partial \Delta \mu_e}{\partial x} \right)$$

The chemical potential of the neutral species $\Delta \mu_{\text{Ni}}$ is related to the chemical potentials of the ions and electrons by the defining equation

$$\Delta \mu_{\text{Ni}} = \Delta \mu_{\text{Ni}^{2+}} + 2 \Delta \mu_e \quad \text{and} \quad \frac{\partial \Delta \mu_{\text{Ni}^{2+}}}{\partial x} + z_i \frac{\partial \Delta \mu_e}{\partial x} = \frac{\partial \Delta \mu_{\text{Ni}}}{\partial x}$$

Hence

$$J_{\text{Ni}^{2+}} = - \frac{1}{2F^2} \cdot \frac{\sigma_i \sigma_e}{\sigma_i + \sigma_e} \cdot \frac{\partial \Delta \mu_{\text{Ni}}}{\partial x}$$

Because of the greater mobility in nickel oxide, the partial conductivity σ_e, due to the counter-migration of electrons with positive holes, is very much larger than σ_i which results from Ni^{2+} diffusion. Hence, in this equation

$$\frac{\sigma_i \sigma_e}{\sigma_i + \sigma_e} \cong \sigma_i$$

The differential equation for the ion flux, and hence the rate of the growth of the oxide may now be separated and integrated with respect to x and $\Delta \mu$.

$$J_{\text{Ni}^{2+}} \int_0^x dx = - \frac{\sigma_i}{2F^2} \int_{\Delta \mu_{\text{Ni}} \text{ (metal–oxide interface)}}^{\Delta \mu_{\text{Ni}} \text{ (oxide–gas interface)}} d\Delta \mu_{\text{Ni}}$$

Since the flux of ions fixes the rate of growth of the oxide, then the equation above can be restated in the simple parabolic form

$$x \frac{dx}{dt} = k$$

where k is now equal to the right-hand side of the equation above which involves the integral of $\sigma_i \, d\Delta \mu_{\text{Ni}}$ across the layer of oxide.

It follows from this analysis of the parabolic reaction mechanism that, when the oxide layer is a semiconductor, the migration of ions through the product layer is rate-determining. The temperature coefficient of the reaction rate will therefore be the same as that for the diffusion coefficient of the rate-determining species. In the specific example which was discussed above, the activation enthalpy for the diffusion of nickel in NiO will therefore be equal to the activation enthalpy for the oxidation of nickel metal.

THE LINEAR RATE-LAW

If the area remains relatively constant but the product layer cracks during growth and so does not impede the access of the gaseous reactants and products to the reacting solid, then the linear rate law is observed

$$\frac{dx}{dt} = k' \qquad x = k't.$$

The circumstances under which the linear law will apply to a given reaction have not been very clearly elucidated. Any mechanical or thermal shock to the product layer which can cause the rupture of the layer will lead to a departure from the parabolic law, but unless the cracking occurs continuously, the linear law need not be observed. Other possibilities such as the sealing of cracks during propagation may lead to other time dependencies such as the logarithmic law. This behaviour is observed when the rate-controlling step in a reaction is transport via the gas phase across sealed-off pores in the product layer.

The two extremes of chemical reaction are represented by the parabolic law in which a constant chemical potential gradient for the reacting species is maintained across the growing product layer, and the linear law in which the chemical potential gradient occurs suddenly at an interface of constant surface area. This situation is most closely approximated when the product layer is porous to about the same extent during growth and the gaseous reactants and products can have unrestricted or constantly restricted movement across the width of the product layer. Transport of gaseous species will therefore determine the rate of the process.

OXIDATION OF COMPLEX SULPHIDES—COMPETITIVE OXIDATION OF CATIONS

Most sulphide minerals contain more than one metal, e.g. chalcopyrite has the formula $CuFeS_2$ and pentlandite $(Fe,Ni)_9S_8$. Thornhill and Pidgeon (1957) have shown semi-quantitatively how such compounds behave during oxidation roasting by means of a metallographic study of the roasted powder specimens.

Although there exists no direct experimental evidence at present, it is probable that the diffusion coefficients of both metallic species are about the same, and both are very much larger than that of sulphur. It should then follow that the metal which undergoes the greater reduction in chemical potential by oxidation, i.e. forms the more stable oxide, will be preferentially removed. Thus, FeO is considerably more stable than Cu_2O and so iron should be preferentially oxidized from chalcopyrite. The resulting copper sulphide after a period of oxidation of $CuFeS_2$ was shown by the authors to give the X-ray pattern of digenite Cu_9S_5. The acid soluble oxide layer which had been formed on the surface was iron oxide.

TABLE IX. Roasting of 30–40 mesh $CuFeS_2$ at 550°C

Time (minutes)	Sulphide Analysis		Phases Present
	Wt.% Cu	Wt.% Fe	
Zero	35·3	30·6	Chalcopyrite
20	60·6	8·3	Mauve digenite
35	68·0	zero	Blue digenite (Covellite)

The difference in stability between FeO and NiO is not so marked as that between iron and copper oxides, and so the preferential oxidation of iron is not so marked in pentlandite. Furthermore, the nickel and iron monoxides form a continuous series of solid solutions, and so a small amount of nickel is always removed into the oxide phase.

TABLE X. Roasting of 65–80 mesh $(Fe, Ni)_9S_8$ at 600°C

Time (min)	Sulphide Analysis		Phases present
	Wt.% Ni	Wt.% Fe	
Zero	35·1	32·3	Pentlandite
65	42·2	22·0	Pyrrhotite type (c/a ratio = 1·603)

The kinetics of the processes of oxidation of these complex sulphides have not been established quantitatively, but the rate of advance of the oxides into sulphide particles of irregular shapes were always linear. This suggests that the oxide films were ruptured during growth thus permitting the gas phase to have relatively unimpeded access to the sulphide–oxide interface in all cases.

THE KINETICS OF SULPHATION ROASTING—DIMINISHING
CHEMICAL POTENTIAL BUT ENHANCED DIFFUSION

A few studies have been made of the rates at which sulphates can be formed on oxides under controlled temperatures and gas composition. The mechanism changes considerably from one substance to another, and there is a wide variability in the rates.

It can be seen by reference back to the data for the thermodynamics of sulphates that the dissociation pressures of a number of the sulphates of the common metals, iron, copper, nickel, etc., reach one atmosphere at quite low temperatures, less than 1000°C. At around 600°C, most of these sulphates have very low dissociation pressures. Thus, $NiSO_4$ has a dissociation pressure of 10^{-5} atm at this temperature. It follows that when a study of the kinetics of sulphation of these oxides is carried out over this temperature range 600–1000°C, the SO_3 pressure exerted at the oxide–sulphate interface will change by five orders of magnitude if local equilibrium prevails. At the same time, the diffusion processes through the sulphate product layer will increase with increasing temperature over this same interval, following a normal Arrhenius relationship between diffusion coefficients and the temperature. Under the right circumstances, it could, and in some instances does, happen that the overall rate of the process would be seen to pass through a maximum somewhere in the temperature interval 600–1000°C because the rate is dependent on the flux of particles across the product, and hence on the chemical potential gradient multiplied by the diffusion coefficient.

The sulphation of cobalt oxide, CoO, which follows the parabolic law up to 700°C and above 850°C, proceeds by outward diffusion of cobalt and oxygen ions through a sulphate-layer which is coherent up to about 700°C, breaks off above a limiting thickness at intermediate temperatures and is coherent again above 850°C. Alcock and Hocking (1966). The rate of sulphate formation passes through a maximum in the intermediate temperature zone probably because the diffusion coefficients are low at low temperatures whilst the chemical potential gradients across the sulphate are high, whereas the converse applies at high temperatures. It is observed that the rate law is the linear law in the intermediate, high velocity, region, and the sulphate layer is seen to be cracked. In the upper and lower temperature regions, where the reaction is parabolic, the sulphate layer is smooth and uncracked.

The sulphation of cuprous oxide in the same temperature interval occurs with the formation of an intermediate basic sulphate $CuO \cdot CuSO_4$ layer below the surface layer which is composed of $CuSO_4$. Once the sulphate layer has been formed at temperatures around 750–800°C, it can be seen to break and a liquid phase exudes from the substrate to solidify as columns. These are found on subsequent sectioning to consist of copper basic sulphate which is clad in a thin layer of cupric sulphate. The formation of a liquid phase in the copper–oxygen–sulphur system at low temperatures has not been substantiated under equilibrium conditions to date, but it is suspected, as a result of the sulphation studies, that the liquid forms at the oxygen and sulphur potentials which are found below the surface of the sulphated oxide Since the growth law is parabolic in the major part, these potentials must be intermediate between those of the gas phase and the $CuO/CuO.CuSO_4$ equilibrium.

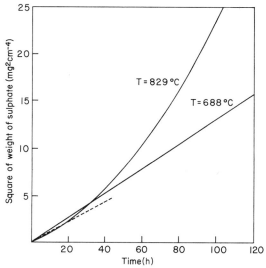

Fig. 26. Weight change of CoO samples as a function of time during sulphation roasting showing parabolic kinetics at 688°C and linear kinetics at 829°C after a short parabolic region.

The overall reaction mechanism can be described by the equation

$$Cu_2O + \tfrac{1}{2}SO_2 + \tfrac{3}{4}O_2 \rightarrow CuO + \tfrac{1}{2}(CuO.CuSO_4)$$

and the mobile species are cuprous ions and electrons. These move out of the Cu_2O core, leaving behind CuO and migrate to the gas–solid interface.

The formation of nickel sulphate on nickel oxide proceeds at widely differing rates, depending on the manner of preparation of the oxide. Pure oxide which is formed by oxidizing a sheet of spectrographically pure nickel, sulphates extremely slowly. NiO prepared by decomposing $NiSO_4$ at 1000°C for one hour in air sulphates rapidly. The rates of sulphation of oxide samples which are prepared by ignition of sulphate at longer times in air, sulphate more slowly. It appears that a dilute solution of nickel sulphide can be formed in NiO during ignition of the sulphate, and that the nickel diffusion coefficient is drastically increased by the presence of small amounts of sulphur in the oxide. This factor is responsible for the rapid corrosion of nickel metal in SO_2–O_2 mixtures. On further heating in air, the sulphur is eliminated and the sulphation rate of the oxide decreases correspondingly.

THE KINETICS OF OXIDATION OF ZINC SULPHIDE—SURFACE ADSORPTION CONTROL

Ong, Wadsworth and Fassell (1956) made a study of the rate of oxidation in oxygen of rectangular samples of sphalerite in the temperature range 700–870°C. The weight change of each sample was measured continuously during each experiment, and the linear law fitted all of the results. It can, therefore, be concluded that the product of oxidation, ZnO, was formed as a porous layer over the sulphide. In a few experiments, a thermocouple was fixed to one of the faces of the sample and a measurable temperature rise was observed at the beginning of the oxidation. The results which are quoted show that this temperature increase was as high as 60°C above the furnace temperature initially, decreasing to zero as the reaction front moved into the sample and away from the position of the thermocouple.

The free energy change for the reaction

$$ZnS + \tfrac{3}{2}O_2 \rightarrow ZnO + SO_2$$

is given by

$$\Delta G° = -108,000 + 18T \text{ cal}$$

and it can be seen that a large heat change is involved in the process. This value is not untypical, and thus it would seem that the anticipated difficulty of heat transfer away from the reaction site plays an important role as in the case of most of the roasting reactions of sulphides.

One further aspect of these reactions which was brought out by this study is the effect of the oxygen pressure on the oxidation rate. It was found, by passing gas mixtures of oxygen and nitrogen at a number of ratios over samples of the sulphide at a fixed temperature, that increasing oxygen pressure only increased the rate of oxidation at low p_{O_2} up to about 20%, but a saturation effect occurred at the higher oxygen pressures. This was ascribed to the saturation of surface sites on which oxygen is absorbed before reaction (Fig. 27). The process of oxidation can be written in a number of elementary steps, thus

$$ZnS + \tfrac{1}{2}O_2(g) \rightarrow ZnS \ldots [O] \text{ adsorbed} \rightarrow ZnO + [S] \text{ adsorbed}$$

$$ZnO + [S] \text{ adsorbed} + O_2 \rightarrow ZnO + SO_2 \text{ desorbed}$$

Since there are a limited number of surface sites at which the oxygen molecules can be adsorbed, it follows that a Langmuir adsorption expression
must be included in the overall reaction rate equation to take this fact into
account.

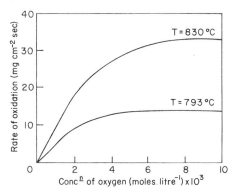

Fig. 27. Kinetics of oxidation of ZnS as a function of the oxygen partial pressure. (After Ong *et al.*)

The adsorption isotherm is of the general form

$$K_L = \frac{\theta}{(1 - \theta)p_{O_2}}$$

where θ is the fraction of the surface sites which are occupied. The isotherm is
deduced by equating the number of atoms desorbed per unit time to the
number which are adsorbed under surface–gas local equilibrium conditions.
The number which is desorbed in unit time is proportional to the number of
atoms already sorbed on the surface sites

$$\dot{n}_{\text{desorbed}} = k\theta$$

The rate of condensation is proportional to the oxygen pressure and the
fraction of surface sites which are unoccupied

$$\dot{n}_{\text{adsorbed}} = k'(1 - \theta)p_{O_2}.$$

At equilibrium

$$\dot{n}_{\text{desorbed}} = \dot{n}_{\text{adsorbed}}.$$

Hence

$$K_L = \frac{k'}{k} = \frac{\theta}{(1 - \theta)p_{O_2}} : \qquad \theta = \frac{K_L p_{O_2}}{1 + K_L p_{O_2}}.$$

The overall rate of the oxidation reaction is governed by the fact that the
reaction can only proceed via the adsorption of oxygen on the surface of the

sulphide and hence the rate will be proportional to the vacant sites concentration which is a function of the oxygen pressure. Rearranging the equilibrium constant for the Langmuir isotherm and substituting into a general rate equation

$$\text{Rate} = k''\theta = \frac{k'' K_L p_{O_2}}{1 + K_L p_{O_2}}.$$

This function will reach a constant value when all of the surface sites are occupied. This occurs when

$$K_L p_{O_2} \gg 1.$$

At higher temperatures, Denbigh and Beveridge (1962) found that the vaporization of zinc sulphide as zinc atoms and sulphur molecules S_2, plays an

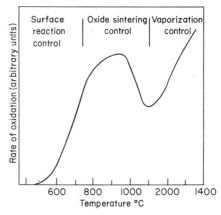

Fig. 28. Variation of the rate constant for the oxidation of ZnS with temperature. (After Denbigh and Beveridge.)

important part in the oxidation mechanism. Above 1000°C, the experimental rate of oxidation of the sulphide decreased with increasing temperature. This was ascribed to an increasing effect of sintering of the oxide product with increasing temperature which impeded the access of oxygen to the sulphide. Above 1200°C, the oxide layer formed as a strong, dense, shell around the smaller zinc sulphide kernel which remains. The inside diameter of the oxide shell was approximately the same as the original diameter of the sulphide sphere. The oxidation of the remaining, smaller, sulphide particle was effected in the gas phase within the oxide shell. Finally, at temperatures above 1350°C, the main quantity of oxide no longer surrounded the sulphide but was distributed around the reaction tube. This product was clearly the result of a vapour phase reaction between air and vaporized ZnS (Fig. 28).

The equation for the vapour pressure of ZnS is

$$\log p \, (\text{atm}) = \frac{-13,980}{T} + 7.09$$

and hence the sulphide exerts a pressure of 10^{-3} atm at 1112°C and at 1350°C where the vapour phase reaction becomes predominant the pressure is 3×10^{-2} atm. Hence, we may conclude that the formation of a cohesive oxide skin around a sulphide will cease to occur when the vapour pressure of the sulphide exceeds 10^{-2} atm.

THE REDUCTION OF IRON OXIDES—REACTION V. TRANSPORT CONTROL

The reduction of iron ore to metallic sponge is carried out industrially on a substantial scale and the factors which affect the reduction kinetics are of some significance. In the iron–oxygen system, there are three oxides to be considered. The two higher oxides, haematite and magnetite, have only a narrow range of non-stoichiometry, whereas wustite has a readily measurable range from $Fe_{0.95}O$ on the metal-rich side to $Fe_{0.88}O$ at the equilibrium with magnetite at 1000°C. The principal defect in this oxide is a vacant site on the cation sublattice, the concentration of defects increasing with increasing oxygen pressure. This defect structure accounts for the fact that the diffusion coefficient of iron in wustite is much greater than that of oxygen, and thus the oxygen ions can be regarded as virtually immobile during the formation of wustite on iron by the oxidation of the metal. In the highest oxide, the relative mobilities are in the opposite order, and oxygen is more mobile than iron. The results for iron diffusion in these oxides were obtained by Himmel, Birchenall and Mehl (1953), who also showed by marker measurements that iron diffuses much more rapidly in magnetite than oxygen does. When a sample of wustite was oxidized to magnetite in a controlled atmosphere, an

TABLE XI. Diffusion coefficients in iron oxides at 1000°C and 800°C

	D_{cation} (cm^2 s^{-1})	D_{oxygen} (cm^2 s^{-1})
$Fe_{0.90}O$	10^{-7} at 1000°C	—
	10^{-8} at 800°C	—
Fe_3O_4	10^{-9} at 1000°C	—
	10^{-11} at 800°C	—
Fe_2O_3	10^{-14} at 1000°C	10^{-13} at 1000°C
	10^{-17} at 800°C	—

inert marker which had been placed on the surface of the oxide before the oxidation was found three-quarters of the way through the magnetite layer, away from the wustite–magnetite boundary. The oxidation process can thus be represented by the equations:

$$12FeO \rightarrow 3Fe_3O_4 + 3Fe^{2+} + 6e^-$$

at the wustite–magnetite boundary and

$$3Fe^{2+} + 6e^- + 2O_2 \rightarrow Fe_3O_4$$

at the magnetite–gas boundary.

Thermodynamic data for the oxide equilibria

In the temperature range which is usually employed for oxide reduction, 500–1000°C, the two-phase equilibria of the iron oxides can be represented by the following free energy changes:

$$2Fe + O_2 \rightarrow 2FeO \qquad \Delta G^\circ = -124,100 + 29 \cdot 9 T \, cal$$

$$6FeO + O_2 \rightarrow 2Fe_3O_4 \qquad \Delta G^\circ = -149,200 + 59 \cdot 8 T \, cal$$

$$4Fe_3O_4 + O_2 \rightarrow 6Fe_2O_3 \qquad \Delta G^\circ = -119,200 + 67 \cdot 3 T \, cal$$

If these free energy equations are combined separately with that for water formation from H_2 and O_2, the equilibrium H_2/H_2O ratios for the two phase systems Fe–FeO, FeO–Fe_3O_4 and Fe_3O_4–Fe_2O_3 can be calculated. The results are as follows:

		500°C	1000°C
Fe–FeO	$H_2/H_2O =$ 10/1*	2/1
FeO–Fe_3O_4		10/1*	1/10
Fe_3O_4–Fe_2O_3		1/10⁵	3/10⁵

These figures clearly demonstrate that over the whole temperature range, the degree of utilization of hydrogen as a reducing agent can be virtually complete before the equilibrium gas for the Fe_3O_4–Fe_2O_3 system is reached. Therefore, even in a stagnant gas, the reduction of Fe_2O_3 to magnetite will practically always go to completion. At the other end of the scale, as soon as the gas contains in excess of 40% water vapour, the reduction of wustite to iron would not be possible, at 1000°C.

The corresponding values for the CO/CO_2 equilibria show much the same effect, the CO/CO_2 ratio being extremely small at Fe_3O_4–Fe_2O_3 equilibrium and being about 2/1 at Fe/FeO and 1000°C.

* Note that FeO is just metastable with respect to disproportionation $4FeO \rightarrow Fe + Fe_3O_4$ below 580°C.

It thus appears, from the thermodynamic data alone, that partial reduction is easily accomplished, but the final reduction to iron necessitates the rapid replenishment of the gas phase with the reducing gas at the solid–gas interface.

The mechanisms of reduction of the separate oxides

When the monoxide, wustite, is reduced to iron either by H_2 or CO reduction, the initial phase of the reaction involves the removal of oxygen atoms from the oxide/gas interface and the excess iron ions and electrons diffuse through the oxide, establishing a higher metal/oxygen ratio in the non-stoichiometric composition range of wustite than that of the original sample. Clearly, this process may develop until the sample is at the metal-rich limit of FeO, about $Fe_{0.95}O$. The iron ions and electrons which are subsequently produced by further reaction will then nucleate to form islands of iron metal, to which other ions and electrons diffuse. In this process of nucleation of the metallic phase, it is most probable that the initial growth will occur at dislocations or grain boundaries in the oxide phase, close to the oxide–gas interface. These islands coalesce, and the large ones grow at the expense of the small nuclei, because of the higher surface energy content per unit volume of the small particles. This process is known as Ostwald ripening. The islands which coalesce to form a bulk phase will probably retain a significant amount of porosity, so that the access of the gas to the oxide–metal interface, and the diffusion of the water vapour or CO_2 away from the reaction site is not severely restricted.

Depending on the temperature at which the reaction is being carried out, the metal layer will remain porous or, at high temperatures, the solid will sinter to produce a layer which will impede the diffusion of water or CO_2 away from the interface, but which will probably not restrict the movement of hydrogen seriously. It is thus frequently found that the rates of such reduction processes go through a minimum value after a small increase of temperature (Fig. 29); the rapid sintering of the metallic layer at the high temperatures slows the reaction down more than the increased mobilities can speed it up.

The reduction mechanism of magnetite to wustite has one very interesting feature. It has been shown by means of X-ray diffraction studies, that the arrangement of iron atoms in Fe_3O_4 is not significantly disturbed during the reduction to FeO. Such a topochemical reaction leads to the formation of a fairly linear, smooth interface and a fairly dense uniform product.

The corresponding reduction of haematite to magnetite could not proceed topochemically because of the difference in crystal structure between these oxides, and so the magnetite layer is found to be porous, containing many fissures. The transport of gases through the produce layer is, therefore, very much easier in the reduction step $Fe_2O_3 \rightarrow Fe_3O_4$ than in the step $Fe_3O_4 \rightarrow FeO$.

The general rates of reduction of the oxide ore, haematite, by H_2 or CO are not only complex because of the number of oxides which can be formed as intermediate phases on the way to the production of metallic iron. It can be seen from the discussion which has been presented above that such factors as the sites for nucleation of the metallic phase in relation to the diffusion paths of iron ions in the oxides, as well as the degree of sintering of the products and the available gaseous diffusion paths, complicate the kinetics of the reduction process.

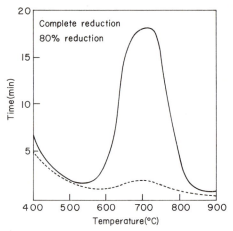

Fig. 29. Schematic representation of the time to achieve 80% (broken line) and 100% (solid line) reduction of haematite powder as a function of temperature.

The empirical studies of the kinetics of reduction show that the fractional reduction equation is nevertheless of a relatively simple form.

McKewan (1958, 1960) initially accounted for his experimental results for the reduction of spherical haematite samples by means of an equation which can be derived on the assumption that the rate depended on the interfacial area between the reduced and unreduced material only.

We will use the symbols W_0 as the original weight of the sphere of radius r_0, W_∞ as the weight after complete reduction and W as the weight at time t when the radius of the unreduced sphere is r.

If the weight loss per unit time is proportional to the interfacial area

$$-\frac{dW}{dt} = kA = 4\pi kr^2$$

$$W = \tfrac{4}{3}\pi r^3 \rho_0 + \tfrac{4}{3}\pi(r_0^3 - r^3)\rho,$$

where ρ_0 and ρ are the densities of the unreduced and reduced products respectively. We will set

$$\rho_0 - \rho = \Delta\rho.$$

Then

$$-\frac{dW}{dt} = -\frac{dW}{dr}\frac{dr}{dt} = -4\pi r^2 \Delta\rho \frac{dr}{dt}; \qquad -\Delta\rho \frac{dr}{dt} = k.$$

Upon integration and evaluation of the integration constant

$$\Delta\rho(r_0 - r) = kt.$$

Now

$$\frac{r}{r_0} = \left(\frac{W - W_\infty}{W_0 - W_\infty}\right)^{\frac{1}{3}} = (W)^{\frac{1}{3}}.$$

Hence

$$(W)^{\frac{1}{3}} = 1 - \frac{kt}{\Delta\rho r_0} \qquad \text{[Reaction control]}.$$

If the reaction rate had been controlled by the diffusion of the gaseous reaction products away from the interface, the weight loss equation would be

$$-\frac{dW}{dt} = \frac{4\pi D(p - p_0)}{(1/r - 1/r_0)} = -4\pi r^2 \Delta\rho \frac{dr}{dt}.$$

Here D is the diffusion coefficient in the gaseous phase, and p and p_0 are the partial pressures of the gaseous products at the reaction interface and surface of the sphere at r_0 respectively.

Hence

$$r^2 \frac{dr}{dt}\left[\frac{1}{r} - \frac{1}{r_0}\right] = -\frac{D}{\Delta\rho}(p - p_0)$$

Upon integrating and substituting for the integration constant

$$\frac{r^2}{2} - \frac{r^3}{3r_0} - \frac{r_0^2}{6} = -\frac{Dt}{\Delta\rho}(p - p_0)$$

$$3(W)^{\frac{2}{3}} - 2(W) = 1 - \frac{6Dt}{\Delta\rho r_0^2} \qquad \text{[Diffusion control in the gas phase]}$$

Although the two equations, for a chemically-controlled reaction and a diffusion-controlled process, look mathematically different from one another, in most experiments in which small spheres are reduced in the laboratory, the difference is close to the experimental error. It is thus very difficult to distinguish between these two alternative reaction mechanisms.

If the equations for the two alternative mechanisms are written as algebraic formulae, thus

$$1 - x^{\frac{1}{3}} = Ay \qquad \text{[Reaction control]}$$

$$1 - 3x^{\frac{2}{3}} + 2x = By \qquad \text{[Gas diffusion control]}.$$

it can be seen that the factors A and B are merely scale factors for these two functions. This means that the values of A and B merely determine the extension of the functions of x along the ordinate y in a graphical plot of the experimental results, and the values of the scale factors are deduced from the results. These scale factors involve an unknown chemical reaction rate constant in A and an unknown effective diffusion coefficient in B. It should be noted that the effective diffusion coefficient involves here the normal volume diffusion coefficient within the homogeneous gaseous phase multiplied by a tortuosity factor. The former is readily obtained or can be closely estimated from existing data for gaseous systems. The latter, the tortuosity factor, takes into account the pore structure of the overlaying product phase through which the gaseous reactants and products must diffuse. As shown earlier, the processes of sintering and stress cracking will probably both be occurring in this layer and thus, not only is the value of a tortuosity factor difficult to estimate, but this value may change significantly during the course of a single experiment.

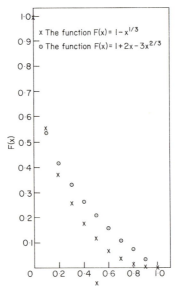

Fig. 30. Functional relationships between weight change and time in reduction kinetics.

The choice between one mechanism or the other on the basis of experimental data only, cannot be reasonably obtained from estimation of the scale factors and is normally made on the basis of the *shape* of the function of x, or W in the real case.

The two functions are shown plotted for values of x in the field $0 \leqslant x \leqslant 1$ which covers this experimental situation. It can be seen that in order to make a clear choice very accurate experimental results are needed over the whole period of each experiment and then the choice might be possible from the ratios of the values of x in the middle of the reduction, at around $x = 0.5$ when compared with the values at the earlier stages $x = 0.1–0.2$.

A very significant study in connection with the reduction of haematite by hydrogen was made by Olsson and McKewan (1966) who demonstrated the probable effect of the tortuosity factor on the observed mechanism of this reaction. By following the reduction kinetics of a sample of wustite which was held in a metal canister having one wall containing a porous iron

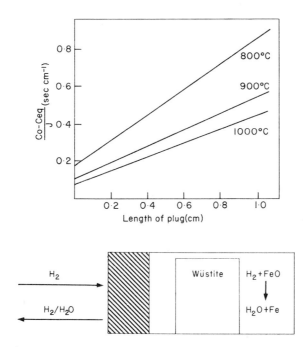

Fig. 31. Effect of a porous plug length on the reduction rate of a sample of wüstite. (After Olsson and McKewan.)

C_0 = conc. of H_s in the gas entering the reaction vessel;
C_{eq} = conc of H_2 in the gas–solid equilibrium mixture;
J = reaction rate.
Schematic representation of the reaction vessel for the porous plug experiment.

plug, and which would be weighed continuously, results were obtained which showed the effect of the length of the plug of iron on the overall reduction rate. In each case, the iron plug was prepared by reducing a pellet of haematite. It was found that the longer the plug the slower the reduction rate of the contained wustite sample, and hence by implication the more the restriction to the counter-diffusion of hydrogen and water vapour. If the rate of the reaction had been controlled by the rate of the chemical reaction at the wustite reaction interface, then the length of the iron plug, which merely connected the gas stream to the wustite sample, should have had no effect.

The experimental temperature coefficient of these measurements which yields the effective diffusion coefficient contains the temperature effect on the tortuosity factor. This latter was found to be about 12 kcal which is too high for the "activation energies" for gaseous diffusion, but must include the temperature coefficient related to the sintering process.

The balance of evidence therefore is that the reduction of iron oxides occurs in a very complicated manner from the point of view of the physical behaviour of the reaction products, but it appears that the reaction rate is controlled by the counter-diffusion of the gaseous reactants and products through the porous product. At the true reaction interface, that is at the surface of the unreduced volume of the specimen, local equilibrium prevails between the gaseous and solid phases.

REFERENCES

Alcock, C. B. and Hocking, M. G. (1966). *Trans. I.M.M.* **75**, C27. Sulphation of CoO-kinetics.

Belle, J. (1969). *J. Nuclear Materials* **30**, 3 .General review of U and O diffusion in $UO_{2+\delta}$.

Darken, L. S. and Gurry, R. W. (1945). *J. Amer. Chem. Soc.* **67**, 1398. Thermodynamics of the Fe–O solid system.

Denbigh, K. G. and Beveridge, G. S. G. (1962). *Trans. Inst. Chem. Engrs.* **40**, 23. Kinetics of ZnS oxidation.

Hawkins, R. J. and Alcock, C. B. (1968). *J. Nuclear Materials* **26**, 112. Diffusion of uranium in $UO_{2+\delta}$.

Himmel, L., Birchenall, E. C. and Mehl, R. F. (1953). *Trans. AIME* **197**, 827. Diffusion of Fe ions in iron oxides.

Hocking, M. G. and Alcock, C. B. (1966). *Trans. AIME* **236**, 635. Sulphation of Cu_2O-kinetics.

Johnson, D. L. and Cutler, I. B. (1963). *J. Amer. Ceram. Soc.* **46**, 541. Models for the sintering of powder compacts.

Kuczynski, G. C. (1949). *Trans. AIME* **185**, 169. Model systems for the initial stages of sintering.

Lay, K. W. and Carter, R. E., (1969). *J. Nuclear Materials* **30**, 74. Sintering study of $UO_{2+\delta}$.

McKewan, W. M. (1958). *Trans. AIME* **212**, 791; **218**, 2 (1960), Kinetics of the reduction of iron oxides.

Niwa, K., Wada, T. and Shiraishi, Y., *Trans. AIME* **209**, 269 (1957). Oxidation of FeS.

Norris, L. F. and Parravano, G. (1963). *J. Amer. Ceram. Soc.* **46**, 449. Sintering of ZnO and NaCl via the vapour phase.

Olsson, R. G. and McKewan, W. M. (1966). *Trans. AIME* **236**, 1518. Diffusion of H_2 and H_2O through porous iron formed by the reduction of haematite.

Ong, J. N., Wadsworth, M. E. and Fassell, W. M., Jr. (1956). *Trans. AIME* **206**, 257. Kinetics of ZnS oxidation.

Thornhill, P. G. and Pidgeon, L. M. (1957). *Trans. AIME* **209**, 989. Roasting of sulphide minerals.

6

HEAT TRANSFER IN GAS–SOLID REACTIONS

The considerations of model systems which have been presented so far are only related to isothermal systems. It is important to realize that during an exothermic reaction, such as the oxidation of a sulphide, it is necessary for the gas phase to transport the heat which is generated by the reaction away from the solid phase. As heat is usually generated at the solid–gas interface where the reaction proceeds, the reaction will only continue at constant temperature if the rate of supply of gas provides a large sink in which the heat generated at the reacting interface can be rapidly dissipated. If this condition is not met, then the reaction cannot be analysed in the simple terms appropriate to an isothermal system.

When the heat dissipation is not sufficient, the temperature of the sample increases and the reaction tends towards the state of ignition. The heating of the solid, due to poor heat transfer to the gaseous phase, can obviously be minimized by making the surface/volume ratio of the solid sample as small as possible and thus minimizing the reaction rate per mole of solid. In this way, the heat capacity of the solid acts as a thermal regulator for a limited period of time. In most practical roasting operations, rapid reaction is the normal objective.

The rate of heat generation during a reaction which obeys the parabolic law is given by

$$\frac{dQ}{dt} = A\frac{dx}{dt} \cdot \frac{\rho}{M} \cdot \Delta H^\circ = \frac{Ak}{x}\frac{\rho}{M} \cdot \Delta H^\circ,$$

where A is the area of interface, ρ is the density of the solid product of molecular weight M and heat of formation $\Delta H^\circ\,mol^{-1}$, k is the rate constant for the reaction and x is the thickness of the product layer.

The precise formulation of the transfer of heat to the moving gas phase from the solid or from the surface of the solid into its interior is complicated and depends very much on such details as the state of flow of the gas, whether it is streamline or turbulent, and the shape of the solid specimen. We can gain a first approximation to the important factors by considering the situation where there is streamline flow in the gas phase, and the solid sample temperature remains constant during the reaction time.

The transfer of heat to the gas phase can be represented by means of the simple linear equation

$$-\frac{dQ'}{dt} = Ah(T_2 - T_1),$$

where h is the heat transfer coefficient to the gas, T_2 is the temperature of the surface of the solid, and T_1 is the temperature of the bulk gas phase.

The conduction of heat away from the surface of the solid and into the interior is

$$-\frac{dQ''}{dt} = k_s A \frac{dT}{dx},$$

where k_s is the thermal conductivity of the solid.

The heat generated in the product layer passes to both the gaseous and solid phases, and in the steady state

$$-\frac{dQ}{dt} = \frac{dQ'}{dt} + \frac{dQ''}{dt}$$

It follows that the solid will only heat up slowly when

$$\frac{dQ''}{dt} \ll \frac{dQ'}{dt} \qquad \text{or} \qquad \frac{h}{k_s}(T_2 - T_1) \text{ is large.}$$

Now the criterion for this partition of heat between the solid and the gas is also the criterion for the possibility of measuring the reaction rate at a reasonably well-defined temperature, and it must be obvious that if the solid consists of small particles, then even a small value of dQ''/dt within the solid could lead to a significant change in temperature.

In the case of a gas streaming slowly over a solid (free or natural convection) a typical value of h is found to be 10^{-3} cal cm^{-2} s^{-1} deg^{-1} which is about the same as the thermal conductivity of a typical non-metallic solid (see Fig. 32). Hence, the temperature gradient in the solid will only remain small when the rate of production of heat is sufficient to produce a small temperature gradient between the solid product layer and the bulk of the gaseous phase, i.e.

$$\frac{1}{h} \cdot \frac{dQ}{dt} = \frac{k}{x} \cdot \frac{\rho}{M} \cdot \frac{\Delta H^\circ}{h}$$

must have a value less than about 10.

The cooling effect of the gas may be improved by passing the gas more rapidly over the solid (forced convection) by means of a pump or by the simple use of compressed gases at the source for the gas mixture which is

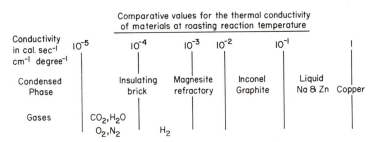

Fig. 32. Comparative values for the thermal conductivity of materials at roasting reaction temperatures.

used. The heat transfer coefficient may then be increased by as much as a factor of ten over the free convection value. This use of the phenomenological transfer coefficient rather than the solution of the appropriate differential equation is usually adopted because of the complex nature of the motion of the gas phase in the region near an interface with a solid. It was pointed out earlier that the heat conduction equation is only readily soluble for well-defined boundary conditions, and when the value of the thermal conductivity, k_g, is a constant. The state of motion of a gas in the vicinity of solid is extremely complex since the mass of moving gas has a velocity profile such that the velocity changes from approximately zero at the gas–solid interface to the bulk velocity at a point somewhat removed from the interface. The velocity change is a relatively smooth function of the distance away from the solid in laminar, or streamline, flow, but it is extremely complicated in turbulent flow. These two distinct conditions of flow must be treated separately in heat flow problems and very sophisticated mathematical techniques must be applied even for the approximate solution of the problem of heat transfer.

The general criterion which indicates whether a gas is in streamline or turbulent flow is based on a consideration of the velocity and of the physical properties of the gas. In order that the criterion shall apply equally to small model systems as to large plants, it is based upon a quantity which has no physical dimensions. The Reynolds number of a gas, for example, is defined by the equation

$$N_{Re} = \frac{u\rho L}{v},$$

where u is the bulk gas velocity, ρ is the density, and v the gas viscosity. The number has a value exceeding 10^3 as the gas motion changes from streamline to turbulent flow, under ideal conditions. L has the dimensions of a length characteristic of the container through which the gas is passing.

HEAT TRANSFER COEFFICIENTS FROM SOLIDS TO GASES

There is no unique solution to the evaluation of the heat-transfer coefficient in solid–gas systems, although it is possible to give approximations for a number of given geometrical situations. If the solid has a flat surface of length L, and the gas is flowing over the solid with a bulk velocity, u, then this is the forced convection situation for which the transfer coefficient at a point x along the length can be given as part of a dimensionless group as follows: for laminar flow

$$\left(\frac{h_x x}{k}\right) = 0.64 \left(\frac{u_x \rho}{v}\right)^{\frac{1}{2}} \left(\frac{c_p v}{k_x}\right)^{\frac{1}{3}}$$

and for turbulent flow

$$\left(\frac{h_x x}{k}\right) = 0.023 \left(\frac{u_x \rho}{v}\right)^{\frac{4}{5}} \left(\frac{c_p v}{k}\right)^{\frac{1}{3}}.$$

In these expressions the symbols represent the following quantities: h_x = heat transfer coefficient at the point x, k = thermal conductivity of the gas, v = the viscosity, c_p = specific heat, and ρ is the density of the gas.

The quantities in brackets are given the following names:

$$\frac{h_x x}{k} = \text{Nusselt number, } N_{Nu} \text{ at the point } x$$

$$\frac{u_x \rho}{v} = \text{Reynolds number, } N_{Re}$$

$$\frac{c_p v}{k} = \text{Prandtl number, } N_{Pr}$$

The first equation, for laminar flow, holds for Reynolds numbers up to 10^3; at values greater than this, the gas phase is probably in turbulent flow, because in most circumstances it is impossible to preserve laminar flow at high values of the Reynolds number. The value of the heat transfer coefficient varies with x, and thus the heat transmitted to the surface per unit area, also changes along the length of the flat surface. For a given length of the surface, L, the average heat transfer coefficient is therefore given by

$$\bar{h} = \frac{1}{L} \int_0^L h_x \, dx$$

where h_x denotes the value of the coefficient at x.

The numerical values of some of the dimensionless groups involved in the heat-transfer coefficient calculations do not change very markedly for a given state of motion from one temperature to another or from one gas to another. The viscosities of gases range between 100 and 500 micropoise under most circumstances (masses up to SO_2 and temperatures up to 1000°C) the thermal conductivities vary from about 10^{-5} to 10^{-3} cal cm^{-1} s^{-1} deg^{-1}, densities are from 10^{-4} to 10^{-2} g cm^{-3} and specific heats are from 10^{-1} to 1 cal g^{-1} deg^{-1} (see Table VII). By manipulating these figures, it can readily be seen that the Prandtl number has a value normally between 0·5 and 1·0 and in the equations for the Nusselt number which involve $(N_{Pr})^{\frac{1}{3}}$, as in the forced convection of gas over a flat bed, this factor may be satisfactorily equated to unity. The Reynolds number is therefore the most important figure in determining the Nusselt number for these situations, and it is the difference between the equations for N_{Nu} for laminar or turbulent flow in respect of the constant and the exponent of N_{Re} which sets the relative values of the Nusselt number for these two regimes of flow. Generally speaking, because of the roughness of the solid surfaces which are involved in roasters, the turbulent condition is more likely to apply than the laminar flow condition even at low values of the Reynolds number, such as 10^2.

For a value of 10 of the Reynolds number, where streamline flow should prevail, the Nusselt number takes the value of approximately two according to the equation given above and, considering the typical value of k, this gives a value of h_L equal to about 10^{-4}. For the Reynolds number 10^4, and hence turbulent flow, the Nusselt number has the value approximately twenty. Hence, there is only a difference of a factor of ten between these two extreme situations.

HEAT TRANSFER WITH NATURAL CONVECTION

When natural convection occurs from a vertical hot surface, the gases rising from the hot surface may, again, be either in laminar flow or in turbulent flow. The criterion for the state of flow which is now used is the value of the Grashof number, N_{Gr}, which is given at a distance x above the surface by

$$N_{Gr, x} = \frac{g \; x^3 \beta (T_s - T_g)}{v^2}$$

In this dimensionless quantity

$$\beta = -\frac{1}{\rho} \left(\frac{\partial \rho}{\partial T} \right)_P$$

is the coefficient of thermal volume expansion, $(T_s - T_g)$ is the temperature difference between the solid and the bulk of the gas and g is the acceleration due to gravity.

When the Grashof number is less than 10^8 the laminar flow equation applies. At higher values, the flow is turbulent. The equations for the Nusselt number are now

$$N_{Nu, x} = \frac{hx}{k} = \frac{0 \cdot 51 \, N_{Gr, x}^{\frac{1}{4}} . N_{Pr}^{\frac{1}{2}}}{0 \cdot 95 + N_{Pr}^{\frac{1}{4}}}$$

for laminar flow, and

$$N_{Nu, x} = \frac{0 \cdot 03 \, N_{Gr, x}^{\frac{2}{5}} . N_{Pr}^{\frac{1}{2}}}{(1 + 0 \cdot 5 \, N_{Pr}^{\frac{3}{4}})^{\frac{2}{5}}} ; \quad N_{Gr, x} > 10^8.$$

for turbulent flow.

Such complicated expressions as are shown above for the evaluation of the heat transfer coefficient are even so to some extent empirical, and the precise value of the exponents of the dimensionless quantities depends on the nature of the gas which is acting as a transfer medium. For ready use, Coulson and Richardson (1956) have proposed some approximate expressions for air as the transfer medium, and it is probably not inappropriate to most metallurgical systems to use the following expressions for any situation when natural convection occurs.

It should be noted again that the Prandtl number has little effect in the Nusselt number for free convective heat transfer, and the number which defines the state of motion of the gas has the predominating effect.

TABLE XII. The heat-transfer coefficient to air through convection

Configuration of the solid	Laminar flow	Turbulent flow
Vertical plate of length L	$h = 2 \cdot 0 . 10^{-5} \left(\dfrac{\Delta T}{L} \right)^{\frac{1}{4}}$	$4 \cdot 8 \times 10^{-5} (\Delta T)^{\frac{1}{4}}$ cal s^{-1} cm^{-1} deg^{-1}
Horizontal surface facing upwards	$h = 2 \cdot 0 . 10^{-5} \left(\dfrac{\Delta T}{L} \right)^{\frac{1}{4}}$	$5 \cdot 7 \times 10^{-5} (\Delta T)^{\frac{1}{4}}$ cal s^{-1} cm^{-1} deg^{-1}

E

HEAT TRANSFER TO SINGLE PARTICLES—FLUOSOLIDS ROASTING

In some gas–solid reaction situations such as flash roasting, or fluosolids roasting, individual particles of sulphide come into contact with the gas phase in isolation from other particles. The Nusselt number, hd_p/k, is then a function of position along the surface of the particle in the direction of gas flow. Ranz and Marshall (1952) have given an approximate average value for N_{Nu} for the case of a gas flowing past a sphere

$$N_{Nu} = 2 \cdot 0 + 0 \cdot 6 N_{Re_p}^{\frac{1}{2}} N_{Pr}^{\frac{1}{3}}$$

N_{Re_p} is the particle Reynolds number which is defined by the equation

$$N_{Re_p} = \frac{d_p u \rho}{v}$$

where d_p is the particle diameter, and the other terms are as previously defined.

This expression yields an average value of the Nusselt number of 4, which is typical of experimental results for small fluosolids roasters when N_{Re_p} is about 10. The minimum value of the flow velocity of gas through a fluidized bed roaster is determined by the pressure drop across the bed which must be sustained in order to expand the solid column into the fluidized state. In this state, the solid particles are separated from one another by a layer of gas, and gas–solid interaction is maximized since the interfacial area is maximized. The condition for fluidization therefore sets the lower limit of velocity.

The maximum velocity of the gas through a fluidized bed is determined by the acceptable level of loss of material which is swept away by the gas. A real assembly of solid particles has a particle size distribution, and at a given gas velocity where the smaller particles in the bed will tend to be swept away (elutriation), the heavier particles will be in the fluidized condition. The calculation of the optimum velocity of gas through the bed must be made in the light of the particle size distribution and the amount of material which can be allowed to pass out of the bed by elutriation.

A good approximation to the gas velocity at which elutriation of a solid particle will occur is to equate this velocity to the terminal velocity of descent of the particle in still air. This velocity can be derived by the use of Stokes' equation for the viscous drag F

$$F = 3\pi d v u_t$$

d is the diameter of the falling particle, u_t is the terminal velocity and v is the viscosity of the gas. Equating this force to the gravitational force we have

$$3\pi d v u_t = \frac{\pi d^3}{6}(\rho_s - \rho_g)g$$

$$u_t = \frac{d^2 g(\rho_s - \rho_g)}{18v}$$

ρ_s and ρ_g are the densities of the solid and gaseous phases respectively. This equation is satisfactory for low values of the Reynolds number, $N_{Re} < 2$, but fails at higher values. In the region of higher values, up to a few hundred, the terminal velocity is expressed by the function

$$u_t = 0.0178 \left[\frac{g^2(\rho_s - \rho_g)^2}{\rho_g v}\right]^{\frac{1}{3}} d.$$

but at Reynolds numbers up to 10 the simpler expression only leads to an error of about a factor of two.

In metallurgical operations, the particle size distribution of the solids which are to be roasted is determined by the milling operation; a typical rod mill product would have an average particle size 100–200 microns with about 30% of the product less than 50 microns. The terminal velocity of 50 micron particles is given by Stokes equation to be

$$u_t = \frac{(50 \times 10^{-4})^2 \times 981}{18} \frac{(\rho_s - \rho_g)}{v}$$

$$= 1.36 \times 10^{-3} \frac{\rho_s - \rho_g}{v} \text{ cm s}^{-1}.$$

A value of 200 micropoise is typical of a gas and the density of an oxide or sulphide may be taken as 3 cm^{-3}; hence we find the value of 20 cm s^{-1} for the terminal velocity of this smallest particle under average conditions. This velocity would be typical for the gas in a fluo–solid roaster and the value of the Reynolds number for the average particle is then

$$N_{Re_p} = \frac{(100 \times 10^{-4}) \times 20 \times 10^{-4}}{200 \times 10^{-6}} = 10^{-1}$$

if $\rho_g \cong 10^{-4}$.

The particle Nusselt number according to the Ranz–Marshall correlation will therefore be 2, and the heat transfer coefficient to the gas would be given by $2 \cong hd/k$ where d is the particle diameter. The value of the coefficient would therefore be 10^{-2} cal cm^{-2} s^{-1} deg^{-1} for a particle of 100 micron diameter in a gas of thermal conductivity equal to 5×10^{-5} cal cm^{-1} s^{-1} deg^{-1}.

HEAT LOSSES BY RADIATION EMISSION

We now have typical expressions for convective heat transfer in terms of dimensionless quantities for most of the physical situations which can arise in the roasting of sulphides. The other important source of heat loss in high temperature systems is loss by the emission of radiation. This loss of energy is calculated according to the Stefan–Boltzmann law which can be written in the form

$$\phi_{\text{radiation}} = \varepsilon s T^4$$

The energy lost by a perfectly emitting substance per unit area in unit time is proportional to the absolute temperature of the substance raised to the fourth power, the Stefan-Boltzmann constant being

$$s = 1.355 \times 10^{-12} \text{ cal s}^{-1} \text{ cm}^{-2} \text{ deg}^{-1}$$

The emissivity, ε, has a value of approximately 0·1–0·3 for metallic surfaces, but approaches the upper limit of unity for non-metallic solids.

The energy emitted by a plane solid at a temperature T_s to a wall of the same area at the lower temperature T_w is given by

$$\phi = Fs(T_s^4 - T_w^4)$$

if both behave as black bodies ($\varepsilon = 1$). The new term, F, in this equation is the "view factor" which represents the solid angle which each surface subtends to the other (Fig. 33).

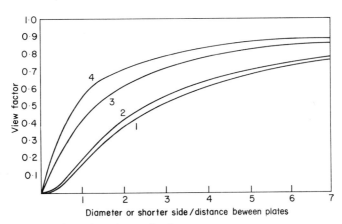

Fig. 33. View factors for various shapes of radiators.
Curve 1. Disks.
     2. Square plates.
     3. Rectangular plates with length $\geqslant$ width.
     4. Square plates connected by an adiabatic surface of emissivity 1.

It is instructive to compare the heat which would be lost to cold surroundings $(T = 300 \text{ K})$ by convective transfer and by radiation from a solid surface at 800 K using the relevant equations from above. The turbulent natural convection transfer from a horizontal bed, according to Coulson and Richardson's equation for the heat–transfer coefficient is

$$\phi_{\text{convection}} = 5\cdot7 \times 10^{-5} (500)^{\frac{5}{4}} \ \{\phi = h(T_s - T_w)\}$$
$$= 0\cdot13 \text{ cal cm}^{-2} \text{ s}^{-1}$$

By radiative loss from a black body with $F = 1$

$$\phi_{\text{radiation}} = 1\cdot355 \times 10^{-12} \ \{(800)^4 - (300)^4\}$$
$$= 0\cdot5 \text{ cal cm}^{-2} \text{ s}^{-1}$$

The radiation loss is therefore greater according to this calculation. In the real case, the gas phase could not be maintained so cold in a roasting furnace at 800 K as the 300 K temperature which has been used. The conditions used in the calculation therefore represent a much greater convection transfer than would normally be the case. It may be assumed, then, that the radiation losses are very important when the hot solid can "see" low temperatures.

Even if the temperature gradient is reduced to 100 degrees (800–700 K), the radiation loss is still by some ten times the larger quantity.

RECOMMENDED READING

Szekely, J. and Themelis, N. J. (1971). "Rate Phenomena in Process Metallurgy". Wiley-Interscience. A general account of the process engineering of extractive metallurgy with many examples.

Bosworth, R. C. L. (1952). "Heat Transfer Phenomena". J. Wiley. A short but comprehensive survey.

Bird, R. B., Stewart, W. E. and Lightfoot, E. N. (1960). "Transport Phenomena". J. Wiley. A classical account from the chemical engineer's standpoint.

Coulson, J. M. and Richardson, J. F. (1956). *Chemical Engineering* 1, 176 Pergamon Press, London.

Ranz, W. E. and Marshall, W. R. (1952). *Chem. Eng. Progress* **48**, 141, 173.

7

INDUSTRIAL ROASTING MACHINES

The types of equipment which are used in industry for the large scale roasting of sulphides were developed some time before the present state of scientific understanding of roasting reactions had burgeoned. With what we now understand of this science, it is informative to review the operation of these machines in terms of the physico-chemical factors which determined the process kinetics.

The Herreshoff or multiple hearth roaster consists of a number of circular stages mounted one above the other on a rotating axle. The stages are enclosed in a circular box and ore particles are admitted from the top and are slowly moved across each stage during their descent by the action of a system of stationary rakes or "rabble arms". The supply of air is made from the bottom of the containing cylinder or at the level of each stage through doors (Fig. 34).

The particles lie in a bed a few inches deep which is slowly turned over by the action of the rabble arms. During this time, the oxidation rate of the sulphides, which are charged, will be determined by the diffusion of the oxidizing gaseous species through the gas phase, which is rising through the machine by natural convection, and by subsequent counter-diffusion of the gaseous oxidation reactants and products through the oxidized layer on each particle. It will be remembered that most practical oxidizing reactions for sulphides obey the linear rate law, and hence solid state diffusion will not play a part in this process.

Such a machine can hold a large amount of material at any given time and the particle size distribution can include both very large, relatively speaking, and very small particles. This is because the gas velocity is low throughout the system, and the dangers of particle elutriation are minimal. The shortcomings of the machine are that gas mixing is poor and the intimacy of contact between individual particles and oxidizing gases can be small

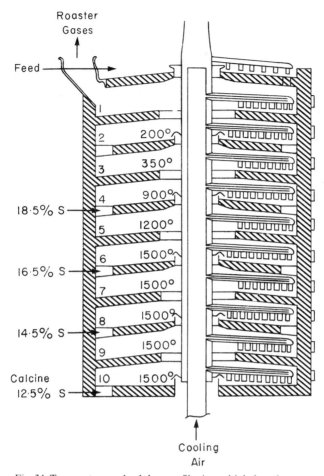

Fig. 34. Temperature and sulphur profiles in multiple-hearth roaster.

because of gas stagnancy and any deficiency in the action of the rabble arms. In fact, it has been found more effective to reduce the number of stages in order to increase the free fall time of particles between stages. This of course will improve the gas–solid contact. Such a change in practice must effect a compromise between a sufficient gas–solid interaction and a sufficiently long residence time of the sulphide particles within the machine to effect the desired degree of oxidation.

The machine which operates at the other end of the spectrum of roasting variables from the Herreshoff roaster is the Fluosolids roaster. Here each particle is separated from the others and can interact with the rapidly moving gas phase (Fig. 35). The problem of elutriation now becomes important and careful control of the ingoing particle size distribution must be exercised.

It has been concluded from consideration of conduction and radiation heat transfer that the heat loss by a particle will be more readily effected by radiation heat loss than by conduction. Hence, in the oxidation of a sulphide, which is usually an exothermic process, the individual particles are likely

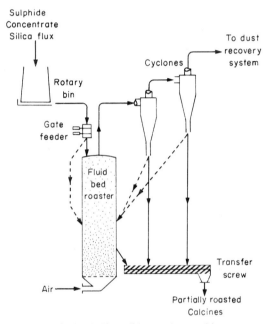

Fig. 35. A Fluosolids roasting machine.

to heat up as a result of oxidation when the radiation loss is minimized, rather than be cooled down by the rapid passage of a cool, but oxidizing, gas. In the fluidized bed of a fluosolids reactor, particles within the main body of the bed are protected from radiation energy loss by being surrounded by the other particles which are undergoing oxidation. Those particles which can suffer heat loss in this way are the ones close to the stationary walls of the reactor. Since there is continuous stirring of the bed by the action of the fluidizing gas stream, no single particle will remain at the walls for a long time, and thus the rate of heat loss to the walls is determined by the ratio of the time a particle spends at the periphery of the fluidized bed to the time it spends within the main volume of the bed. This fraction is usually quite small.

The reactor is therefore very effective as a means for bringing about gas–solid reactions with particles of a well-controlled size distribution.

Apart from this drawback, the only other factor would seem to be that the gas phase does not have a long residence time within the reactor, and it is quite probable that solid–gas thermal equilibrium is never reached. This effect can always be overcome to some extent by pre-heating of the ingoing oxidizing gas, but this adds a thermal penalty to the operating costs of the machine.

A compromise between these two modes of sulphide oxidation is the "flash-roaster" in which the sulphide particles are allowed to fall under gravity through a long cylinder up which the oxidizing gas is moving through convection. A rotary kiln in which the residence time of the solid is increased by setting this cylinder at a low angle to the horizontal and rotating the whole is another industrial compromise which provides a gas–solid contact efficiency somewhere between that of the multiple-hearth and the Fluosolids roaster without requiring the relatively closer particle size control which is necessary for the operation of the latter.

There are thus a number of alternative procedures which may be used all of which derive their simplicity from the fact that sulphide roasting is normally an exothermic process, a so-called "autogenous" roasting reaction.

For a reaction which is practically athermic, such as the reduction of ferrous oxide with hydrogen or CO, there is practically no heat of reaction available, and heat must be supplied to the reacting system. Furthermore, sintering effects become significant in this process, and some fairly vigorous stirring of the reactant–product mixture is called for. It would appear to be for these reasons that the multiple-hearth and the Fluosolids roaster do not appear to be in favour in industry today for the reduction of haematite in the solid state to produce solid iron powder. Some heat is provided by the reduction of haematite to wustite in the overall reduction reaction, but that is all, and in the rotary kilns which are now coming into use, a supply of heat either by preheating of the gaseous phase, or via the walls of the reactor appears necessary. The principal disadvantages of the rotary kiln would appear to be the mechanical stresses which act on the system at operating temperatures and the erosion which takes place at the inner walls of the kiln due to the scouring action of the solid reactant and product particles.

RECOMMENDED READING

"Forum on Roasting". (1961). *Trans. C.I.M.M.* **64**, 315. A survey of current Canadian practice in a number of typical industries.
Boldt, J. R. (1967). "The Winning of Nickel". Longmans. A general account of sulphide ores in the nickel industry.
Newton, J. (1959). "Extractive Metallurgy". J. Wiley.
Bailey, A. R. (1960). "Textbook of Metallurgy". Macmillan.

Part II

Metal Extraction Reactions which produce Liquid or Gaseous Metals

8

THERMODYNAMICS OF METAL EXTRACTION REACTIONS

INTRODUCTION

Because of the many kinetic restrictions to metal production in the solid state, the mainstream of development in metal extraction processing is centred around the formation of liquid products. The metal compound, sulphide, oxide or chloride is acted upon by a reducing agent which produces either a liquid or in a few cases, a gaseous metallic product together with the corresponding compound of the reducing agent in liquid or gaseous form.

By carrying out the reaction in this régime, the separation of the desired metal, in a relatively pure form, from the compound of the reducing agent may be achieved more rapidly than is the case when the products are all solid.

One disadvantage of operation with liquid products is that substances tend to mix more readily in the liquid state than in the solid state. An impurity metal which is co-produced during the reduction of the primary metal is thus more likely to form an alloy with the major element than would be the case where solid products are obtained.

The simplest requirement for the practical reducibility of a compound of one metal MX by a reducing agent R, is the necessity for the standard free energy change of the extraction reaction to be negative at the extraction temperature. The sequence of stabilities of metal oxides which is displayed in an Ellingham free energy diagram (Fig. 4) gives the broad perspective of the possible extraction reactions which can be employed in winning metals from their oxides, and in the industrially significant cases, from the sulphides also.

It can be seen that the free energy of formation of carbon monoxide is more negative than those of iron, tin, lead, copper and zinc oxides above 1000°C; the extraction reaction

$$MO + C \rightarrow M + CO$$

can therefore be employed in this case.

This reaction is usually carried out on the industrial scale in a vertical shaft furnace through the base of which air is forced. This furnace, the blast furnace, produces heat by the aerial oxidation of carbon in the form of metallurgical coke. The burden of the furnace consists of the ore of the desired metal mixed with coke and limestone. The precise mechanism of the reduction reaction probably involves the solid-gas reaction sequence

$$C + \tfrac{1}{2}O_2 \rightarrow CO$$

$$CO + MO \rightarrow M + CO_2$$

$$CO_2 + C \rightarrow 2CO$$

rather than the simple solid-solid reaction which is indicated above. The overall effect is the same, however, from the thermodynamic point of view.

The reaction of sulphur with oxygen has a more negative standard free energy change than that of copper with oxygen, and therefore the direct oxidation of the sulphide

$$Cu_2S + O_2 \rightarrow 2Cu + SO_2$$

is a reaction which is used in copper production. This reaction is practically unique, apart from minor exceptions.

For the reduction of the oxides of intermediate stability, Cr_2O_3 and MnO, clearly some strongly reducing agent must be found to produce the metal. In these cases, the aluminothermic reaction is used, e.g.

$$Cr_2O_3 + 2Al \rightarrow 2Cr + Al_2O_3$$

Finally, it can be seen that special methods must be used to reduce the very stable oxides MgO and CaO, since no simple reduction reaction is indicated by the Ellingham diagram.

The consideration of the standard free energy change indicates the circumstances under which reaction will be feasible, but it becomes immediately apparent from the further consideration of the equilibrium constant that a reaction does not proceed to completion unless one product can be removed, whilst a reactant is supplied to the reacting system. In a closed system which approaches equilibrium the constant for the reduction reaction

$$MO + R \rightarrow M + RO; K = \frac{a_M \cdot a_{RO}}{a_{MO} \cdot a_R}$$

is related to the standard free energy change by the equation

$$-\Delta G^{\circ} = RT \ln K$$

The activities which are calculated from the equilibrium constant must be considered together with the mixing properties of the primary and reducing metals to form alloys (M + R) and of their oxides to form molten salt mixtures, usually known as slags, (MO + RO) in order to determine the *compositions* of the product phases. In the simple case where this two-phase system, a metallic phase (M + R) and a slag phase (MO + RO), is at equilibrium, it follows that the oxygen potentials for the two sub-systems M—MO, R—RO which form the total system M–R–O will also be equal

$$M + \tfrac{1}{2}O_2 \rightarrow MO; RT \ln p_{O_2} = 2\{\Delta G^{\circ}_{MO} + RT \ln a_{MO} - RT \ln a_M\}$$

$$R + \tfrac{1}{2}O_2 \rightarrow RO; RT \ln p_{O_2} = 2\{\Delta G^{\circ}_{RO} + RT \ln a_{RO} - RT \ln a_R\}$$

During the reduction process, more reductant oxide and more primary metal will continuously be formed and thus the activities of these components will increase while those of the reactants will decrease. The effect of the progress of the reaction is therefore to raise the oxygen potential of the sub-system R–RO whilst lowering that of the sub-system M–MO until equilibrium is achieved (Fig. 36).

It is obvious that if some other chemical substance which can alloy with the metallic phase, or which can dissolve in the slag, is added to the system, then the compositions of the final metal and slag phases can be substantially altered. For example, the addition of a very stable oxide which cannot be reduced will provide a diluting substance which can dissolve in and alter the composition of the slag phase and therefore affect the activities of the oxides in that phase. The distribution of M and R between the metal and slag phases can therefore be manipulated through the slag composition by the addition of compounds which do not enter the main reduction reaction. Such additions can affect differentially the mole fractions of M and R in the two phases. Consider the equilibrium constant for the metal extraction reaction in the form

$$K = \frac{\gamma_{[M]}\gamma_{\{RO\}}}{\gamma_{[R]}\gamma_{\{MO\}}} \frac{X_{[M]}X_{\{RO\}}}{X_{[R]}X_{\{MO\}}} \qquad \begin{array}{l} [\ ] = \text{metallic phase} \\ \{\ \} = \text{slag phase} \end{array}$$

It can be seen that the effect of some non-reactive slag component AO which changes the slag composition could also change the activity coefficients of MO and RO in the complex slag phase MO–RO–AO from the corresponding values of activity coefficients in the simple binary slag MO–RO.

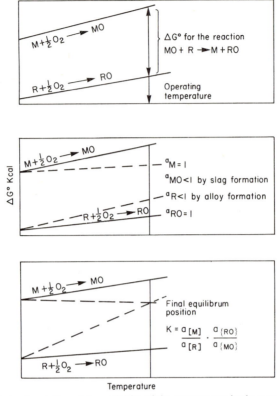

Temperature

Fig. 36. The effects of variation of the activities of the components in the reaction MO + R → M + RO on the oxygen potentials of the sub-systems M–MO and R–RO.

Similarly the presence of an oxide BO which could be co-reduced with MO would provide a metallic constituent B which could affect the activity coefficients of M and R in the equilibrium metallic phase.

As examples of these chemical effects of ternary additions, we may take the presence of calcium oxide in the slag phase and of carbon in the metal phase during the blast furnace reduction of iron oxide. In this process, silicon is found in the metal and slag phases as a result of the presence of siliceous gangue in the furnace charge. On reduction of the iron oxide to produce liquid iron, the metal becomes saturated with carbon due to the contact between the metal phase and coke during the passage of the metal to the bottom of the furnace, the hearth. The final distribution of silicon between the resulting metal and the oxide slag phases results from the effects of carbon on the activity coefficient of silicon dissolved in iron and of lime on the activity coefficient of silica dissolved in the slag phase.

In this example, carbon raises the activity coefficient of silicon in solution in liquid iron, and an addition of lime lowers the activity coefficient of silica dissolved in an oxide slag. Hence, carbon and lime will act in concert to shift the silicon from the metal to the slag phase as equilibrium is achieved.

The manipulation of the chemical composition of metal and slag phases to alter the metal-slag distribution of elements could be achieved by an informed use of knowledge of the thermodynamic properties of complex product phases, and more often than not, operating with complex systems is also forced on the extraction metallurgist by the nature of his starting materials. Thus, impurity oxides, known as "gangue" material, are invariably present in the mineral systems with which he must work, and these impurities are distributed between the metal and slag phases. They are usually removed from the crude metal which is obtained from the primary extraction reaction in a subsequent metal-refining stage. The consideration of metal refining will be left to Part III of this work. Additions are frequently made to the furnace charge with the intention of affecting the compositions of the equilibrium phases by altering the activity coefficients of the products which are dissolved in the metal and slag phases.

METAL-SLAG DISTRIBUTION AT EQUILIBRIUM

It has been already stated that in many extraction reactions liquid metal is produced together with a molten slag phase which consists mainly of mixtures of oxides or halides. A chemical function of the slag is in removing some of the impurities from the metallic phase, and the efficiency of this operation depends on the slag-metal thermodynamic equilibria and the rates at which these equilibria can be achieved.

The thermodynamics of the refining behaviour of a slag phase with respect to the liquid metal is a function of temperature and of the compositions of the metal and slag phases. In the removal of an impurity from a metal during the reduction of oxides, the equilibrium distribution of the impurity (G) between metal [ ] and slag { } is determined by the values of the activities in the equilibrium constant for the reaction

$$[G] + n/2O_2 \rightarrow \{GO_n\} \qquad K_G = \frac{a_{\{GO_n\}}}{a_{[G]}p_{O_2}^{n/2}}$$

One important limitation on the extent to which the absorption of the impurity from the metal and into the slag can be affected by control of the slag composition is the corresponding reaction for the principal metal to be extracted, M.

$$[M] + x/2O_2 \rightarrow \{MO_x\} \qquad K_M = \frac{a_{\{MO_x\}}}{p_{O_2}^{x/2}}$$

Here it is assumed that conditions are adjusted so that the metal M is extracted in a reasonably pure state (say, better than 95% pure), and hence the activity of M can be assumed to be unity.

Since the same oxygen partial pressure applies to the transfer of G and of M from the metallic phase to the slag, the two equations above can be brought together to yield the displacement reaction

$$[G] + n/x\{MO_x\} \rightarrow \{GO_n\} + n/x[M]; \quad K_{G-M} = \frac{a\{GO_n\}}{a_{[G]}a_{\{MO_x\}}^{n/x}}$$

As one means of increasing the transfer of G to the slag, the activity coefficient of G in the metal phase could be raised by the addition of another impurity to the metal phase which displaces element G to the slag phase. This would merely replace one impurity in the metal by another. Hence, the important consideration in the elimination of the impurity into the slag phase without adding an impurity to the metal is changing the chemistry of the slag phase. Consider the quotient

$$Q = \left(\frac{a_{\{GO_n\}}}{a_{\{MO_x\}}^{n/x}}\right)_{a_G} = \left(\frac{\gamma_{\{GO_n\}} \cdot X_{\{GO_n\}}}{\gamma_{\{MO_n\}}^{n/x} \cdot X_{\{MO_x\}}^{n/x}}\right)_{a_G}$$

This quotient is clearly a constant for the system for a given value of $a_{[G]}$ and of temperature, and hence to optimize the transfer of this impurity into the slag phase, we need only consider the ways in which it is possible to raise $\gamma_{\{MO_x\}}$ with respect to $\gamma_{\{GO_x\}}$ in the slag phase. If no obvious way appears for making this change in activity coefficients, then the value of Q must be accepted as limiting the efficiency which can be achieved in a given process. Of course, an increased elimination of an unwanted impurity can always be achieved by sacrificing an increased amount of the major metal, M, to the slag phase, i.e. by raising $a_{\{MO\}}$ through an increase of the oxygen potential of the system.

The heat change for the displacement reaction will determine the variation of the equilibrium constant of this reaction K_{G-M} with respect to temperature according to the Gibbs–Helmholtz equation

$$\frac{\partial \ln K_{G-M}}{\partial(1/T)} = \frac{-\Delta H°_{G-M}}{R}$$

but the variation of the value of Q with temperature will be determined not only by this effect, but also by two other factors. Firstly, the change of a_G for a given composition of the metal phase with temperature which is given by the equation

$$\left(\frac{\partial \ln a_{[G]}}{\partial(1/T)}\right)_{X_G} = \frac{\Delta \bar{H}_{[G]}}{R}$$

where $\Delta \bar{H}_{[G]}$ is the partial heat of solution of G in the metallic phase.

Secondly, the displacement of M by G in the slag phase will change with temperature according to the equation

$$\left(\frac{\partial \ln Q}{1/T}\right) = \frac{\Delta \bar{H}_{\{GO_n\}} - \Delta \bar{H}_{\{MO_x\}}^{n/x}}{R}$$

at constant a_G, X_{GO_n}, X_{MO_x}

There are clearly, then, a number of factors which are involved in the effect of temperature variation of the metal–slag equilibrium for the elimination of a given impurity (Fig. 37). The first of these is the standard heat change of

Slag phase:
$$\left.\left(\frac{\partial \ln a_{GO_n}}{\partial 1/T}\right)\right|_{X_{GO_n}} = \frac{\Delta \bar{H}_{GO_n}}{R}$$
$$\left.\left(\frac{\partial \ln a_{MO_x}}{\partial 1/T}\right)\right|_{X_{MO_x}} = \frac{\Delta \bar{H}_{MO_x}}{R}$$

These quantities are sensitive to slag composition,

Pure phase reaction: $G + n/x \, MO_x \rightarrow GO_n + n/xM$

$$\frac{\partial \ln K_{G-M}}{\partial 1/T} = -\frac{\Delta H^\circ_{G-M}}{R}$$

Metal phase:
$$\left.\left(\frac{\partial \ln a_G}{\partial 1/T}\right)\right|_{X_G} = \frac{\Delta \bar{H}_G}{R}$$
Probably large for a minor component.

$$\left.\left(\frac{\partial \ln a_M}{\partial 1/T}\right)\right|_{X_M} = \frac{\Delta \bar{H}_M}{R}$$
Small for a major component.

Fig. 37. The thermodynamic quantities which are involved in the temperature coefficient of the general reduction reaction of the metal impurity G, and its distribution between the metal and slag phases.

the displacement reaction ΔH°_{G-M}. This can be assessed quite closely using the two free energy lines on the Ellingham diagram for the oxides, from the respective intercepts at the absolute zero (Fig. 4). The second contributory factor, the partial heat of solution of the impurity in the metal, can only be obtained from separate thermodynamic studies of the M–G metallic alloys. If this has not been measured, then only very crude guesses can be made since no quantitative theory for alloy thermodynamics is available.

The final quantity, the difference between the partial heats of solution of GO_n and MO_x in the slag phase can be approximately calculated if there is information available about the thermodynamics of the solution of these oxides, separately, in the slag. The interactions between metal oxides which are dissolved in the slag can be ignored to a fair degree of approximation because usually only dilute solutions of these oxides in the slag are involved. Unless there is a large difference between the affinities of the oxides GO_n and MO_x for the slag phase, which would yield a large value for the difference

$$\Delta \bar{H}_{GO_n} - \Delta \bar{H}^{n/x}_{MO_x}$$

this factor can probably be ignored in comparison with the first factor in the temperature coefficient, ΔH°_{G-M}.

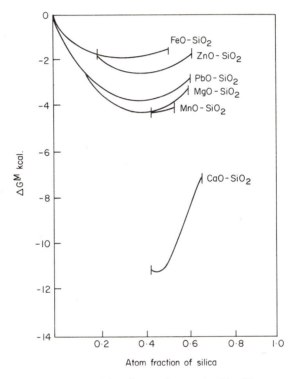

Fig. 38. Integral free energy of mixing diagram for metal oxide–silica systems at 1600°C.

A very important slag constituent in the extraction of metals from their oxides which does usually occur at high concentration is silica. The slag phase usually attains a temperature exceeding 1400°C before fusion occurs and the system becomes sufficiently fluid to enable a ready separation of the metal from the slag. The thermodynamic data for binary liquid systems formed by metal oxides and silica at temperatures above 1400°C are therefore of great importance to the understanding of metal extraction processes. These results show that many metal oxides, such as FeO, ZnO, MnO, MgO and PbO, do not have a strong interaction with silica, the strongest inter-action by a large factor being that with lime, which is added in the burden as limestone, and silica (Fig. 38). Because of this strong interaction, the integral free energy of mixing in the liquid lime–silica system has a maximum negative value, at approximately the equimolar composition, of about $-12\ \text{kcal mol}^{-1}$ at slag-making temperatures and a rapid change occurs around this composition of the chemical potentials of lime and silica. On the lime-rich side, the lime activity is close to unity, and the silica activity is very low, and on the silica-rich side the silica activity is nearly unity and the lime activity is very low. Now, it is because this interaction dominates the chemical behaviour of slags that the terms "basic" for high-lime low-silica activities, and "acid" for the converse situation, has been applied to the description of slag composition and behaviour. Typical experimental data for this system are shown in Table XIII.

TABLE XIII. Experimental results for CaO–SiO$_2$ at 1500°C

X_{SiO_2}	$\Delta\mu_{SiO_2}$ cal	X_{CaO}	$\Delta\mu_{CaO}$	ΔG^M
0·413	-18338	0·587	-7798	$-12,153$
0·457	-11140	0·543	-13453	$-12,396$
0·480	-7918	0·520	-16452	$-12,303$
0·500	-4713	0·500	-18852	$-11,783$
0·520	-2656	0·480	-20908	$-11,416$

In general, we expect the CaO/SiO$_2$ ratio to control the activity coefficients of dilute solutions of other metal oxides in liquid slags, and thus it is mainly through this ratio that any effect can be made on the value of Q in any given situation. The experimental results show that the effects of changing this ratio are not very profound for the metallic oxides in the slag phase, such as FeO, MnO, ZnO, etc. The general effect of increasing lime content is to raise the activity coefficients of these oxides by a small amount, usually less than a factor of two overall. The distributions of non-metallic elements, such as sulphur and phosphorus, are quite sensitive to this ratio, the transfer from metal to slag being enhanced considerably by high lime activity in both cases.

It is therefore to be anticipated that the difference between the heats of formation of metallic oxides ΔH°_{G-M} will determine the distribution of metallic impurities between metal and slag, but the distribution of non-metallic impurities can be affected significantly through the lime-silica ratio in the slag phase.

EXPERIMENTAL STUDIES OF SIMPLE SLAG SYSTEMS

Chipman and his co-workers (1961) have made a study of silica activities in blast furnace slags by determining the silicon distribution between Fe–Si–C and slags containing SiO_2. The two phase system was held in a graphite crucible, and under 1 atm. pressure of CO, 150 g of slag and 300 g of metal were typical samples, and the system was stirred with a graphite stirrer at 100 rev min^{-1}. Runs usually lasted 10 h before slag-metal equilibrium, as indicated by the Si content of the metal, was reached. The calculation of activities in the slag were made with the use of the standard free energy change

$$SiO_2 + 2C \rightarrow Si(1) + 2CO \qquad K = \frac{a_{[Si]} \cdot p_{CO}^2}{a_{\{SiO_2\}} \cdot a_C^2}$$

$$\Delta G^\circ = 161,500 - 87\cdot4\, T\, cal$$

Since graphite and CO were at unit activity, the only other piece of information which was required was the activities of silicon in the Fe–Si–C system as a function of composition. These were obtained from a separate study of the distribution of silicon between Fe–Si–C and molten silver in which silicon forms almost Raoultion solutions.

Baird and Taylor (1958) also measured the activities of SiO_2 in slags by the direct measurement of the CO pressure which was evolved in the equilibrium between slag, carbon and silicon carbide

$$3C + \{SiO_2\} \rightarrow SiC + CO.$$

It was considered for some time that the results which were obtained from these different techniques were not in agreement. The source of the discrepancy was finally traced to an error in the accepted heat of formation of SiO_2. After a thorough re-examination of the calorimetric procedure, results were obtained for this quantity which brought the results for the activities of silica in the slag systems which were obtained by these two methods into agreement.

The free energy of formation of SiC was obtained by Grieveson and Alcock (1961) who measured the silicon pressure above the carbide in equilibrium with carbon by the Knudsen weight-loss technique. The vapour pressure of pure silicon was established in separate measurements by means of the transportation technique in which $MoSi_2$ was employed as a container material.

Abraham, Davies and Richardson (1960) established the pattern of activities in the binary $MnO–SiO_2$ system and the ternary system $MnO–CaO–SiO_2$ in the temperature range 1500–1650°C. The technique for the measurement of the MnO activities involved equilibrium of the melt with platinum wire under a fixed oxygen partial pressure. From earlier studies, the manganese contents of platinum wires which were equilibrated with pure MnO in a fixed oxygen pressure had been established in the same temperature interval. The MnO activities in the slags were thus obtained by direct comparison with the results for pure MnO.

$$[\text{Mn}]_{\text{Pt}} + \tfrac{1}{2}O_2 \rightarrow \{\text{MnO}\}_{\text{slag}}$$

The results showed that the MnO activities in the MnO–SiO$_2$ system were markedly higher than the CaO activities in the corresponding CaO–SiO$_2$ system, thus indicating a lower affinity for SiO$_2$. In the ternary MnO–CaO–SiO$_2$ system, the MnO activity coefficient increased with increasing CaO/SiO$_2$ ratio. The activity of MnO at an atom fraction of 0·2 was increased from 0·1 to 0·6 when the CaO/SiO$_2$ ratio changed from 3:7 to 6:4. The effect is thus similar to that of CaO on the FeO activity in FeO–SiO$_2$ mixtures, but more marked.

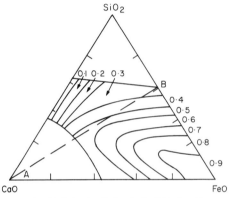

Fig. 39. Isoactivity contours for FeO in the FeO–CaO–SiO$_2$ liquid ternary systems at 1600°C. (After Elliott, loc. cit.)

Schuhmann and Ensio (1951) obtained the activities of FeO in FeO–SiO$_2$ slags which were held in iron crucibles by equilibration with CO/CO$_2$ gas mixtures in the temperature range 1250–1400°C. The composition range which could be achieved in this system was from pure FeO to silica saturation at 40 wt % SiO$_2$. The activities of FeO were obtained by the use of Darken and Gurry's results (1946) for the binary Fe–O system, and it was found that the activity coefficients varied between 1·1 and 0·7 over the composition range. The temperature coefficients of activities were also very small, and thus to a good approximation the system may be considered as conforming to Raoult's Law.

Elliott (1955) surveyed the data for the FeO–CaO–SiO$_2$ system, and presents iso-activity curves for the liquid system at 1600°C (Fig. 39). The effect of CaO in increasing the activity coefficient of FeO can be clearly seen in the diagram. The iso-activity contours for FeO cut across the lines of constant

FeO/SiO_2 ratios with the addition of CaO. The line AB shows the effect of adding CaO to the composition which is saturated with silica in the binary $FeO-SiO_2$ system.

Finally, Charette and Flengas (1968) have studied the activities of PbO in $PbO-SiO_2$ solutions in the temperature range 720–1100°C. The experimental method which was used was electrochemical, the oxygen potential of the slag in contact with liquid lead being measured in order to get PbO activities.

Pb, PbO–SiO$_2$	ZrO–CaO Electrolyte	Pt, O$_2$ (1 at.)

In this cell, the oxygen pressure of the right-hand side electrode is fixed merely by the gas composition, and the platinum-electrolyte contact which is surrounded by the gas phase functions as an oxygen electrode. The oxygen pressure of the left-hand electrode is obviously determined by the equilibrium

$$Pb + \tfrac{1}{2}O_2 \rightarrow \{PbO\}_{slag}$$

The results showed that the integral heat of mixing was -1 kcal g mol^{-1} and the integral entropy of mixing was 1·6 cal g mol^{-1} degree^{-1} at the equimolar composition. The lead silicate slags are about as stable as the $MnO-SiO_2$ solutions, and therefore, once again the $CaO-SiO_2$ system is much more stable by comparison.

There is a marked similarity between the order of stability of the liquid metal oxide–silica systems and the heats of formation of the corresponding crystalline silicates. The solid compounds are the metasilicates $MSiO_3$ and the orthosilicates M_2SiO_4 where M is a divalent cation. The crystal structures of these two groups of compounds are different in a very important way in that the anionic structure of the metasilicates consists of SiO_4 units arranged in chains and in the orthosilicates these units are separated. The

TABLE XIV. Heats of formation of crystalline metal silicates from metal oxides

Cation (M^{2+})	$\Delta H°$ metasilicate Kcal mol^{-1}	$\Delta H°$ orthosilicate Kcal mol^{-1}
Iron	—	3·0
Lead	2·5	7·0
Zinc	—	7·0
Manganese	5·9	11·8
Magnesium	8·7	15·1
Calcium	21·5	30·2
Barium	38·0	64·5
Sodium (Na$^+$)	55·5	74·9

comparative stabilities of compounds among a number of cations neverthe-less follow the same sequence as the results in Table XIV show. It would therefore seem reasonable to estimate the stabilities of other liquid metal oxide–silica systems from data for the crystalline compounds in an analogous fashion. As examples, there do not appear to be reliable data for liquid barium and sodium silicates at the present, but the results for the formation of the crystalline silicates suggest that these should be even more stable than those of the lime-silica system.

THE THERMODYNAMICS OF LIQUID ALLOY SYSTEMS

Theoretical considerations

The basic model of liquid alloys which has seen most use in the correlation of the thermodynamic properties is a very simple random distribution model. It is assumed that the bonds between metal atoms have a characteristic value of bond energy for a given pair of atoms which are in contact. This bond energy is unchanged when either atom makes contact simultaneously with other atoms, no matter what these are. This description of interaction in metallic systems is called the "constant pairwise bonding" model. In order to calculate the total bonding energy of an assembly of atoms, it is assumed that the coordination number of each atom in a liquid alloy, Z, is the same. When two metals A and B are mixed, then following the Born–Haber cycle procedure, A–A bonds and B–B bonds must be broken during the atomization of the pure components and some A–B bonds must be formed in the condensed alloy. If the distribution of atoms in the alloy is random, then $\frac{1}{2}(N_0 Z)(X_A + X_B)$ bonds are broken in the first step and $\frac{1}{2}(N_0 Z)(2X_A X_B + X_A^2 + X_B^2)$ bonds are formed in the alloy.

If the energies of the pairwise bonds are written as E_{AA}, E_{BB} and E_{AB}, the energy of formation of the randomly-mixed alloy from the pure components with atom fractions X_A and X_B is given by

$$\Delta E^M = \frac{N_0 Z}{2}\left(E_{AB} - \frac{E_{AA} + E_{BB}}{2}\right) X_A X_B$$

A randomly assembled mixture of particles of two species has an entropy of mixing

$$\Delta S^M = -R[X_A \ln X_A + X_B \ln X_B].$$

An alloy system which has these properties of pairwise bond energies, a constant coordination number and a random distribution is called a "strictly regular" system, and it can be seen from the model that the heat of mixing has a parabolic dependence on alloy composition.

The activity coefficient, γ, of a component is given by

$$RT \ln \gamma_A = \Delta \overline{H}_{A(A+B)} = \text{constant} \cdot X_B^2$$

where the constant is equal to

$$\frac{N_0 Z}{2} \left(E_{AB} - \frac{E_{AA} + E_{BB}}{2} \right).$$

This symmetrical form of the heat of mixing curve has been found for a number of binary alloy systems, usually those in which both metals are found in a given group of the Periodic Table of the elements, e.g. Ag + Cu, Zn + Cd, Ga + In and Pb + Sn alloys.

The term "regular solutions" was originally applied to liquids in which the entropy of mixing had the ideal form of the strictly regular liquids, but the heat of mixing was not necessarily of the symmetrical form. This generalization does not seem to be appropriate to metallic liquid systems because the departure from symmetry of the heat of mixing curve which is obtained most frequently when the elements in a binary alloy are not drawn from the same group of the Periodic Table, is usually found to be accompanied, when a detailed study is made of the system, by departures from the simple form of the entropy of mixing equation. There are a number of reasons why this may be, and a few possibilities are the following:

(1) The coordination number probably changes with alloy composition and hence the number of contacts and therefore of bonds may change. If this is because of an important size difference between the species, then an entropy term must be included to allow for this. The coordination numbers of many liquid elements have been obtained through X-ray and neutron diffraction studies and Table XV presents some typical results.

TABLE XV. Coordination numbers for liquid metals

Element	Coordination number Crystal structure of the solid	Liquid
Sodium	b.c.c.	9·5
Potassium	b.c.c.	9·5
Copper	f.c.c.	11·5
Silver	f.c.c.	10·0
Gold	f.c.c.	8·5
Zinc	h.c.p.	10·8
Cadmium	h.c.p.	8·3
Gallium	Orthorhombic	11·0
Tin	Tetragonal (white tin)	8·5
Lead	f.c.c.	8·0

f.c.c. = face centred cubic   b.c.c. = body centred cubic
h.c.p. = hexagonal close packed

(2) The entropy of mixing can contain, in principle, not only the positional entropy of which the random mixing equation is one form, but non-configurational terms also. The Kopp–Neumann rule requires that the heat capacity change on formation of the alloy from the elements is zero. If the heat capacity of the alloy is not given by this simple form

$$Cp_{alloy} = X_A Cp_A + X_B Cp_B$$

then a non-configurational term in the entropy of mixing

$$\Delta S = \int_0^T \Delta Cp \, d \ln T$$

must be present.

(3) If the unlike-atom energy is much greater or much less than the arithmetic mean of the like-atom bond energies, then short range order or like-atom clustering respectively, may occur in the liquid. There is some X-ray diffraction evidence, for example, that in the gold-tin system, because of the strong interaction between gold and tin atoms, a short range ordering similar to that in the solid intermetallic compound AuSn occurs in the liquid alloys. This interpretation is by no means the only possible one for the X-ray results, but the strongly exothermic nature of the alloys could be used to give a convincing picture along such lines.

In metals where there is a large endothermic heat of mixing, the phase diagram usually shows some immiscibility in the liquid state, and this may be used as evidence to support the idea that even in the homogeneous alloys there is some tendency for the like atoms to cluster together.

Both of these effects should cause a departure from the ideal entropy of random mixing which the strictly regular liquids would possess. The experimental evidence now begins to suggest that all of these traits can be found in real systems, i.e. departure from randomness and non-configurational contributions to the entropy from heat capacity effects. From the thermodynamic data alone it is almost impossible to distinguish between these two causes of departure from regularity.

The "sub-regular" equation, which does not have a well-defined basis in atomic model systems, gives a sufficient account of liquid alloy systems to be of practical value. This equation requires that the excess free energy of mixing divided by the product of the atom fractions is a linear function of composition

$$\Delta G^{XS} = \Delta G^M - RT(X_A \ln X_A + X_B \ln X_B)$$

$$\frac{\Delta G^{XS}}{X_A X_B} = (A + BX_A) \quad \text{(A strictly regular solution has B = 0)}$$

If it is assumed that the unlike atom bond energy E_{AB} is a function of composition, then the heats of solution at infinite dilution of A in pure B and of B in pure A can be expressed by the equations:

$$\Delta \bar{H}^{\infty}_{A(B)} = \frac{N_0 Z}{2} \left\{ E_{AB} - \left(\frac{E_{AA} + E_{BB}}{2} \right) \right\}.$$

$$\Delta \bar{H}^{\infty}_{B(A)} = \frac{N_0 Z}{2} \left\{ E'_{AB} - \left(\frac{E_{AA} + E_{BB}}{2} \right) \right\} \qquad E_{AB} \neq E'_{AB}$$

In the simple situation where E_{AB} is a linear function of composition, the integral heat of mixing can be represented by the equation

$$\Delta H^{M} = \frac{N_0 Z}{2} \{ X_A \Delta \bar{H}^{\infty}_{B(A)} + X_B \Delta \bar{H}^{\infty}_{A(B)} \} X_A X_B$$

This equation is identical in form with the sub-regular equation.

The equation can be used for alloy systems involving metals from neighbouring groups of the Periodic Table, but important departures are often found when the alloys involve the elements of Group V, arsenic and antimony. This may be because these alloys have a significant polar contribution to the atomic bonding which leads to short range order.

The subject of electron transfer in alloy systems is still very far from being clarified. There are results for alloy systems which indicate that electron transfer from s bands to holes in d bands can occur in some alloys of the Group VIII elements but precisely what quantitative effect this has on the energy of mixing is unknown. There appears to be growing evidence that electron transfer from one atomic species to the other will occur in a binary system when the elements have a large difference in electronegativity. Again this idea is very difficult to quantify, but it would account for the much greater stabilities of gold alloys, for example, when compared with the corresponding copper or silver-containing alloys. Recent theoretical treatments suggest that a wave mechanical basis can be found for the use of Pauling's electronegativity table in alloy systems, although the values were originally obtained from gaseous diatomic molecules.

Another wave mechanical contribution to the theory of alloys which has proved a useful guide, springs from the calculations first made by Friedel (1954) concerning solute-solute interactions in moderately dilute alloys. These show that a solute atom having a different valence from the solvent species will have the excess nuclear charge screened by the conduction band electrons. When two solute atoms come close together, as will happen more frequently with increasing concentration, there will be a repulsive interaction if the solute has a greater valency than the solvent, and an attractive interaction if the solute has a lower valency than the solvent. This effect can also account for the Hume–Rothery rule which shows that the phase diagrams for the systems Cu or Ag plus Zn, Cd, Ga, In, Ge and Sn can be superimposed when electron concentrations are used rather than atom fractions for the compositional abcissa.

The dilute alloy problem has not yet been satisfactorily resolved theoretically, but the range of conformity to Henry's Law, usually from 0 to one atomic percent concentration will be dealt with empirically in Part III.

The measurement of activities in liquid alloys

The vapour pressures of the components of a binary alloy provide one of the simplest bases for the measurement of thermodynamic activities. In the most primitive form of measurement, when the vapour pressure of one component is very much less than that of the other, the total weight loss during Knudsen evaporation or as the result of inert carrier gas transport can be used to establish the vapour pressure of the more volatile component as a function of alloy composition. The activity of the second, non-volatile component can be obtained in principle by application of the Gibbs–Duhem relationship. To provide sufficient data for this calculation, the activity of the volatile component must be measured over the range of composition such that the activity obeys Raoult's or Henry's law at each extreme. In most systems, the dilute solution limiting law only applies for mole fractions from 0 to 0·01 for the dilute constituent. It follows, therefore, that the vapour pressure of the volatile component should be at least 10^4 times greater than that of the non-volatile component to allow a simple weight loss technique to be applicable to a binary system which obeys Raoult's law across the composition range. Then, even when the volatile component is present at an atom fraction 0·01, the contribution to the weight loss from the almost pure non-volatile component would only account for one percent of the total weight loss.

This difference between the vapour pressures of two elements would largely arise from the difference in the values of the heats of vaporization, since according to Trouton's rule, the entropy of vaporization is approximately constant for all metals at about 23 entropy units. The difference in free energy in vaporization between two elements at 1000°K to make a difference of 10^4 between the vapour pressure is simply obtained thus:

$$\Delta G^\circ_{evap} = \Delta H^\circ_{evap} - T\Delta S^\circ_{evap} = -RT \ln p^\circ$$

$$\delta \Delta G^\circ_{evap} \cong \delta \Delta H^\circ_{evap} \cong RT \ln \frac{p^\circ_1}{p^\circ_2} = 4{,}575 \times 4$$

$$= 18\,300 \text{ cal mol}^{-1} \text{ at } 1000 \text{ K}.$$

Since we now have reliable information for the free energy of vaporization of the major common elements, this criterion can be used to assess the applicability of the simple weight loss technique for binary alloys between these elements.

TABLE XVI. Vapour pressure data for the commoner metals

Metal	M.pt.(°C)	B.pt.(°C)	Temp. at which $p = 1$ mm.(°C)	$\Delta G^{\circ}_{\text{vaporization}}$ (cal mol^{-1})
Sodium	98	882	439	$26\,440 + 5\cdot40T\ \log T - 39\cdot44T$
Potassium	64	779	341	$21\,820 + 6.27T\ \log T - 39\cdot80T$
Magnesium	650	1105	621	$31\,400 - 22.80T$
Calcium	843	1483	800	$37\,300 - 21\cdot25T$
Aluminium	659	2450	1284	$72\,800 - 27\cdot18T$
Copper	1083	2570	1628	$73\,110 - 25\cdot70T$
Silver	961	2200	1357	$62\,590 - 25\cdot69T$
Zinc	420	907	487	$28\,340 - 24\cdot04T$
Cadmium	321	765	394	$24\,400 - 23\cdot53T$
Tin	232	2750	1492	$71\,830 - 25\cdot03T$
Lead	327	1740	973	$44\,610 - 22\cdot42T$
Iron	1536	3070	1787	$84\,440 - 26\cdot38T$
Nickel	1455	2810	1810	$94\,600 - 30\cdot31T$
Manganese	1244	2010	1292	$55\,500 - 24\cdot00T$

The vapour pressure data which are gathered in Table XVI are for the common metals in the liquid state. The free energy equations have the gas at one atmosphere pressure as the standard state, and hence the pressures which will be obtained from the equation

$$-\Delta G^{\circ} = RT \ln p^{\circ}$$

are for the pure liquid elements and are in atmospheres as units. The equations for most of the metals have been given in the Ellingham two-term form since this is a good approximation within the normal experimental error for measurement of the free energy of vaporization (± 200 cal). The equations for sodium and potassium, however, include a term for the heat capacity change on vaporization which is important in these cases.

The heat capacities of the monatomic vapour species which predominate in all of the cases cited in the Table have the constant value derived from the equipartition law, viz. 5 cal mol^{-1} deg^{-1}, but the values for the condensed phases vary significantly with temperature, especially in the temperature range up to 1000 K (Fig. 40). This effect can be seen clearly in the curves for Na and K, and the value for tin has a marked dependence on temperature between the melting point and 600 K. However, the vapour pressure of tin is far too low to be measurable at this temperature and so the inclusion of

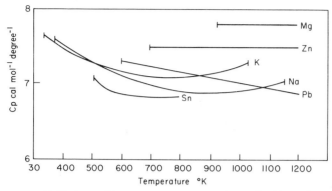

Fig. 40. Heat capacities at constant pressure of some liquid metals.

the complication of the heat capacity change term in the free energy equation for vaporization is not justified in the present instance.

The Knudsen effusion and gas transportation techniques function most satisfactorily in the pressure range 10^{-8}–10^{-4} atm. It can be seen from the vapour pressure data which are presented in Table XVI, that this technique has a wide applicability to alloys involving the common metals in the liquid range. In those binary alloys where the vapour pressures of the two components are very similar, it is necessary to incorporate a gas sampling device to separate the contributions of the two metal vapours to the total weight loss. Zellars, *et al.* (1959) condensed the exit vapours from their gas transport studies of iron–nickel liquid alloys and analyzed chemically for each component. The consistency of the results for each component could be checked for accuracy by the interrelation between the two activities through the Gibbs–Duhem equation (Fig. 41).

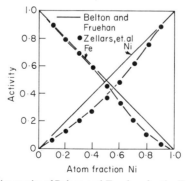

Fig. 41. A comparison of the results of Belton and Fruehan for the Fe + Ni system with those of Zellars *et al.*

Belton and Fruehan (1967) obtained results which are in very good agreement with the gas transport studies on the iron-nickel alloys by the use of a mass spectrometer and a novel form of the Gibbs-Duhem integration. In the spectrometer, the ion current intensities I^+ can be measured corresponding to the ionized beam which is obtained by electron bombardment of the mixed molecular beam of atoms Fe° and Ni° effusing from the Knudsen cell. These measured ion intensities are proportional to the vapour pressures according to the equation

$$I_i^+ = \frac{p_i \sigma_i \beta_i}{CT}$$

σ_i is the ionization cross section of the species i
β_i is the relative detector efficiency for the ions I_i^+
C is an instrumental constant containing the difference between the energy of the bombarding electrons and the critical energy for the ionization process

$$M^\circ + e^- \rightarrow M^+ + 2e^-$$

This constant also contains geometrical factors related to the position of the Knudsen cell with respect to the electron beam.

The activity of a component is related to the ion current thus

$$a_i = \frac{p_i}{p_i^\circ} = \frac{I^+ C}{\sigma_i \beta_i} \cdot \frac{1}{p_i^\circ}$$

where p_i° is the vapour pressure of the pure component. For the iron-nickel system, according to the Gibbs–Duhem equation

$$X_{Fe} \, d \ln a_{Fe} = -X_{Ni} \, d \ln a_{Ni}$$

and adding $X_{Ni} \, d \ln a_{Fe}$ to each side and re-arranging

$$d \ln a_{Fe} = -X_{Ni} \, d \ln \frac{a_{Ni}}{a_{Fe}} \, (X_{Fe} + X_{Ni} = 1)$$

Substituting for the activities on the right hand side in terms of ion currents

$$d \ln a_{Fe} = -X_{Ni} \, d \ln \frac{I_{Ni}^+}{I_{Fe}^+} \quad (p_{Fe}^\circ \,\&\, p_{Ni}^\circ \text{ are constants at a given temperature})$$

hence,

$$\ln a_{Fe} = -\int_{X_{Fe}=1}^{X_{Fe}=X_{Fe}} X_{Ni} \, d \ln \frac{I_{Ni}^+}{I_{Fe}^+}$$

and

$$\ln \gamma_{Fe} = -\int_{X_{Fe}=1}^{X_{Fe}=X_{Fe}} X_{Ni} \, d \left[\ln \frac{I_{Ni}^+}{I_{Fe}^+} - \ln \frac{X_{Ni}}{X_{Fe}} \right]$$

Results have been obtained using this procedure by the author and co-workers for a number of liquid copper alloys, $Cu + Sn$, $Cu + Ga$ and $Cu + Ge$ which have not been measured by means of the gas transport method. These results all show a characteristic asymmetry in the free energies and heats of mixing which is anticipated by the free electron theory of alloy thermodynamics of systems formed between elements of Groups I and IV (Fig. 42).

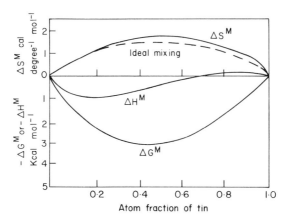

Fig. 42. Thermodynamic properties of the liquid $Cu + Sn$ system obtained from a number of mass spectrometric studies.

When the vapour pressure of one component is higher than 10^{-4} atmos. in the temperature range of interest, the Knudsen and gas transport techniques can no longer be satisfactorily used. The technique which has been employed by Eldridge *et al.* (1966) seems to present an elegant solution under these circumstances. This isopiestic technique where for example liquid magnesium at a variable but known temperature is equilibrated with $Mg + Sn$ alloys at another fixed temperature can be used in suitable circumstances over a wide range of composition and temperature. In this instance, the system $Mg + Sn$ was studied from 990–1290°K and from $X_{Sn} = 0.05$–0.90. The principal source of experimental difficulty in this method is that inert container materials are needed for the volatile metal vapour phase. These workers used an evacuated and sealed titanium tube as a container for the system and held the alloys in carbon crucibles. The results for this system which show strong negative deviations from Raoult's law are typical of what might be expected in the liquid alloys of a system in which quite stable intermetallic compounds can be formed in the solid alloys.

Electromotive force techniques have found some application for alloy systems, but generally these have not proved so reliable as the vapour pressure methods. To employ a fused salt electrolyte in the study of a liquid

F

metal system, it is obvious that if a current through the electrolyte is carried by a cationic species, the species should belong to one of the metals in the alloy only. This usually requires that the free energies of formation of the corresponding compounds should be widely different. Laurie *et al.* (1966) studied the silver-tin system by the use of a molten electrolyte containing 58 mole per cent LiCl–42 mole percent KCl and a small amount of stannous

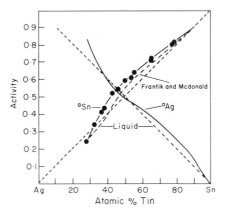

Fig. 43. A comparison of results for the liquid Ag + Sn system.

-------- = Results of Laurie, Morris and Pratt.
— — — = Results of Frantik and McDonald, *Trans. Electrochem. Soc.* **88**, 243, 253 (1945).

chloride. The heat of formation of $SnCl_2$ is $-83.6 \, kcal \, mol^{-1}$ and that of AgCl is $-30.3 \, kcal \, mol^{-1}$ and the entropies of formation are roughly the same. The displacement reaction in the electrolyte

$$2[Ag] + \{SnCl_2\} \rightarrow [Sn] + 2\{AgCl\}$$

has an equilibrium constant which does not allow significant substitution of silver for tin in this electrolyte until the activity of tin is quite low. The range of composition which was studied by this technique was from nearly pure tin down to $X_{Ag} = 0.265$ in the liquid system, and the activities are approximately Raoultian (Fig. 43).

One other source of uncertainty in the EMF technique is that the assumption must be made that only one cationic species occurs in the electrolyte, e.g. only Sn^{2+} not accompanied by Sn^{4+}. This is because the activity of tin in this instance is obtained from the thermodynamic relationship

$$E = \frac{RT}{nF} \ln a_{Sn}$$

where $n = 2$ for Sn^{2+}. There is therefore a second cation-displacement reaction to be considered in this system

$$2\{Sn^{2+}\} \rightarrow [Sn] + \{Sn^{4+}\}$$

which can be evaluated from thermodynamic data for $SnCl_2$ and $SnCl_4$ together with the activities of tin in the alloy system.

Yang, Hudson and Chien (1961) studied the EMFs of cells involving elements which form a number of chlorides in a study of the effective valency of the cation in the molten salt phase. The cells had a Ag–AgCl reference and the molten salt solvent was the LiCl–KCl liquid eutectic.

$$
\begin{array}{c|c|c|c|c}
M & MCl_n \text{ in LiCl–KCl} & \dfrac{\text{Pyrex}}{\text{membrane}} & \dfrac{\text{LiCl–KCl–AgCl}}{\text{eutectic}} & Ag
\end{array}
$$

It was found that the valency 3 predominated for uranium and niobium at temperatures between 500 and 600°C and that the higher valencies of these metals were not present in the melt to any significant extent. The difference between the free energies of formation of the chlorides which might be found in these molten salt electrolytes may be assessed approximately from the respective heats of formation of the solids.

TABLE XVII.

Compound	$\Delta H^{\circ}_{298}/\text{mol Cl}_2$
$U\,Cl_3$	142
$U\,Cl_4$	126
$U\,Cl_5$	105
$U\,Cl_6$	90
$NbCl_3$	93
$NbCl_4$	83
$NbCl_5$	76

It is apparent from these data that quite a small difference in stability is sufficient to ensure that only one cationic species is present in the melt in equilibrium with a polyvalent metal at these relatively low temperatures. In the tin chloride system, the difference in heat of formation per mole of chlorine of $SnCl_2$ and $SnCl_4$ is about 15 kcal mol^{-1}, and this would seem to be sufficient to yield only Sn^{2+} in the salt phase system unless the activity of tin in the alloy system takes on a very low value. This range was not covered by Laurie et al. in their study of the Ag + Sn system.

The volatility of the halides which makes them so important in vapour transport reactions for metal systems places an upper limit on the temperature of molten salt EMF measurements. In the light of earlier comments, it is obvious that the fluorides are least susceptible to vaporization as electrolyte systems.

A method which for obvious reasons has found very little application is the measurement of the partition of an element between two immiscible metals, as a means for studying alloy systems. One very important exception to this has been the use made by Chipman and co-workers (1961) in the determination of the activity of silicon in solution in liquid iron. Iron and silver are virtually immiscible in the liquid state, and the silicon activity coefficient in Ag + Si alloys has a value close to unity, and hence the partition of silicon between liquid iron and silver may be used to obtain the activity coefficient of silicon in solution in liquid iron. Furthermore, the interaction of carbon and silver is also very small and so the activities of silicon in liquid Fe + Si + C alloys were also obtained in this manner. This information has very practical application in the analysis of the iron blast furnace.

THE MEASUREMENT OF HEATS OF FORMATION

When compared with the heats of formation of the oxides, sulphides and halides, the heats of formation of liquid alloy systems are usually quite small. Thus, integral heats of mixing of the order of one kcal g atom^{-1} are typical of most systems of industrial significance. No large thermal effects are to be anticipated in metal-winning reactions in the metallic phases, the main interest being in the effects of the heats of solution on the activity coefficients of the components, especially in the dilute solution range.

One of the most significant contributions to the knowledge of heats of mixing is that of Kleppa (1958) who developed the reaction calorimeter in which the heat of mixing was measured directly. His studies of zinc and cadmium alloys with a number of elements gallium, indium, tin, lead and bismuth gave the first experimental indication of the importance of the electron/atom ratio in determining the asymmetry of the heat of mixing curve as a function of composition. Heat effects on mixing between 50 and 300 cal could be measured to $\pm$ one percent when 5 to 25 cc of liquid metal were mixed (Fig. 44).

Very sensitive calorimeters have been developed from the work of Calvet by Laffitte and co-workers in which a multi-junction thermopile is used to detect very small heat effects. These highly sensitive devices make it possible to obtain accurate values of heats of mixing over a range of temperatures and recently the departures of real alloy systems from the Kopp–Neumann law have been clearly demonstrated in this way. A good example is in the results for the Ag + In system which were obtained by Castenet, Claire and Laf-

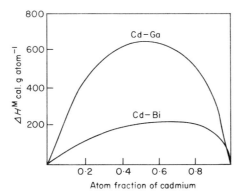

Fig. 44. Heats of mixing in the liquid Cd + Ga, Cd − Bi systems as a function of composition. (After Kleppa, loc. cit.)

fitte (1970) where there is a difference of 135 cal g atom^{-1} for a mean value of 410 cal g atom^{-1} in ΔH^M at $X_{Ag} = 0.25$ (Fig. 45). Such a variation acts as a word of caution to the use of heat of mixing and free energy of mixing data to provide entropies of mixing. Quite false values may be obtained when the calorimetric and the activity data are for widely differing temperature ranges especially when the alloying metals are widely separated in the periodic Table.

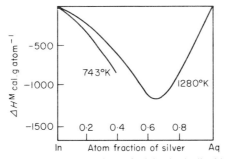

Fig. 45. The temperature dependence of the heat of mixing in the liquid Ag + In system. (After Castenet et al., loc. cit.)

REFERENCES

Abraham, K. P., Davies, M. W. and Richardson, F. D., (1960). *J. Iron Steel Inst.* **196**, 82, Activities of MnO in silicate melts.

Alcock, C. B., Sridhar, R. and Svedberg, R. C., (1970). *J. Chem. Thermodynamics* **2**, 255, Activities in Cu + Ga and Cu + Ge liquid alloys.

Baird, J. D. and Taylor, J., (1958). *Trans. Faraday Soc.* **54**, 527, Measurement of SiO$_2$ activities.

Belton, G. R. and Fruehan, R. J., (1967). *J. Phys. Chem.* **71**, 1403, Mass spectrometric study of Fe + Ni alloys.

Castanet, R., Claire, Y. and Laffitte, M., (1970). *Journal de Chim. Phys.* **67**, 789, Heats of mixing in the Ag + In system.

Charette, G. G. and Flengas, S. N., (1968). *Canadian Met. Quarterly* **7**, 191, Activities in $PbO-SiO_2$ liquid slags.

Chipman, J., (1961). 'Thermodynamics of Blast Furnace Slags In "Physical Chemistry of Process Metallurgy", Part 1, Interscience Publishers Inc. A review of experimental results.

Darken, L. S. and Gurry, R. W. (1946). *J. Amer. Chem. Soc.* **68**, 798. Thermodynamics of the iron–oxygen system.

Eldridge, J. M., Miller, E. and Komarek, K. L., (1966). *Trans. AIME* **236**, 114, Isopiestic measurements on Mg + Sn alloys.

Elliott, J. F., (1955). *Trans. AIME* **203**, 485, Activity data for CaO–FeO–SiO₂ slags.

Friedel, J. (1954). *Adv. Phys.* **3**, 446. Wave mechanics of dilute metallic solutions.

Grieveson, P. and Alcock, C. B. (1961). "Special Ceramics", p. 183, Heywood and Co., London. Thermodynamics of silicon carbide by Knudson effusion.

Hultgren, R., Orr, R. L., Anderson, P. D. and Kelley, K. K., (1965). "Selected Values of Thermodynamic Properties of Metals and Alloys", John Wiley, London.

Kleppa, O. J., (1958). *Acta Met.* **6**, 225, Heats of mixing in Zn + Cd, Zn + Ga, Zn + In and Zn + Sn alloys; ibid. **6**, 233, Results for cadmium alloys.

Kubaschewski, O., Evans, E. Ll. and Alcock, C. B., (1967). "Metallurgical Thermochemistry", Pergamon Press, London. A general source for techniques and data.

Laurie, G. H., Morris, A. W. H. and Pratt, J. N., (1966). *Trans. AIME* **236**, 1390, Activities in Ag + Sn alloys.

Schuhmann, R. and Ensio, P. J., (1951). *J. Metals* **3**, 401, Activities in the FeO–SiO₂ system.

Waghorne, R. M., Rivlin, V. G. and Williams, G. I., (1967). "The properties of Liquid Metals", (Ed.) P. D. Adams, H. A. Davies and S. G. Epstein, Taylor and Francis Ltd., London. Coordination number in liquid metallic systems.

Yang, L., Hudson, R. G. and Chien, C. Y., (1961). "Physical Chemistry of Process Metallurgy", Part 2, p. 925, Interscience Publishers. Equilibria between metals and LiCl–KCl eutectic metals.

Zellars, G. R., Payne, S. L., Morris, J. P. and Kipp, R. L., (1959). *Trans. AIME* **215**, 181, Activities by gas transport technique.

9

KINETICS OF METAL–SLAG SYSTEMS

THE SEPARATION OF SLAG FROM LIQUID METAL

The two condensed phases which are formed during the metal extraction reaction are usually in the form of two liquids in a varying state of mutual dispersion. These liquids frequently have significantly different physical properties and it is these properties which determine the ease of separation.

The densities of liquid metals are usually considerably higher than those of the liquid oxide slags, and hence separation under gravity is nearly always possible. Table XVIII shows some typical values which represent the metals of major importance in industry. The metals of Groups I–IIIA have lower densities than the commoner non-ferrous metals, but they are generally extracted from their halides. A consideration of these systems is given later in this Part.

The rate at which the separation of metal and slag can be achieved depends on the degree of aggregation of the two phases. A very typical situation is that of metal spheres or droplets falling through a layer of liquid slag. The descent is usually quite slow and so it is appropriate to calculate the rate of fall of a metallic sphere of mass m and radius r through the liquid slag layer of viscosity v by means of Stokes' equation. This shows that the limiting constant velocity, V, of descent of the liquid metal sphere which is eventually reached is given by the equation

$$V = \frac{(m - m_0)\, g}{6\pi v r}$$

Here, m_0 is the mass of liquid slag which is displaced by the metal sphere and g is the gravitational constant. The function $6\pi v r V$ is the frictional force which opposes the fall of the metal droplet.

TABLE XVIII. Densities of typical metals, oxides and sulphides in the liquid state

Metals	Density ($g\ cm^{-3}$)	Temperature (°C)
Aluminium	2·357	700
	2·304	900
Zinc	6·58	500
	6·22	800
Tin	6·92	300
	6·34	1200
Iron	7·03	1550
	6·80	1700
Copper	7·99	1100
	7·66	1500
Lead	10·56	400
	10·17	700
Oxides		
Ferrous oxide	4·55	1410
Boric oxide	1·49	1400
FeO-CaO-SiO_2 slags	3·30–3·70	1410
Sulphides		
Lead sulphide	6·47	1200
Antimony sulphide	3·70	800

An oxide slag is more viscous than liquid metallic phases and has a viscosity which at a given temperature can be markedly affected by control of the chemical composition of the slag. The choice of slag-making components to be added to the furnace burden is therefore an important consideration not only from the chemical point of view which has already been discussed but also in connection with the physical properties of the slag.

A high-silica slag will always have a higher viscosity than the corresponding slag with a high metal oxide content. This is because the structure of silica, which is a 3-dimensional network, is broken down by metal oxide addition, and the high viscosity of siliceous slags is a direct result of the presence of the silica network. The retention of the metal, which is to be extracted, in the slag phase as an oxide in not desirable, and so calcium carbonate, which is transformed to CaO at the high temperatures, is added as a slag-forming metal oxide. The addition of calcium carbonate also adds CO_2 to the furnace burden, but this is evolved at quite low temperatures and is merely a source of heat loss due to the endothermic nature of the reaction

$$CaCO_3 \rightarrow CaO + CO_2 \quad \Delta H° = 42·7 \text{ kcal}.$$

Returning to the structures of molten silicates, the discussion of the physical properties of liquid slags is largely based upon information which has been drawn from the corresponding solids. It is known that the structure of crystal-line quartz shows a fourfold coordination of silicon with oxygen. Each oxygen atom, or ion, is shared between two silicon atoms giving the stoichiometry SiO_2. The crystal structure is therefore made from oxygen tetrahedra with all corners shared.

In the ortho-silicates, e.g. $2CaO . SiO_2$ or Ca_2SiO_4, there are separated tetrahedra having no corners shared, with metal ions, in this example Ca^{2+}, regularly distributed between the tetrahedra. The addition of CaO to SiO_2 thus leads to an increasing separation of the basic tetrahedral structural units. Discrete anionic species $Si_2O_7^{6-}$, $Si_3O_9^{6-}$, etc. have been detected in naturally occurring silicates, and as well as these discrete anions, there are others of a chain-like nature, e.g. SiO_3^{2-} in which each tetrahedron is joined to its neighbours in the chain at one corner. Ring structures have also been found in some minerals, e.g. the $(Si_6O_{18})^{12-}$ structure which is found in beryl.

These large units are reduced in size by metal oxide additions to yield elementary silicate tetrahedra, and it is believed that it is occurrence of the same degradation in the liquid slag phase which accounts for the decrease in viscosity as the metal oxide/silica ratio is increased. Other minor anion-forming species phosphorus and aluminium occur in oxide ion tetrahedra which are derived from PO_4^{3-} and AlO_4^{5-} if we use a purely ionic description of the bonds.

A typical example of the drastic changes which occur in the viscosities of liquid slags as a function of composition is shown by the comparison in Table XIX.

TABLE XIX. Viscosity of liquids which occur in slags

	Temperature	Viscosity
SiO_2	1940°C	$1·5 \times 10^5$ poise
$CaO:2SiO_2$	1800°C	100 ,,
$CaO:SiO_2$	1800°C	0·7 ,,
$CaO:Al_2O_3$	1900°C	4·0 ,,
$CaO:Al_2O_3:SiO_2$	1900°C	4·0 ,,

The viscosities of liquid oxide slags have temperature coefficients which follow the equation characteristic of an activated process, the Arrhenius equation. Typical values of the activation energy for viscous flow are between 40 and 60 kcal, thus there is a substantial change with temperature, the viscosity decreasing markedly as the temperature increases

$$v = v_0 \exp \Delta E/RT.$$

For comparison with liquid silicates, it should be noted that water has a viscosity of 5×10^{-3} poise at 50°C with an activation energy of 3·5 kcal,

TABLE XX. Viscosities of liquid metals just above the melting
point

Metal	Melting point °C	Viscosity (centipoise)
Sodium	97·8	0·68
Lead	327	2·56
Zinc	420	3·79
Silver	961	3·88
Iron	1535	4·95

whilst a typical liquid metal has a viscosity between 1–10 centipoise with a
similarly small value of ΔE (5–10 kcal).

THE MEASUREMENT OF VISCOSITY

The viscosities of the metal and slag phases which are encountered in
extractive metallurgy range from 10^{-2} to 10^5 poise and the measurement of
viscosity must therefore be carried out with very flexible techniques to cover
such a range. Two techniques have now found considerable application, the
electromagnetic rotating crucible viscometer of Bockris, Lowe and Mackenzie (see Mackenzie, J. D. (1959)), and the oscillating cylinder method used
by Towers and Kay see McKenzie (1959).

In the rotating crucible method, the crucible containing a sample of the
liquid to be studied is rotated at constant velocity, and a measurement must
be made of the torque which is exerted on a cylinder immmersed from a
torsion wire suspension within the volume of the liquid. In the electromagnetic variant of this procedure, the torque acting on the suspension because
of the viscous drag of the experimental liquid was counterbalanced by an
electro-magnetic torque which was exerted between electromagnetic coils
mounted on the suspension wire, and permanent magnets placed around the
coils. The furnace which was used to heat the sample was molybdenum wound
and capable of operating up to 1800°C.

The basic equation which is applicable to this experimental procedure
shows that the torque exerted by the viscous drag of the liquid upon the
immersed cylinder is given by

$$\tau = 4\pi v \Omega \frac{a^2 b^2}{b^2 - a^2} (L + E)$$

v is the viscosity of the liquid
Ω is the angular velocity of rotation of the crucible
a is the radius of the immersed cylinder which is immersed to a depth L
    in the liquid
b is the inner radius of the crucible

E is a correction for end effects arising from the finite length of the suspended cylinder

Another method for the measurement of viscosity which has found a good deal of value at high temperatures is the oscillating bob or cylinder method. In the latter procedure, a cylinder is suspended from a torsion wire or tape in the liquid under investigation, and the decrease of amplitude of oscillations of the cylinder while immersed in the liquid is compared with the damping effect of liquids of known viscosity. The unknown viscosity is obtained from the use of the equation

$$v = \frac{(DI)^{\frac{1}{2}}}{K}.\lambda.$$

where D is torque constant of the suspension

I is the moment of inertia of the oscillating cylinder

K is a calibration constant which is obtained from measurements on liquids of known viscosity

λ is the logarithmic decrement of the oscillations of the cylinder

The decrement is defined as the ratio of the amplitudes of two successive oscillations. This technique is used for viscosities down to 0·5 poise and up to about 50 poise. Its range of application is therefore much more restricted than that of the electromagnetic viscometer which has a range 0·05 to 10^5 poise.

THE SEPARATION OF METALS FROM HALIDE SLAGS

Because of their greater simplicity of structure, halide slags have very much lower viscosities than oxide slags, a typical value being that for NaCl at 900°C of one centipoise. This value is in the same range as the viscosities of liquid metals. The densities of molten halides increase broadly with the size of the cation for a given anion. Some examples are shown in the accompanying table.

TABLE XXI. Densities of some typical molten chlorides.

Salt	Density (g cm^{-3})	Temp. (°C)	Salt	Density	Temp. (°C)
LiCl	1·420	800	$CaCl_2$	2·074	800
NaCl	1·550	800	YCl_3	2·472	800
RbCl	2·176	800	$MnCl_2$	2·289	800
CsCl	2.629	800	AgCl	4·57	800
			$CdCl_2$	3·20	800
			$PbCl_2$	4·67	700

It will be clear from a comparison of these data with the values for the elements of Groups I–III that the separation of these metals from their molten salts will not occur very readily because of the similarity in density. Thus, magnesium metal has a density of $1 \cdot 55 \, cm^{-3}$ at 800°C and magnesium chloride a density of $1 \cdot 66 \, cm^{-3}$ at the same temperature. The alkali metals all have lower densities than their chlorides but the separation of these metals from the molten chlorides is a more complicated subject than can be discussed in simple terms of density differences. This is because of the mutual metal–metal halide solubilities which will be discussed later.

KINETICS OF TRANSFER OF ELEMENTS FROM METAL TO SLAGS

In liquid metallic systems, the diffusion coefficients of most atomic species lie between 10^{-6} and $10^{-4} \, cm^2 \, s^{-1}$ with a very small change in the values with temperature. In slag phases, there is a greater amount of variation, the values lying between 10^{-8} and $10^{-4} \, cm^2 \, s^{-1}$. Since the mean displacement, $\bar{x}$, of an atom with diffusion coefficient D in a time t is given by

$$\bar{x}^2 = 2Dt$$

it is clear that the transfer of an atom over 1 mm length by diffusion only would require 10^4 and 10^2 s. at each limit of the range of values for D in metal-making liquids.

A widely used technique for the measurement of self-diffusion coefficients in liquid systems is that originally devised by Anderson and Saddington (1949). A capillary of uniform diameter with one end open is filled with the liquid which contains a radioactive isotope of the element under study at a concentration C^*. The capillary is immersed in a large volume of liquid of the same composition but with no radioactive tracer; concentration C_0. The larger volume is well stirred to ensure homogeneity and thus the concentration of the radioisotope has a negligible value within this volume during the course of the experiment. After a time t of inter-diffusion of the liquid in the capillary with the large volume, the concentration of the isotopic species at a distance x from the open end C_x is given by

$$\frac{C^* - C_x}{C^* - C_0} = \left[1 - \text{erf} \frac{x}{2(Dt)^{1/2}} \right]$$

Provided that the length of the capillary is long when compared with the mean displacement of the isotope during the experiment, the average concentration of the isotope in the capillary after an experiment, $\bar{C}$, is given by

$$\frac{C^* - \bar{C}}{C^* - C_0} = \frac{2}{l}\left(\frac{Dt}{\pi}\right)^{\frac{1}{2}} \quad (l \text{ is the length of the capillary})$$

Finally, when

$$\frac{C_x}{C^* - C_0} = \frac{1}{2}, \quad \frac{x}{(Dt)^{\frac{1}{2}}} = 0 \cdot 954.$$

This technique may also be used for the study of the diffusion coefficient of a dilute solute in a liquid alloy if the capillary holds the liquid solvent containing a known concentratjon of the dilute solute, and the large volume holds the pure liquid solvent. C^* in the equations above is now the initial concentration of the solute in the capillary, and C_0 is equal to zero. The requirement for the use of this technique in alloy systems is that the diffusion coefficient of the solute must not vary between C^* and zero concentration for the solute.

Most situations in extractive metallurgy require the exchange of matter between phases of much greater thickness than a few millimeters, and therefore the phases must be well stirred in order for a reasonable rate of transfer to occur. The ideal sitation for a metal + slag system would be that the metal and slag would be so well stirred that any concentration gradient which occurs anywhere within the bulk of the liquids will soon be removed. The rate-determining step for transfer would then be the exchange of atoms across the interface between the two phases.

Unless evidence to the contrary is found for a specific system, it may generally be assumed that local equilibrium prevails at the interface between two liquid phases. The transfer of atoms to and from the interface therefore depends in part on the bulk states of motion of the two liquid phases. The transfer of matter within each phase can be approximately calculated in terms of the physical properties of the phase in a manner analogous to the calculation of heat transfer. The flux of atoms is now related to a mass transfer coefficient, k, and a concentration difference of the moving species between the bulk and the interface, thus

$$J = k(C_{\text{bulk}} - C_{\text{interface}}) \qquad \{C_{\text{bulk}} > C_{\text{interface}}\}$$

The dimensionless groups which were used to calculate the heat transfer coefficient through the gaseous phase also have analogues in the liquid phase mass transfer calculation.

For mass transfer from a solid flat surface to a liquid in laminar flow which is in a state of forced convection over the surface

$$\frac{kx}{D} = 0 \cdot 332 \left(\frac{U_{\text{bulk}} \rho x}{v}\right)^{\frac{1}{2}} \left(\frac{v}{D\rho}\right)^{\frac{1}{3}}$$

where

$\left(\dfrac{kx}{D}\right)$ is the Sherwood number at the point x, symbol N_{Sh}

$\left(\dfrac{U_{bulk}\,\rho x}{v}\right)$ is the Reynolds number at x

$\left(\dfrac{v}{D\rho}\right)$ is the Schmidt number of the liquid, symbol N_{Sc}

Thus, the Sherwood and Schmidt numbers replace respectively the Nusselt and Prandtl numbers of the heat transfer equations.

Also for natural convection from a vertical plate

$$N_{Sh} = \frac{0\cdot902\,N_{Sc}^{\frac{1}{2}}}{(0\cdot861 + N_{Sc})^{\frac{1}{4}}}\left(\frac{N_{Gr}}{4}\right)^{\frac{1}{4}}$$

Because of the marked variation of the viscosity with composition for silicate slags, it follows that the Schmidt number would also have a wide variation in these phases as the composition changes. In liquid metal systems, however, the viscosity does not vary widely usually taking a value around one centipoise. A typical value of the Schmidt number for metallic liquids is therefore 10^2 ($v = 10^{-2}$ poise, $\rho = 5\ \mathrm{g\ cm^{-3}}$, $D = 2 \times 10^{-5}\ \mathrm{cm^2\ s^{-1}}$), and as with the Prandtl number in the heat transfer equations for gases, the Schmidt number does not effect the value of the mass transfer coefficient for laminar flow in metals in a highly variable manner. The broad effect is to multiply the Reynolds number by a constant factor around 5. It is the Reynolds number which has the major effect on the variation of the mass transfer coefficient in metallic systems through the bulk velocity which appears in the Reynolds number.

For most practical applications in extractive metallurgy it is not necessary to proceed to the calculation of the mass transfer equation by means of dimensionless groups, and very simple models of the real situation can be used to predict the major effects.

Two simple idealized models have been used as a basis for a semi-quantitative calculation for the mass transfer process, the stagnant boundary layer method of Nernst, and the surface regeneration method. In the former, it is assumed that there exists a stagnant layer at the edge of each phase across which diffusion can occur according to Fick's first law, that is to say with a constant flux. The flux of a given species across the layer, J, is then given by

$$J = \frac{D}{\delta}\cdot(C_1 - C_2)$$

where δ is the thickness of the boundary layer, and C_1, C_2 are the constant, concentrations at each edge of the boundary layer.

Jackson and Grace (1961) made use of this simplified analysis in a study of dissolution of zinc cylindrical rods in liquid bismuth as a function of the rotation speed of the rod and the temperature of the liquid solvent. The equilibrium solubility of zinc in bismuth in the temperature range of the investigation 300–400°C is 4·5–12 wt% and thus the diffusion coefficient of zinc at these relatively low concentrations into the liquid bismuth through the boundary layer may be assumed to be independent of composition. The results of this study could be interpreted to show that the rate constant for the dissolution, j, changed from 4×10^{-3} to 16×10^{-3} cm s^{-1} as the speed of rotation was changed from 80 to 400 revolutions per minute. Since

$$j = D/\delta$$

and D is approximately 10^{-5} cm^2 s^{-1} it follows that δ has a value of about 10^{-3} cm for this system. This value is found to be typical for a wide range of systems and can always be used as a first approximation. It is quite wrong to think that experiments such as these have established the reality of a boundary layer. It can be clearly seen in model systems that the state of motion of two liquids which are stirred in contact with one another presents a velocity profile which changes continuously over a measurable distance from the interface. The measured properties of the system are such that these behave *as if* the profile extended over a smaller distance, the boundary layer, and varied in a simple linear manner over the thickness of the boundary layer (Fig. 46). The same remarks can be applied to the concentration gradient in real systems and the corresponding approximation of a concentration boundary layer.

When two liquids, a metal M and a slag S are in contact two simultaneous equations can be written for the flux from metal to slag of a dilute solute across the intervening boundary layers.

$$J^M = \frac{D^M}{\delta^M} \cdot (C_B^M - C_E^M): \qquad J^S = \frac{D^S}{\delta^S} \cdot (C_E^S - C_B^S)$$

where $C_B^{M,S}$ corresponds to the bulk concentration

$C_E^{M,S}$ corresponds to the interfacial concentrations.

There is no accumulation of matter at the interface, where it is assumed that local equilibrium prevails.

Thus $C_E^M/C_E^S = K_{M-S}$ the metal–slag partition coefficient and when a steady state is reached

$$J^M = J^S \text{ and} \frac{D^M}{D^S} \cdot \frac{\delta^S}{\delta^M} = \frac{(C_E^S - C_B^S)}{(C_B^M - C_E^M)}$$

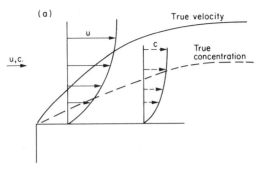

Fig. 46a. True velocity and concentration distributions normal to a fixed bed between when a diffusion-controlled transfer is occurring between the solid and liquid.

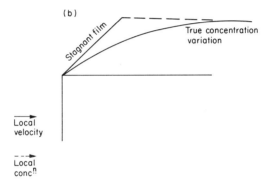

Fig. 46b. The boundary layer ("stagnant film") approximation to the true concentration variation.

If C_B^S is very small compared with C_E^S, and showing the ratio D_i/δ_i, which has the same dimensions as the mass-transfer coefficient, by the symbol, j_i, then

$$\frac{j^S}{j^M} = \frac{(C_B^M - C_E^M)}{C_E^S} = K_{M-S}\frac{(C_B^M - C_E^M)}{C_E^M}$$

$$= K_{M-S}\left[\frac{C_B^M}{C_E^M} - 1\right]$$

On re-arrangement of this equation, we obtain

$$1 + \frac{j^S}{j^M \cdot K_{M-S}} = \frac{C_B^M}{C_E^M} = G^M$$

Clearly, the value of G^M is high compared with unity when

$$j^S > j^M \cdot K_{M-S}$$

and if the value of K_{M-S} is of the order of unity or greater, then mass-transfer through the metal phase is the slower step. Conversely, if

$$j^S < j^M \cdot K_{M-S}$$

and K_{M-S} is greater than unity, the mass transfer through the slag is the slower step, and G^M will be close to unity.

When C_B^S is no longer negligible when compared with C_E^S.

$$\frac{j^S}{j^M} \cdot (C_E^S - C_B^S) = C_B^M - C_E^M$$

hence

$$\frac{j^S}{j^M} \cdot \left(\frac{C_E^M}{K_{M-S}} - C_B^S\right) = C_B^M - C_E^M$$

$$\left(1 + \frac{j^S}{j^M \cdot K_{M-S}}\right) C_E^M = C_B^M + \frac{j^S}{j^M} \cdot C_B^S = \left[\frac{K_{M-S} \cdot j^M + j^S}{K_{M-S} \cdot j^M}\right] C_E^M$$

Since $J^M = j^M (C_B^M - C_E^M)$, it follows that

$$J^M = \frac{j^M j^S}{K_{M-S} \cdot j^M + j^S}(C_B^M - C_B^S \cdot K_{M-S})$$

The rate of transfer of material in the steady state can now be calculated in terms of the bulk concentrations, the partition equilibrium constant and the boundary transfer coefficients j^M and j^S. These latter are analogous to the electrical conductivities in the equation for the transport of electricity and the inverse functions are sometimes referred to as "boundary layer resistances".

The steady state equation deduced above can also be written to bring out this similarity, by a simple rearrangement. Then

$$(C_B^M - C_B^S \cdot K_{M-S}) = J^M \left(\frac{1}{j^M} + \frac{K_{M-S}}{j^S}\right)$$

which is analogous to Ohm's law with

$$C_B^M - C_B^S \cdot K_{M-S} \equiv \text{Potential drop}$$
$$J^M \qquad\qquad \equiv \text{Current}$$

and

$$\frac{1}{j^M} + \frac{K_{M-S}}{j^S} \qquad \equiv \text{Two resistances in series}$$

This form of the mass transfer equation brings out very clearly the fact that the two resistances

$$R_M = \frac{1}{j^M} \text{ and } R_S = \frac{K_{M-S}}{j^S}$$

must be compared in order to determine which is the greater and therefore which will be rate-determining under steady state conditions of mass transfer.

Returning now to the calculation of the mass transfer coefficient in terms of dimensionless groups, we can also evaluate the ratio of the two boundary layer resistances shown above in these terms if we write the following equations:

For laminar flow and using the symbolism { } for the slag phase and [ ] for the metal phase,

$$\frac{\{N_{Sh}\}}{[N_{Sh}]} = \frac{k^S}{k^M} \cdot \frac{D^M}{D^S} = \frac{\{N_{Re}\}^{\frac{1}{2}} \{N_{Sc}\}^{\frac{1}{3}}}{[N_{Re}]^{\frac{1}{2}} [N_{Sc}]^{\frac{1}{3}}}$$

$$= \frac{\{u_{bulk}\}^{\frac{1}{2}} [\bar{v}^{\frac{1}{6}} D^{\frac{2}{3}}]}{[u_{bulk}]^{\frac{1}{2}} \{\bar{v}^{\frac{1}{6}} D^{\frac{2}{3}}\}}$$

$\bar{v} = v/\rho$, the kinematic viscosity.
Through the equivalence

$$k \equiv j = \frac{D}{\delta}$$

It follows that

$$\frac{\delta^S}{\delta^M} = \frac{[u_{bulk}]^{\frac{1}{2}} \{\bar{v}^{\frac{1}{6}} D^{\frac{2}{3}}\}}{\{u_{bulk}\}^{\frac{1}{2}} [\bar{v}^{\frac{1}{6}} D^{\frac{2}{3}}]}$$

and therefore

$$\frac{R^S}{R^M} = K_{M-S} \frac{j^M}{j^S} = K_{M-S} \frac{D^M}{D^S} \cdot \frac{\delta^S}{\delta^M}$$

$$= K_{M-S} \frac{[u_{bulk}]^{\frac{1}{2}} \{\bar{v}^{\frac{1}{6}} D^{-\frac{1}{3}}\}}{\{u_{bulk}\}^{\frac{1}{2}} [\bar{v}^{\frac{1}{6}} D^{-\frac{1}{3}}]}$$

Since the bulk velocities will not differ greatly in an experimental study and the viscosity only appears as the one-sixth power, then the diffusion coefficients largely determine the mass transfer resistance ratio thus

$$\frac{R^S}{R^M} \simeq K_{M-S} \left(\frac{D^M}{D^S}\right)^{\frac{1}{3}}$$

Usually $D_S < D^M$ and hence $R^S > K_{M-S} R^M$, and $j^M > j^S$. The transfer in the slag phase is therefore generally rate-determining in the transfer of a solute such as sulphur, from metal to slag.

One further simplification which has been introduced in this procedure is the assumption that D^M and D^S are constant quantities no matter what the concentration distribution within the boundary layers. It is well-known that diffusion coefficients within a binary system do vary with concentration. The more correct flux equation should therefore be used which employs the

TABLE XXII. Diffusion coefficients in liquid iron and liquid $CaO-Al_2O_3-SiO_2$ slags

Metallic phase	Element	Concentration	$D \times 10^5$ $(cm^2 s^{-1})$	Temperature
	Sulphur	0·96–1·24 wt%	4·5	1550
	Silicon	1·5–2·0 wt%	10	1560
	Manganese	0·8 wt%	20	1600
	Carbon	2·1 wt%	5·9	1550
Slag Phase	Sulphur	1·6–1·8 wt%	0·15	1515
	Silicon	$\begin{cases} 40\text{--}20\text{--}40 \text{ slag} \\ CaO-Al_2O_3-SiO_2 \end{cases}$	0·01	1430
	Iron	$\begin{cases} 40\text{--}20\text{--}40 \text{ slag} \\ CaO-Al_2O_3-SiO_2 \end{cases}$	0·4	1500

thermodynamic activity gradient rather than the concentration gradient as the driving force for diffusive flow

$$J_i = \frac{B_i c_i}{N} \frac{\partial \Delta_{\mu_i}}{\partial x}$$

where B is the mobility coefficient. It has been pointed out earlier that

$$D_i = B_i kT \left[1 + X_i \frac{d \ln \gamma_i}{d \ln X_i} \right]$$

where γ_i and X_i are the activity coefficient and mole fraction of the ith species. Hence, it is only when the activity coefficient is independent of composition that there is any likelihood that the use of a constant value of D_i could be justified. This is so for dilute solutions which obey Henry's law. At present, it is not possible to make a rigorous calculation of the variation of B_i, the mobility coefficient, with composition, and hence the extension of the use of the simple equations for mass transfer to the concentrated solutions could not be made precisely, even if the values of δ_i could be calculated from a fundamental mathematical treatment.

The alternative approach to mass transfer rates across interfaces, the surface replenishment model, assumes that packets of bulk phase are propelled by convection to the equilibrium interface and here the composition is readjusted to the equilibrium value. The difference of approach is that the extent of exchange across an interface at any given instant depends upon the fraction of the interfacial area which contains the elements of the bulk phase. The extent to which the composition of any packet can be altered clearly depends upon the time of residence of the packet at the interface, during which diffusional transfer can occur, and on the diffusion coefficients in metal and slag phase. The general form of the mass transfer equation

$$J_i = j_i (C_B^i - C_E^i)$$

can still be applied, but the mass transfer coefficient j_i must now contain a time of residence and a fractional occupation of the interfacial area by packets of bulk phase.

It is clear that this model is no more simple to apply to the problem of metal-slag equilibration kinetics than is the boundary layer theory. The general mass transfer coefficient can be used as a phenomenological number only, without too detailed an enquiry into the exact details of the transfer process. As a result of a wide variety of studies, a typical value of δ of 10^{-3} cm can be applied as a good approximation in most circumstances.

THE MECHANISM OF DIFFUSION IN LIQUID SYSTEMS

The results of the early experimental studies of self-diffusion in liquid metals were interpreted in much the same way as were those for solid metals. The variation of the diffusion coefficient with temperature was fitted to an Arrhenius equation and an activation enthalpy for diffusion, usually around 5 kcal was obtained. The Arrhenius expression applies when a rate process occurs by means of a number of identical unit steps; thus in the solid state an atom moves to a vacant site in the lattice, leaving behind an identical vacancy to replace the filled one. It is easy to see why the unit step should always be the same for solid systems where the structure is a regular three-dimensional lattice. The same argument cannot be transposed to liquid systems mainly because this long-range structural order is not a property of liquid systems. In these, it is more probable that thermal vibrations will generate unoccupied volumes in the liquid into which atoms may flow by a diffusive step, but that the length of these translations could well vary over a wide range of values. The regular fixed step of the solid is replaced by a step of varying length in the liquid. The activation enthalpy for the formation of these holes in the structure and the enthalpy for the movement of atoms into them will also vary in the liquid.

A system in which a number of processes occur with a number of activation enthalpies does not fit the Arrhenius equation, but tends to have a low apparent activation enthalpy at low temperatures and a higher value at higher temperatures. In the case of liquid diffusion, this would suggest that the average diffusion steps grows longer as the temperature increases. Swalin (1959) was the first to show conclusively that the apparent fit to an Arrhenius expression for most previous work on liquid metal diffusion was probably due to the narrow range of temperatures which had been employed for any single system. When a wider range of temperature could be achieved, then significant departures were found from the Arrhenius equation. Instead of the relationship

$$D = D_0 \exp - (\Delta H/RT)$$

Swalin found for liquid tin and indium that the simpler equation which he had derived

$$D = \frac{1 \cdot 29 \times 10^{-8}}{A} T^2$$

fitted the experimental results adequately. The constant A which has the dimensions cal cm^{-2} mol^{-1} can be calculated approximately from structural data for the force constants α, of a particular liquid metal, and the heat of vaporization per g atom ΔH_v. Swalin assumed that metal atoms interact in a pairwise fashion according to a Morse function. The energy to increase the distance between two neighbouring atoms by an amount x, which is related to the energy to form a vacancy for diffusion, is given by

$$E = E_D \left[1 + \exp(2\alpha x) - 2 \exp(-\alpha x) \right]$$

E_D is the energy required for the breaking of the chemical bond between the atoms and this is obtained in good approximation by the equation

TABLE XXIII. Self diffusion coefficients of liquid metals expressed by an Arrhenius equation.

Metal	Diffusion coefficient $\times 10^4$	Temperature range (°C)
Copper	$14\cdot6 \exp(-9700/RT)$	1140 1260
Silver	$7\cdot10 \exp(-8150/RT)$	1000–1100
Zinc	$8\cdot2 \ \exp(-5090/RT)$	450–600
Mercury	$0\cdot85 \exp(-1000/RT)$	0–98
Gallium	$1\cdot07 \exp(-1120/RT)$	30–100
Indium	$2\cdot89 \exp(-2430/RT)$	170–750
Tin	$3\cdot24 \exp(-2760/RT)$	300–600
Lead	$9\cdot15 \exp(-4450/RT)$	600–930
Sodium	$11\cdot0 \ \exp(-2450/RT)$	100–220
Potassium	$16\cdot0 \ \exp(-2500/RT)$	340–490

$$E_D = \frac{2\Delta H_v}{ZN_0} \text{ per bond,}$$

where Z is the coordination number of each atom and N_0 is Avogadro's number. The constant A is then given by the relationship

$$A = \Delta H_v \alpha^2.$$

The Morse function which is given above was obtained from a study of bonding in gaseous systems, and this part of Swalin's derivation is the weakest. The general idea of a variable diffusion step in liquids which is more nearly akin to diffusion in gases than the earlier treatment which was based on studies carried out in solids remains as a valuable suggestion.

THE MECHANISMS OF METAL-SLAG TRANSFER REACTIONS

The transfer of an element from the metal to the slag phase is one in which the species goes from the free-electron metallic phase to an essentially ionic medium in the slag. It follows that there must be some electron redistribution accompanying the transfer in order that electroneutrality is maintained throughout. A metallic atom must be accompanied by oxygen which will absorb the electrons which are released in the formation of the metal ion thus

$$[Mn^\circ] + \tfrac{1}{2}O_2 \rightarrow \{Mn^{2+}\} + \{O^{2-}\}$$

$[Mn^\circ]$ indicates an atom of manganese in solution in a metallic phase. Another possibility is that electroneutrality can be maintained by the exchange of particles across the slag–metal interface thus

$$\{O^{--}\} + [S] \rightarrow \{S^{--}\} + [O]$$

For any given chemical species, there are probably many ways in which the transfer across the slag–metal boundary can be affected under the constraint of conservation of electric charge. King and Ramachandran (1958) have shown that sulphur transfer from liquid iron saturated with carbon to slag can occur by any of the following processes:

$$[S] + \{O^{--}\} + [C] \rightarrow \{S^{--}\} + CO(g)$$
$$[S] + [Fe] \rightarrow \{Fe^{++}\} + \{S^{--}\}$$
$$2[S] + [Si] \rightarrow \{Si^{4+}\} + 2\{S^{--}\}$$

The experimental arrangement consisted of a graphite crucible containing molten iron, which, of course, was saturated with carbon, in which was dissolved a fixed initial weight of sulphur. Above this was a liquid lime–

alumina–silica slag (CaO 48%, Al_2O_3 21% and SiO_2 31%), and the whole
system was maintained at 1500°C. The rate of evolution of carbon monoxide
was measured, and by taking slag samples periodically, the sulphur, iron and
silicon contents of the slag were obtained as a function of time. It was found
that the number of equivalents of sulphur which were transferred to the slag
from the metal was at all times equal to the number of equivalents of iron and
silicon transferred plus the number of equivalents of oxygen which were
evolved as carbon monoxide (Fig. 47). This equivalence is the same as that

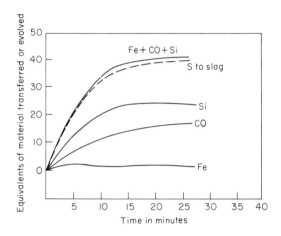

Fig. 47. Rates of transfer of components between metal, slag and gas during the desulphurization
of carbon-saturated liquid iron. (After King and Ramachandran, loc. cit.)

required by electrical charge conservation during the transfer, and hence
suggested the reaction scheme which is shown above.

Because of the requirement for electric charge conservation, it is possible
under suitably chosen conditions of experiment to force the transfer of an
element *up* a chemical potential gradient. Turkdogan and Grieveson (1962)
separated a molten Fe–S–O alloy from a mixture of Mn/MnO/MnS by a
liquid layer of sodium silicate at 1200°C. The sulphur and oxygen pressures
were both initially higher on the iron alloy side of the membrane than on the
manganese side. Nevertheless, because electron transport was negligibly
small in the silicate layer, the only way in which oxygen transfer could occur
from iron to manganese was if sulphur could transfer atom for atom from
manganese to iron. Hence, the sum of oxygen and sulphur remained constant
on either side of the membrane, and electric charge was conserved in the
molten silicate.

$$\text{Mn–MnO–MnS} \quad \underset{\substack{\text{Molten} \\ \text{sodium silicate}}}{} \quad \text{Fe–O–S}$$

$$p_{O_2} = 8{\cdot}7 \times 10^{-22}\,\text{atm} \xleftarrow{\quad O\text{--}\quad} p_{O_2} \text{ initially} = 9{\cdot}6 \times 10^{-14}\,\text{atm}$$

$$p_{S_2} = 7{\cdot}2 \times 10^{-14}\,\text{atm} \xrightarrow{\quad S\text{--}\quad} p_{S_2} \text{ initially} = 10^{-6}\,\text{atm}$$

REFERENCES

Anderson, J. S. and Saddington, K. (1949). *J. Chem. Soc.* **152**, S381, Capillary-reservoir technique for liquid diffusion measurement.

Jackson, J. K. and Grace, R. E. (1961). "Physical Chemistry of Process Metallurgy", Part 1, p. 633, Interscience Publishers. Kinetics of the dissolution of Zn in liquid bismuth.

King, T. B. and Ramachandran, S. (1958). "Physical Chemistry of Steelmaking", (ed.) J. F. Elliott, Technology Press. Slag-metal transfer of sulphur.

Kozakevitch, see Reference 2, p. 97, Viscosity of $CaO–Al_2O_3–SiO_2$ melts between 1600 and 2100°C.

Mackenzie, J. D. (1959). "Physicochemical Techniques at High Temperatures", (ed.) J. O'M. Bockris, J. L. White and J. D. Mackenzie, p. 313, Butterworth. The measurement of viscosity.

Smithells, C. J. (ed.) (1967). "Metals Reference Book", 4th edition, Butterworth. Physical data for metallic systems.

Swalin, R. A. (1959). *Acta Met.* **7**, 736. Fluctuation theory of liquid diffusion.

Turkdogan, E. T. and Grieveson, P. (1962). *Trans. AIME* **224**, 316. Transfer of sulphur and oxygen through an ionic membrane.

Walls, H. A. (1970). "Physicochemal Measurements in Metals Research", (ed.) R. A. Rapp, Vol. 2, Interscience Publishers. Diffusion coefficients in liquid metal systems.

Yang, L. and Derge, G., see reference 2, p. 503, General consideration of diffusion in metallurgical melts.

RECOMMENDED READING

"Physical Chemistry of Melts in Metallurgy", F. D. Richardson. Academic Press, London and New York (1974).

10

THE OXIDATION OF SULPHIDES TO PRODUCE METAL

INTRODUCTION

The few metals which can be obtained by the direct oxidation of their sulphides all have the characteristic property that their oxides are very much less stable than SO_2. When the pure metal is placed in contact with the lowest sulphide of the metal in an atmosphere of pure SO_2, the sulphur dissociation pressure is such that the resulting oxygen pressure of the atmosphere, which is determined by the SO_2 equilibrium

$$K_{SO_2} = \frac{p_{SO_2}}{p_{S_2}^{\frac{1}{2}} \cdot p_{O_2}}$$

is too low to cause oxidation of the metal.

As an example, the system copper in contact with Cu_2S and SO_2 at one atmosphere pressure produces an oxygen pressure which is less than that required to form Cu_2O at all temperatures. This can be seen for the solid system on the Cu–S–O stability diagram which is shown in Part I, Fig. 2 p. 8.

The metals which show this behaviour are silver, copper and mercury, together with the platinum metals. The metal lead can also be extracted by partial oxidation of galena to form a sulphate ("Scotch hearth" or Newnham process). The reaction which produces metal can be written in a very simplified form with the equation

$$PbS + PbSO_4 \rightarrow 2Pb + 2SO_2$$

This description is, however, oversimplified and a more complete picture will be assembled in this section.

THE EXTRACTION OF MERCURY FROM CINNABAR

The mineral cinnabar is a relatively pure form of HgS and the metal may be obtained by direct oxidation according to the reaction

$$HgS + O_2 \rightarrow Hg(l) + SO_2$$

$$\text{or } Hg(g) + SO_2$$

The constituent thermodynamic data for the assessment of the extent to which these reactions occur can be obtained in good approximation by using the values for ΔH°_{298} and ΔS°_{298} to form free energy equations. The following equations were derived from the published data:

$$Hg(l) \; \tfrac{1}{2}O_2 \rightarrow HgO(s) \qquad \Delta G^\circ = -21{,}700 + 25{\cdot}9\,T \text{ cal}$$
$$Hg(l) + \tfrac{1}{2}S_2 \rightarrow HgS(s) \qquad \Delta G^\circ = -29{,}500 + 25{\cdot}9\,T \text{ cal}$$

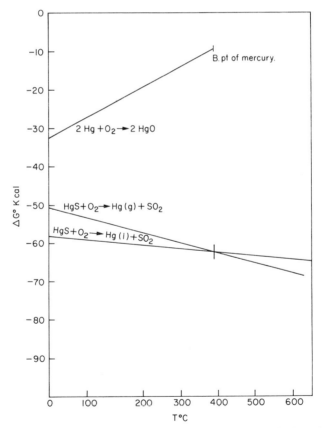

Fig. 48. Ellingham diagram for the production of mercury from HgS.

$$HgS + O_2 \rightarrow Hg(l) + SO_2 \qquad \Delta G^\circ = -57,000 - 8\cdot6\ T\ cal$$
$$Hg(l) \rightarrow Hg(g) \qquad \Delta G^\circ = 14,100 - 22\cdot4\ T\ cal$$
and hence
$$HgS + O_2 \rightarrow Hg(g) + SO_2 \qquad \Delta G^\circ = -42,100 - 31\cdot0\ T\ cal$$

These equations are shown in graphical form in Fig. 48 from which it can be deduced that the formation of elementary liquid or gaseous mercury by the oxidation of HgS is always favoured over the oxidation of the metal. Furthermore, from the fact that the standard free energy change for the principal oxidation reaction is a large negative quantity, it follows that p_{SO_2}/p_{O_2} will always be large at equilibrium, and hence that the utilization of oxygen is very efficient. This is obvious since

$$K = \frac{p_{SO_2}}{p_{O_2}}$$

when $a_{HgS} = a_{Hg} = 1$, and when ΔG° is a large negative quantity, then K is a large positive quantity. Thus at 400°C, one atmosphere of mercury vapour and solid HgS are in equilibrium with p_{SO_2}/p_{O_2} of the value 10^{20} since $\Delta G^\circ = -63\ kcal$.

NEWNHAM HEARTH PROCESS FOR LEAD PRODUCTION

In the Pb–S–O system, there are three basic sulphates besides the normal sulphate, monoxide and monosulphide. The equilibria amongst the various phases have been studied by Kellog and Basu. (1960) The stability diagram which can be obtained from the results of this study (Fig. 49) shows the high stability of the basic sulphates and suggests that the phases which could be in

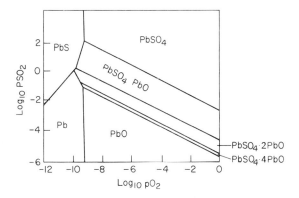

Fig. 49. Stability diagram for the Pb–S–O system at 900°C. (After Kellogg and Basu, loc. cit.)

contact in the hearth process for the production of lead are the sulphide and the basic sulphate, either $PbSO_4.2PbO$ or $PbSO_4.PbO$. When lead sulphide is partially roasted in this manner and metal is produced, SO_2 is readily evolved. The location of the phases which are in equilibrium during metal production must therefore lie near the region where p_{SO_2} is between 10^{-2} and 1 atmosphere.

However, the stability diagram deals in this instance only with condensed phases which contain lead. At metal-making temperatures, reduction of the sulphates could occur by reaction with PbS gas at a pressure lower than that in equilibrium with the solid PbS phase. According to Tuffley and Russell (1964), these reactions may be written thus

(i) $PbS(g) + 10(PbSO_4.PbO) \rightarrow 7(PbSO_4.2PbO) + 4SO_2$
(ii) $2PbS(g) + (PbSO_4.2PbO) \rightarrow 5 Pb + 3SO_2$
(iii) $3PbS(g) + (PbSO_4.4PbO) \rightarrow 8 Pb + 4SO_2$

and at 825°C liquid lead at unit activity would be one of the products. The conditions under which metallic lead can be produced are obtained from consideration of the general reaction

$$\tfrac{1}{2}S_2 + O_2 \rightarrow SO_2$$

in which the substitution is made where

$$K_{SO_2} = \frac{p_{SO_2}}{p_{S_2}^{1/2} \cdot p_{O_2}}$$

$$p_{S_2} = \left(\frac{a_{PbS(g)}}{a_{Pb}}\right)^2 \cdot \frac{1}{K_{PbS(g)}}$$

From the reaction for the decomposition $(PbSO_4.2PbO) \rightarrow 3Pb + SO_2 + 2O_2$ we derive

$$p_{O_2}^{(ii)} = \left(\frac{K' a_{(PbSO_4.2PbO)}}{a_{Pb}^3 \cdot p_{SO_2}}\right)^{\frac{1}{4}} \text{ for Reaction (ii) above}$$

and from the reaction $(PbSO_4.4PbO) \rightarrow 5Pb + SO_2 + 3O_2$

$$p_{O_2}^{(iii)} = \left(\frac{K'' a_{PbSO_4.4PbO}}{a_{Pb}^5 \cdot p_{SO_2}}\right)^{\frac{1}{3}} \text{ for Reaction (iii) above}.$$

If the condensed phases are given unit activity and the partial pressure of SO_2 is made one atmosphere, then the oxygen and sulphur pressures which coexist during the formation of metal are

$$p_{O_2}^{(ii)} = (K')^{\frac{1}{4}}; \qquad p_{O_2}^{(iii)} = (K'')^{\frac{1}{3}}$$

$$p_{S_2} = \left(\frac{p_{PbS}}{p_{PbS}^{\circ}}\right)^2 \frac{1}{K_{PbS}}$$

The values of the equilibrium constants for these reactions can be obtained from the pertinent free energy data of Kellogg and Basu which are as follows:

$PbSO_4 \cdot 2PbO \quad \Delta H_{298}^{\circ} = -324\,500 \text{ cal} \quad \Delta G_{298}^{\circ} = -287\,000 \text{ cal}$
$PbSO_4 \cdot 4PbO \quad \Delta H_{298}^{\circ} = -434\,000 \text{ cal} \quad \Delta G_{298}^{\circ} = -380\,000 \text{ cal}$

These values are the average values from the experimental data and correspond to the reactions for the formation of the compounds from the elements in their standard states at 298 K. Since we are interested in the formation of lead at much higher temperatures, it is an informative exercise to apply the heat capacity data for these phases to calculate the deviations of the free energies of these reactions at high temperatures from these equations for 298 K.

We may write the equations for temperatures around 298 K

$$3Pb(s) + 3O_2 + S(s) \rightarrow PbSO_4 \cdot 2PbO$$
$$\Delta G^{\circ} = -324\,500 + 125 \cdot 84\,T \text{ cal}$$

and $5Pb(s) + 4O_2 + S(s) \rightarrow PbSO_4 \cdot 4PbO$
$$\Delta G^{\circ} = -434\,000 + 181 \cdot 2\,T \text{ cal}$$

together with

$$S(s) + O_2 \rightarrow SO_2 \quad \Delta G^{\circ} = -71\,000 + 2 \cdot 63\,T \text{ cal}$$
$$Pb(s) \rightarrow Pb(l) \quad \Delta G^{\circ} = +1150 - 1 \cdot 92\,T \text{ cal}$$

The following equations are then obtained:

$$3Pb(l) + SO_2 + 2O_2 \rightarrow PbSO_4 \cdot 2PbO$$
$$\Delta G^{\circ} = -256\,700 + 129 \cdot 01\,T \text{ cal}$$
$$5Pb(l) + SO_2 + 3O_2 \rightarrow PbSO_4, 4PbO$$
$$\Delta G^{\circ} = -368\,700 + 188 \cdot 17\,T \text{ cal}$$

The heat capacity data for these phases are now used to calculate the deviations of the free energy at high temperatures from these equations for 298 K (Table XXIV). These data yield the heat capacity change equations for the formation of $PbSO_4 \cdot 2PbO$

$$\Delta C_p = -18 \cdot 89 + 41 \cdot 58 \times 10^{-3}\,T + 6 \cdot 42 \times 10^5\,T^{-2} \text{ cal mol}^{-1} \text{ deg}^{-1}$$

TABLE XXIV

Substance	Heat Capacity
Pb(1)	$7 \cdot 75 - 0 \cdot 74 \times 10^{-3}\,T$ cal mol^{-1} deg^{-1}
SO$_2$	$10 \cdot 38 + 2 \cdot 54 \times 10^{-3}\,T - 1 \cdot 42 \times 10^5\,T^{-2}$
O$_2$	$7 \cdot 16 + 10^{-3}\,T - 0 \cdot 40 \times 10^5\,T^{-2}$
PbSO$_4 \cdot$ 2PbO	$29 \cdot 06 + 4 \cdot 38 \times 10^{-2}\,T + 4 \cdot 2 \times 10^5\,T^{-2}$†
PbSO$_4 \cdot$ 4PbO	$47 \cdot 16 + 5 \cdot 66 \times 10^{-2}\,T + 4 \cdot 2 \times 10^5\,T^{-2}$†

† Estimated values

and for $PbSO_4 . 4PbO$

$$\Delta C_p = -23 \cdot 45 + 54 \cdot 76 \times 10^{-3} T + 6 \cdot 82 \times 10^6 T^{-2} \text{ cal mol}^{-1} \text{ deg}^{-1}$$

The free energy deviation function which is needed can be defined by the equation

$$\text{F.E.D.} = \frac{1}{T}(H_T - H_{298}) - (S_T - S_{298})$$

$$= \frac{1}{T} \int_{298}^{T} C_p . dT - \int_{298}^{T} \frac{C_p}{T} . dT$$

and using the symbol θ to represent $298/T$ K, the general equation for the F.E.D. is

$$\text{F.E.D.} = a(1 - \theta + \ln \theta) + bT \left(\theta - \frac{1}{2} - \frac{\theta^2}{2} \right) + \frac{c}{T^2} \left(\frac{1}{\theta} - \frac{1}{2\theta^2} - \frac{1}{2} \right)$$

for a reaction in which the heat capacity change is

$$\Delta Cp = a + bT + \frac{c}{T^2} \text{ cal mol}^{-1} \text{ deg}^{-1}$$

The functions which are multiplied by a, b and c in the equation for the F.E.D. function apply to all systems for which the heat capacity changes can be written in the three term form and are therefore useful generally. They are shown multiplied by T and plotted together as a function of T K for the relevant temperature range of this calculation 298–1200 K (Fig. 50). The values of the terms in F.E.D. multiplied by the temperature are the correction terms which must be added to the free energy equation for 298 K to arrive at the correct value of the free energy change at the new temperature. At 1000 K the deviation terms for the formation of $PbSO_4 . 2PbO$ are $+9615$, $-10\,395$ and $+1792$ cal respectively giving a total of $+1012$ cal. Such a small deviation can be neglected in comparison with the total experimental errors in all of the contributing data to these calculations and therefore the equation derived for ΔG_{298}° can be used as a good approximation even up to 1000 K. Similarly for the formation of $PbSO_4 . 4PbO$, the deviations are $+11\,936$, $-13\,690$ and $+1910$ cal. respectively giving a total effect of $+156$ cal., which is negligible.

In Fig. 51 the standard free energy changes for the reactions

$$Pb + \tfrac{1}{2} S_2 \rightarrow PbS$$
$$\tfrac{1}{2} S_2 + O_2 \rightarrow SO_2$$

and their combination

$$PbS + O_2 \rightarrow Pb + SO_2$$

are shown with those for the basic sulphates which are now found to be acceptable. It can be concluded from the temperature of intersection of this latter line with those for the basic sulphates that the metal can be obtained by reduction of the basic sulphates with the sulphide at temperatures above 1150 K to produce SO_2 at one atmosphere pressure. If the SO_2 pressure is reduced to 0·01 atmosphere, the intersection of the dashed line shows that the temperature for the production of the metal is reduced by about 150 K. The

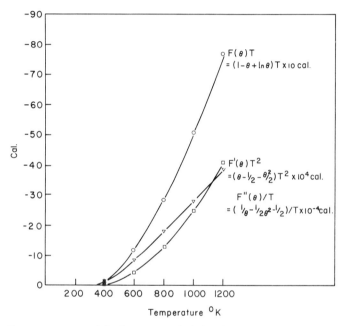

Fig. 50. Component terms of the free energy deviation function. $\odot$ $F(\theta)\,T$ multiply calories ordinate by 10. $\square$ $F'(\theta)\,T^2$ multiply by 10^4. $\triangledown$ $F''(\theta)/T$ multiply by 10^{-4}.

effect of reducing the SO_2 pressure on the free energy of formation of the basic sulphates is not shown, but clearly this will be one third $\Delta\mu_{SO_2}$ for the effect on $PbSO_4 . 4PbO$ formation, and one half $\Delta\mu_{SO_2}$ for the effect on $PbSO_4 . 2PbO$ formation when compared with the effect $\Delta\mu_{SO_2}$ on the reaction

$$PbS + O_2 \rightarrow Pb + SO_2$$

which is shown by the dashed line.

The reduction of the basic sulphates via the gaseous phase can only occur when the vapour pressure of PbS becomes sufficiently large for rapid gas transport to occur. Colin and Drowart (1962) have studied the vaporization of PbS and found two vapour species PbS and Pb_2S_2. The vapour pressure

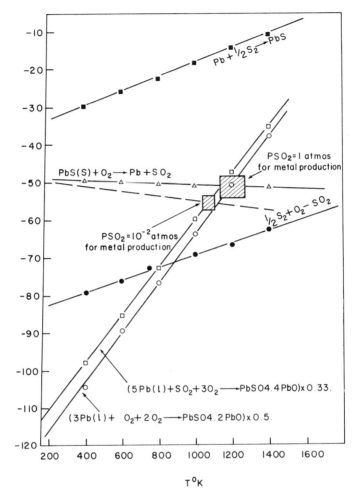

Fig. 51. Ellingham diagram for the production of lead from PbS and lead sulphates.

of the dimer was considerably lower than that of the monomer, and hence Pb_2S_2 will be neglected.

The results for PbS(g) can be fitted to a free energy equation thus

$$PbS(s) \rightarrow PbS(g) \qquad \Delta G° = 55\,700 - 36\cdot4\ T\ cal$$

According to this equation, the vapour pressure of PbS reaches 10^{-3} atm at 1120 K (847°C) and Tuffley and Russell (1964) found empirically that lead was produced at 825°C when synthetic lead sulphide specimens were roasted in a slowly flowing stream of air. Below this temperature the reaction products of roasting for times up to 1 h only contained the basic sulphates. It can there-fore be concluded that the reduction of these sulphates to the metal occurs at temperatures where the removal of oxygen in the form of SO_2 is thermo-dynamically feasible with pSO_2 at 10^{-2} atm or more, and that PbS vapour is very probably the chemical potential transporting medium which acts as a reductant.

COPPER SMELTING

The most important industrial process which depends upon the oxidation of the sulphide is the production of copper from sulphide concentrates. The starting material usually consists of minerals composed of copper and iron sulphides, such as $CuFeS_2$, chalcopyrite, which have a high sulphur pressure at matte smelting temperatures (1200–1400°C). The first stage of a metal-making process therefore consists of a roasting operation in which about one-third of the sulphur is removed as SO_2 followed by melting under a siliceous slag. The molten matte contains essentially a mixture of Cu_2S and FeS. From the sulphide Ellingham diagram (Fig. 4), it can be seen that the sulphur decomposition pressure of these two substances in equilibrium with the pure metals is about 10^{-6} atm. at 1300°C.

THERMODYNAMICS OF MOLTEN MATTES

The molten sulphide phases which are produced during the extraction of copper and nickel can be represented in an idealized manner as mixtures of the sulphides, Cu_2S, FeS, and in the case of nickel mattes, Ni_3S_2. The true matte phases have a sulphur deficit when compared with the idealized mattes, and also contain a small amount of oxygen. However, the representa-tion of the thermodynamic behaviour of real mattes in terms of the idealized mixtures is not so drastic an approximation as to make a useful discussion of the metal extraction reaction impossible in these simple terms.

The experimental study of the binary molten sulphides FeS and Cu_2S were made by Maurer, Hammer and Mobius (1942) and by Schuhmann and Moles (1951) respectively. From these results, it can be concluded that the activities of the sulphides remain fairly constant across a substantial range of compositions around the stoichiometric composition. The separate metal and sulphur activities vary markedly over the same composition range.

$$K_{FeS} = \frac{a_{FeS}}{a_{Fe}p^{\frac{1}{2}}_{S_2}} \quad K_{Cu_2S} = \frac{a_{Cu_2S}}{a^2_{Cu} \cdot p^{\frac{1}{2}}_{S_2}}$$

G

The equation which relates each metallic activity and the corresponding sulphur activity is, of course, the Gibbs–Duhem equation

$$X_{Cu} \, d \ln a_{Cu} = -X_S \, d \ln a_S$$

The activity of sulphur was determined in each case by equilibration of the molten phase with H_2S/H_2 gas mixtures. Then

$$X_{Cu} \, d \ln a_{Cu} = -X_S \, d \ln p \, H_2S/p \, H_2$$

and the activity of the metal is obtained by integration of this equation over the composition range from pure metal to that of the sulphide.

There is a qualitative difference between the copper–sulphur system on one hand and the iron–sulphur and nickel–sulphur systems on the other. The phase diagram for the former system shows a miscibility gap between Cu containing sulphur and Cu_2S containing copper. The phase diagrams for the other two metal–sulphur systems do not show such a gap, and there is a continuous range of liquid solutions from pure metal to the molten sulphides. The sulphur activities at constant temperature as a function of sulphur atom fraction in the Fe–S and Ni–S systems both show that over a broad range of composition the sulphur activity only changes by a small amount. This activity, as in the Cu–S system, climbs most rapidly near to the sulphide composition. The phase diagrams are shown with the corresponding thermodynamic data in Figs 52–57.

Krivsky and Schuhmann (1957) studied the ternary matte phases Cu–Fe–S over the whole liquid range at temperatures between 1150 and 1350°C.

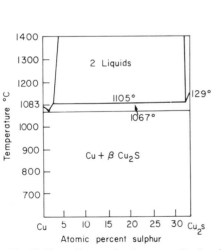

Fig. 52. Phase diagram for the system Cu–S.

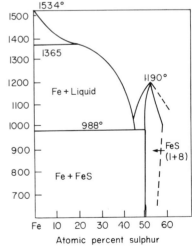

Fig. 53. Phase diagram for the system Fe–S.

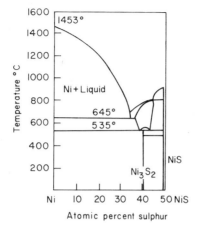

Fig. 54. Phase diagram for the system Ni–S.

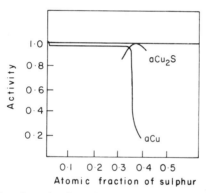

Fig. 55. Activities of metal and metal sulphide in the system Cu–S at 1250°C.

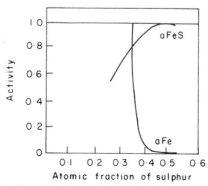

Fig. 56. Activities of metal and metal sulphide in the system Fe–S at 1250°C.

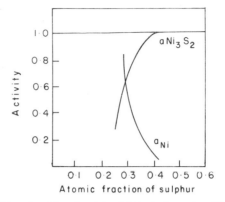

Fig. 57. Activities of metal and metal sulphide in the system Ni–S at 1250°C.

They also measured the sulphur activities by H_2S/H_2 equilibration and then calculated the metal activities by integration of the ternary Gibbs–Duhem equation

$$X_{Fe} \, d \ln a_{Fe} + X_{Cu} \, d \ln a_{Cu} + X_S \, d \ln \frac{p_{H_2S}}{p_{H_2}} = 0.$$

The results of the study showed that the pseudobinary Cu_2S–FeS was fairly close to an ideal Raoultian mixture. The corresponding nickel mattes were studied by Matousek and Samis (1963) at 1200°C, and their evaluation of the results was based on the assumption that Cu_2S–Ni_3S_2 liquid mixtures were ideal also. The general thermodynamic feature which is observed in the binary metal–sulphur systems appears also in the ternary systems; this is the rapid increase of the sulphur pressure around the sulphide composition range. The miscibility gap between the metal-rich and sulphide-rich phases decreases

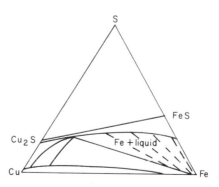

Fig. 58. Phase diagram for the Cu–Fe–S system at 1250°C.

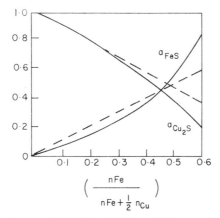

Fig. 59. Activities of sulphides in Cu–Fe–S pseudobinary section at 1250°C. (1957), **209**, 981.)

$— — —$ = Raoultian activity.
$———$ = Measured values.

in width fairly uniformly as iron or nickel are substituted for copper in the ternary systems.

It follows from consideration of all of the ternary metal–sulphur systems that the activities of the metals decrease rapidly from the range of compositions at the miscibility gaps in the ternary systems, *towards* the pseudobinary systems. It thus follows that the activities of the metals remain fairly close to unity over a substantial range of mixtures which are towards the *metal-rich* end of the metal–sulphur binary and ternary systems.

Since the real system in the production of copper from molten matte consists of the liquid sulphide phase in contact with a liquid slag mainly containing FeO and SiO_2, then the industrial liquid mattes will also contain oxygen. Kameda and Yazawa (1961) have studied the solubility of oxygen in liquid Cu_2S–FeS mattes which were equilibrated with magnetite under an atmosphere of nitrogen. They found that the oxygen content was very small in the copper-rich matte but increased in an approximately linear manner with the increasing FeS constant.

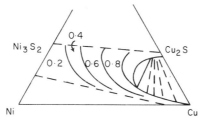

Fig. 60. Isoactivity lines for Cu_2S in the pseudobinary Cu_2S–Ni_3S_2 at 1250°C.

Rosenqvist and Hartvig (1958) made a limited study of quaternary melts in the system Cu–Fe–S–O. They also used a magnetite crucible as a container, and brought the melts to equilibrium with $SO_2 + S_2$ mixtures. The conclusions from this study were that the melts behaved as Raoultian mixtures to a very good approximation in the temperature range 1135–1185°C. In this complex mixture, the partial molar entropies of the molecular species are obtained from the atomic species by the use of Temkin's model. Thus

$$a_{FeS} = \frac{n_{Fe}}{n_{Fe} + n_{Cu}} \cdot \frac{n_S}{n_O + n_S}$$

$$a_{Cu_2S} = \left(\frac{n_{Cu}}{n_{Fe} + n_{Cu}}\right)^2 \cdot \frac{n_S}{n_O + n_S}$$

THE EVALUATION OF EXPERIMENTAL DATA IN TERNARY SYSTEMS

In the studies of ternary matte phases, such as Cu–Fe–S, the experimental technique consisted of a measurement of the sulphur potential across the whole composition range. The potentials of all three components in the matte are interrelated through the ternary Gibbs–Duhem equation

$$X_A \, d\mu_A + X_B \, d\mu_B + X_C d\mu_C = 0$$

It is possible to obtain the activities of two components by calculation using results for the chemical potential of the third component providing that the compositions at which measurements are made in the ternary field are carefully chosen. The method used by Schuhmann and Krivsky will be discussed since this has been most widely applied in the study of molten mattes.

Schumann's method for the solution of the ternary Gibbs–Duham equation

In a multi-component solution, the partial free energies of each component $\Delta \bar{G}_i$ are related to the integral free energy of mixing ΔG^M by the equation

$$\Delta \bar{G}_1 = \left(\frac{\partial \Delta G^M}{\partial n_1}\right)_{n_2, n_3, n_4 \ldots} \quad , \quad \Delta \bar{G}_2 = \left(\frac{\partial \Delta G^M}{\partial n_2}\right)_{n_1, n_3, n_4 \ldots} \quad \text{etc.}$$

where n_i is the number of moles of the ith component in the solution. Now because the integral free energy of mixing is a continuous function of the mole fractions, if a second differential is formed by differentiation with respect to a second variable, the sequence of differentiation is unimportant.

$$\left(\frac{\partial^2 \Delta G^M}{\partial n_1 \, \partial n_2}\right)_{n_3, n_4 \ldots} = \left(\frac{\partial^2 \Delta G^M}{\partial n_2 \, \partial n_1}\right)_{n_3, n_4 \ldots}.$$

Hence

$$\left(\frac{\partial \Delta \bar{G}_1}{\partial n_2}\right)_{n_3, n_4 \ldots} = \left(\frac{\partial \Delta \bar{G}_2}{\partial n_1}\right)_{n_3, n_4 \ldots} \tag{i}$$

In the case of a ternary solution

$$\left(\frac{\partial \Delta \bar{G}_1}{\partial n_2}\right)_{n_1, n_3} = -\left(\frac{\partial \Delta \bar{G}_1}{\partial n_1}\right)_{n_2, n_3} \left(\frac{\partial n_1}{\partial n_2}\right)_{\Delta \bar{G}_1 n_3} \tag{ii}$$

and

$$\left(\frac{\partial \Delta \bar{G}_2}{\partial n_1}\right)_{n_2, n_3} = \left(\frac{\partial \Delta \bar{G}_2}{\partial \Delta \bar{G}_1}\right)_{n_2, n_3} \left(\frac{\partial \Delta \bar{G}_1}{\partial n_1}\right)_{n_2, n_3} \tag{iii}$$

following the fundamental relationships between partial differentials. Hence, using (ii) and (iii) in (i)

$$\left(\frac{\partial \Delta \bar{G}_2}{\partial \Delta \bar{G}_1}\right)_{n_2, n_3} = -\left(\frac{\partial n_1}{\partial n_2}\right)_{\Delta \bar{G}_1, n_3}$$

It is therefore possible to arrive at values of $\Delta \bar{G}_2$ by means of a simple integration thus

$$\left\{[\Delta \bar{G}_2]_{n_2 = n_2}^{n_2 = n_2'} = -\int_{n_2 = n_2}^{n_2 = n_2'} \left(\frac{\partial n_1}{\partial n_2}\right)_{\Delta \bar{G}_1, n_3} d \Delta \bar{G}_1\right\}_{n_2/n_3}$$

It can be seen that results are needed for the slope $(\partial n_1/\partial n_2)$ along the constant activity curve of component 1 at the particular value of n_2/n_3 which is chosen as the path of integration (Fig. 61). This equation is similar in appearance to the simple binary equation

$$[\Delta \bar{G}_2]_{n_2 - n_2}^{n_2 = n_2'} = -\int \frac{n_1}{n_2} d \Delta \bar{G}_1$$

It follows that enough experimental information must be gathered to define the constant activity curves of component 1 before the integration can be carried out along the path of constant n_2/n_3 ratio.

Wagner (1952) also presented a solution for the ternary Gibbs–Duhem equation which was obtained through the introduction of the variables

$$y = \frac{n_3}{n_1 + n_3} \quad \text{and} \quad x = \frac{n_2}{n_1 + n_2 + n_3}$$

which is the atom fraction of component 2.

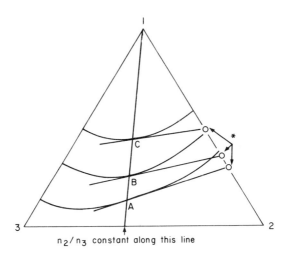

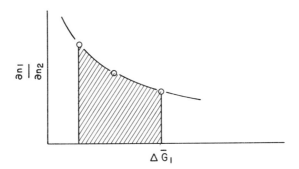

Fig. 61. Graphical data for Schuhmann's method for the integration of the Gibbs–Duhem equation.

The hatched area represents the integral which must be evaluated to find the change in $\Delta \bar{G}_2$ when $\Delta \bar{G}_1$ changes in value from that at point C, above to point A, above.

Values of $\partial n_1 / \partial n_2$ for the points C, B and A on the line with n_2/n_3 constant.

Wagner then finds that for the excess partial molar free energies of components 1 and 3 are given by

$$[\Delta\bar{G}_1^{xs}]_{x=0}^{x=x'} = \int_0^{x=x'} \left[\frac{\Delta\bar{G}_2^{xs}}{(1-x)^2} - y\frac{\partial}{\partial y}\left(\frac{\Delta\bar{G}_2^{xs}}{(1-x)^2}\right) \right]_{n_2/n_3} dx$$

and

$$[\Delta\bar{G}_3^{xs}]_{x=0}^{x=x'} = \int_0^{x=x'} \left[\frac{\Delta\bar{G}_2^{xs}}{(1-x)^2} + (1-y)\frac{\partial}{\partial y}\left(\frac{\Delta\bar{G}_2^{xs}}{(1-x)^2}\right) \right]_{n_2/n_3} dx$$

It follows that fairly complete data for $\Delta\bar{G}_2$ are required to obtain $\Delta\bar{G}_1$ and $\Delta\bar{G}_3$ in order to evaluate the tangent

$$\frac{\partial}{\partial y}\left(\frac{\Delta\bar{G}^{xs}}{(1-x)^2}\right)$$

along each selected value of n_2/n_3 throughout the path of integration (Fig. 62).

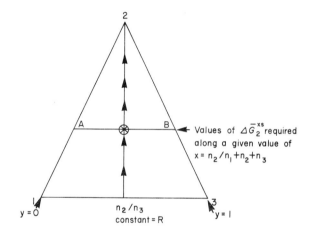

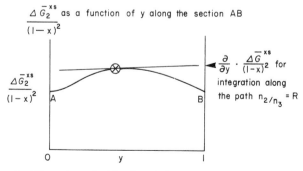

Fig. 62. Graphical data for Wagner's method for the solution of Gibbs–Duhem equation.

THE CONVERTER PRODUCTION OF COPPER METAL

In the industrial process, the liquid matte which is produced by melting down in the reverberatory furnace under a FeO-silica slag is charged into the converter and air is blown through the matte which is covered with a silica slag. The temperature in this process is between 1300–1400°C. Sulphur is evolved as SO_2, and iron is oxidized into the slag, this reaction accounting for practically the complete utilization of oxygen in the ingoing air. Due to the inverse relationship of metal and sulphur activities around the pseudobinary Cu_2S–FeS matte compositions, the tendency of iron or sulphur to oxidize preferentially will change across this range. Thus, on the metal-rich side, iron will oxidize, and on the sulphur-rich side, SO_2 will form preferentially. Hence, the composition of the matte tends to remain along or parallel to the pseudobinary line, and iron and sulphur are eliminated at roughtly equal rates. It follows that, since the sulphur pressure of a matte is roughly 10^{-6} atm, and p_{SO_2} is about one-fifth of an atmosphere during sulphur elimination, then according to the Ellingham oxide diagram, using the line for

$$\tfrac{1}{2}S_2 + O_2 \to SO_2$$

the oxygen potential in the system is roughly -40 kcal. This is too low to oxidize copper, as can be judged from the position of the line for

$$4Cu + O_2 \to 2Cu_2O$$

$\Delta G^\circ_{1573\,K} = -30$ kcal, but can bring about the oxidation of iron since

$$2Fe + O_2 \to 2FeO$$

$\Delta G^\circ_{1573} = -76$ kcal. The iron is, therefore, oxidized to ferrous oxide which reacts with the silica to form a liquid FeO–SiO_2 slag. It will be remembered that this slag is practically Raoultian in its thermodynamic behaviour.

The line for the reaction

$$6FeO + O_2 \to 2Fe_3O_4$$

shows a value of $\Delta G^\circ_{1573\,K} = -54$ kcal. Hence, the activity of FeO in the slag must be kept low so that the liquid is not oxidized to solid magnetite which would make the working of the process very difficult. It follows that

$$\Delta G^\circ = -RT \ln K = -2\Delta\mu_{Fe_3O_4} + 6\Delta\mu_{FeO} + \Delta\mu_{pO_2}$$

and if

$$a_{Fe_3O_4} = 1; \Delta\mu_{Fe_3O_4} = 0.$$

Then

$$6\,\Delta\mu_{FeO} < -54 - (-40, \Delta\mu_{O2})\,\text{kcal}$$

$$\Delta\mu_{FeO} < -2,300 \text{ cal}$$

$$\log a_{FeO} = \frac{-2300}{4 \cdot 575 \times 1573} = -0 \cdot 32$$

$$a_{FeO} < 0 \cdot 48$$

After the virtually complete elimination of iron by oxidation, a fairly pure liquid Cu_2S remains behind. This may be converted to almost pure copper by further oxidation at 1300–1400°C with air to give metal which contains a small amount of residual sulphur and oxygen. It is because the liquid copper which is formed at this stage of the process will sink to the bottom of the reactor that the vessel must be side-blown with air. The cold blast of the gas which is admitted to the matte for oxidation through the tuyère generates enough heat by oxidation of the sulphide to heat the gas and maintain the condensed phases liquid. If the converter were bottom-blown, then liquid copper containing only about 2 atomic percent of sulphur would meet the incoming cold air, and not enough heat would be evolved by sulphur oxidation to heat the gas, and the liquid surrounding the tuyère would freeze. This process is to be contrasted with the Bessemer method for the removal of carbon and silicon by air from liquid iron which is produced by the blast furnace process. Because iron–carbon–silicon alloys have a higher density than the pure liquid iron, the more dense carbon-containing liquid remains around the tuyères at the bottom of the vessel during carbon and silicon elimination. Heat is thus always generated immediately above the tuyères by the chemical reactions

$$2[C] + O_2 \rightarrow 2CO$$

$$[Si] + O_2 \rightarrow \{SiO_2\}$$

The final chemical state of copper which is produced in the converter is determined by the sulphur–oxygen interaction in solution in the liquid metallic phase. As sulphur is removed from this metal by oxidation, the oxygen partial pressure of the equilibrium gas phase will increase, and this increases the oxygen content of the metal. When the process is interrupted, the product metal will contain sulphur and oxygen which can further react on cooling. The SO_2 pressure in equilibrium with these solutions can be calculated to first approximation by combining the free energy equations for the constituent reactions thus

$$\tfrac{1}{2}S_2 \rightarrow [S]_{Cu} \qquad \Delta G° = -29\,200 + 5 \cdot 0\,T \text{ cal}$$

$$O_2 \rightarrow 2[O]_{Cu} \qquad \Delta G° = -41\,080 + 3 \cdot 45\,T \text{ cal}$$

$$\tfrac{1}{2}S_2 + O_2 \rightarrow SO_2 \qquad \Delta G° = -86\,620 + 17 \cdot 31\,T \text{ cal}$$

These equations yield the solution equation

$$SO_2 \rightarrow [S] + 2[O] \quad \Delta G° = +16\,340 - 8\cdot86\,T\ cal$$

In the equation for the dissolution of diatomic sulphur and oxygen in copper, the standard states for the metallic solutions are one atomic percent of these elements dissolved separately in liquid copper. In order that the final equation given above may be used to represent the dissolution equilibrium for SO_2 in liquid Cu, then sulphur and oxygen must not interact with one another in the

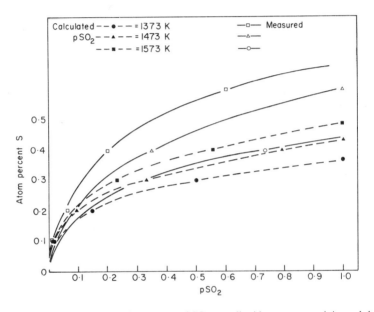

Fig. 63. Measured and calculated pressures of SO_2 over liquid copper containing sulphur and oxygen.

solution. Neither element must therefore affect the activity coefficient of the other element in the ternary solution. This requirement is difficult to test satisfactorily with the available experimental data, but the calculated SO_2 pressures obtained from the equation above which should be exerted as a function of temperature from the given contents of oxygen and sulphur of liquid copper do not coincide with experimental data (Fig. 63). The predicted SO_2 pressures are higher than the observed pressures, thereby indicating a mutual *lowering* of the activity coefficients of oxygen and sulphur in the presense of one another.

The experimental results for studies of SO_2 dissolution in liquid copper in the temperature range 1100–1300°C in which

$$2X_{[S]} = X_{[O]}$$

because of the stoichiometry of the gaseous species SO_2, can be fitted to the free energy equation

$$\Delta G° = 29\,830 - 19·7\,T\ \text{cal}$$

At 1300°C, this equation yields a standard free energy of solution of SO_2 in liquid copper of -1160 cal mol^{-1}. The theoretical equation which was arrived at above, which ignores changes in the activity coefficient of sulphur and oxygen in the presence of one another, yields a value of 2400 cal at this temperature. We may therefore use an equation which corrects for the interaction by subtraction of these two results

$$\delta \Delta G°_{1573} = RT \ln \gamma_{[S]} . \gamma^2_{[O]} = -3560\ cal.$$

From the behaviour of the integral free energy function as a homogeneous function, it follows that the sequence of differentiation with respect to two composition variables is unimportant. Hence,

$$\frac{\partial^2 \Delta G^M}{\partial X_S . \partial X_O} = \frac{\partial^2 \Delta G^M}{\partial X_O . \partial X_S}$$

and

$$\frac{\partial \ln \gamma_O}{\partial X_S} = \frac{\partial \ln \gamma_S}{\partial X_O}$$

which may be written in the shorthand form

$$\varepsilon^S_O = \varepsilon^O_S$$

where ε^O_S is called the "interaction coefficient" for the effect of oxygen on the activity coefficient of sulphur.

If we assume that ε^O_S and ε^S_O are independent of composition in the range under consideration, then the natural logarithms of the activity coefficients of oxygen and sulphur in the presence of one another differ from the values in the separate binary Cu–O and Cu–S systems by $X_S\varepsilon^S_O$ and $X_O\varepsilon^O_S$ respectively. Now because of the stoichiometric relationship between sulphur and oxygen in solution in liquid copper which is saturated with SO_2, and because of the relationship between ε^S_O and ε^O_S it follows that

$$X_S\varepsilon^S_O = \tfrac{1}{2}X_O\varepsilon^S_O = \tfrac{1}{2}X_O\varepsilon^O_S \text{ for the solubility results}$$

Hence

$$RT \ln \gamma_{[S]} \cdot \gamma_{[O]}^2 = RT (X_O \varepsilon_S^O + 2X_S \varepsilon_O^S)$$
$$= 2RT (X_O \varepsilon_S^O) = 4RT \, X_S \varepsilon_S^O$$

If we now use the value of -3560 cal which was calculated above to correspond to the interaction effect, we find

$$\varepsilon_S^O = \frac{-890}{RT \, X_S} = \frac{-0 \cdot 28}{X_S}$$

In the middle of the experimental composition range where X_S is 0·25 atom percent ($X_O = 0 \cdot 5$) the calculated value for the activity coefficient of sulphur is 0·75 using this interaction coefficient. The value for oxygen at the same composition is then 0·75 and the ratio

$$\frac{p_{SO_2} \text{ (measured)}}{p_{SO_2} \text{ (calculated)}} = \gamma_{[S]} \cdot \gamma_{[O]}^2 = 0 \cdot 42$$

for the comparison of the experimental and the measured solubility of SO_2 in liquid copper at 1300°C.

The standard free energy of formation of Cu_2O from liquid copper and oxygen is given, according to Jacobs and Jeffes (1971), by the equation

$$\Delta G° = -45\,900 + 21 \cdot 04 \, T \text{ cal mol}^{-1} \, Cu_2O.$$

The oxygen potential for Cu_2O formation in the presence of oxygen-saturated liquid Cu at 1300°C is equal to $-23\,500$ cal. The oxygen partial pressure in the gaseous mitxture of SO_2 and N_2 which is formed by blowing air through the matte at this temperature can rise to the value which corresponds to this oxygen potential before the formation of a separate phase of Cu_2O will occur ($p_{O_2} \simeq 5 \cdot 4 \times 10^{-4}$ atm). We may calculate the approximate sulphur content of the metal at this stage by using the uncorrected equation which was derived first of all, in a modified form. If we combine the equations

$$\tfrac{1}{2}S_2 \rightarrow [S] \qquad \Delta G° = -29\,200 + 5 \cdot 0 \, T \text{ cal}$$
$$\tfrac{1}{2}S_2 + O_2 \rightarrow SO_2 \qquad \Delta G° = -86\,620 + 17 \cdot 31 \, T \text{ cal}$$

we obtain the free energy change for the reaction

$$[S] + O_2 \rightarrow SO_2 \qquad \Delta G° = -57\,420 + 12 \cdot 31 \, T \text{ cal}$$

Whence

$$\Delta G° = RT \ln [S] + RT \ln p_{O_2} - RT \ln p_{SO_2}$$

and inserting the values $-23\,500$ cal for $RT \ln p_{O_2}$ and $p_{SO_2} = \tfrac{1}{5}$ atm we obtain the final result

$$[S] = 0 \cdot 002 \text{ atom percent at } 1300°.$$

The oxygen content of the metal under conditions of saturation with Cu_2O is obtained from the equation given already

$$\tfrac{1}{2} O_2 \rightarrow [O] \qquad \Delta G^\circ = -20\,540 + 1.72\,T \text{ cal}$$

$$\Delta G^\circ = \tfrac{1}{2} RT \ln p_{O_2} - RT \ln [O]$$

whence on substitution of the value $-23\,500$ cal for $RT \ln p_{O_2}$ we obtain

$$[O] = 7.00 \text{ atom percent at } 1300°C.$$

These calculations indicate that fairly pure copper can be obtained with respect to sulphur content from the converter process, but a subsequent stage of refining is needed to reduce the oxygen content to an acceptable level. This is done in a separate refining furnace where carbon is introduced in the form of wooden poles, traditionally, enabling the reduction of the oxygen content by CO and CO_2 formation. A typical final value which is aimed at for "tough pitch" copper would be 0.1 atom percent. The water vapour which is released when the poles are immersed in the liquid metal probably forms bubbles which can then act as nuclei for the gaseous carbon oxides which are evolved in the reduction process.

THE DIRECT PRODUCTION OF NICKEL FROM LIQUID MATTE

In the nickel-making process, a mixture of nickel–copper–iron sulphides is melted as in the copper process, and again the matte is blown with air to remove the iron sulphide which is preferentially oxidized to form a siliceous slag and SO_2 gas. It will be seen on the Ellingham diagram that nickel is oxidixed at an oxygen potential of -46 kcal at 1300°C and, hence, if the oxygen potential were allowed to rise to the value achieved in copper converting, the nickel would be completely oxidized. Instead of immediate conversion, the matte of copper and nickel sulphides is separated into a nickel-rich and copper-rich fraction in the solid state.

The sulphides Cu_2S and Ni_3S_2 are virtually immiscible in the solid state and their separation can be achieved by slow cooling of the nickel-copper matte through the temperature range where the diffusion-controlled separation in the solid state can occur at a sufficiently rapid rate. This is in the region of 400–600°C. The final solid which contains the separated phases in intimate mixture is then crushed, ground and separated into the two principal components by flotation concentration. The treatment of the copper sulphide proceeds in the manner already described, but the nickel sulphide is normally oxidized to NiO before being reduced to metal.

We may carry out the same approximate calculation for the oxygen and sulphur contents of liquid nickel which would be obtained by converter oxidation as was done above for copper using the equations

$$\tfrac{1}{2} S_2 \rightarrow [S]_{Ni} \qquad \Delta G^\circ = -32\,400 + 3 \cdot 53\, T \text{ cal}$$

$$\tfrac{1}{2} S_2 + O_2 \rightarrow SO_2 \qquad \Delta G^\circ = -86\,620 + 17 \cdot 31\, T \text{ cal}$$

to yield the equation for the reaction

$$[S] + O_2 \rightarrow SO_2 \qquad \Delta G^\circ = -54\,220 + 13 \cdot 78\, T \text{ cal}$$

The oxygen potential which must now be used in solving this equation for the sulphur content of the metal when oxidation to NiO occurs is once more obtained from the Ellingham oxide diagram. Since nickel melts at a much higher temperature than copper, 1452°C rather than 1083°C, it is obvious that a higher final temperature must be achieved in nickel matte conversion to produce the liquid metal than in copper matte conversion.

In the latter case, air is used as the source of oxygen and hence four moles of chemically inactive nitrogen must be heated to converter temperature for every mole of SO_2 which is formed. This inert "ballast" requires more than one half of the thermal energy which is available from Cu_2S oxidation. The final temperature which can be achieved in converting practice by using pure oxygen as the gas phase would obviously be higher than that which can be attained by utilization of air because of the elimination of nitrogen.

The heat capacity of nitrogen gas is given by the equation

$$C_p = 6 \cdot 66 + 1 \cdot 02 \times 10^{-3}\, T \text{ cal mol}^{-1}$$

Hence, the heat required to raise one mole of the gas from room temperature to 1300°C is obtained from the integral of this equation with respect to temperature $\Delta H = 6 \cdot 66\,(T_2 - T_1) + 0 \cdot 51 \times 10^{-3}\,(T_2^2 - T_1^2)$ cal, and this is 9880 cal.

The heat change for the reaction

$$Cu_2S + O_2 \rightarrow 2Cu + SO_2$$

is equal to $-51\,400$ cal at room temperature and, as is usually found to be the case, this value will not be very much different if the reaction is carried out at 1300°C.

The details of the calculation supporting this statement are as follows: the heat capacity data for the solid–gas reaction are given in Table XXV, and we shall assume that the heat capacities of liquid Cu_2S and Cu are the same as those of the corresponding solids.

The heats of fusion of Cu_2S and Cu are 2·6 kcal and 3·1 kcal mol^{-1}, and the equation for the change of heat of reaction between T_1 and T_2 where T_2 is for liquid Cu_2S and Cu is

$$H_{T_2} - H_{T_1} = -6\cdot86\,(T_2 - T_1) + 2\cdot27 \times 10^{-3}\,(T_2^2 - T_1^2) + 3600 \text{ cal.}$$
$$= 129 \text{ cal} \,(T_2 = 1300°C, \, T_1 = 0°C)$$

Hence it can be seen that the heating of the nitrogen during matte conversion with air is a very important "sink" for the chemical energy which is liberated during the oxidation of Cu_2S. We will therefore assume that oxygen is used and that p_{SO_2} may be taken as equal to one atmosphere for this calculation.

TABLE XXV

Heat capacities for the reaction $Cu_2S + O_2 \rightarrow 2Cu + SO_2$

Cu: $C_p = 5\cdot41 + 1\cdot50 \times 10^{-3}\,T$ cal
Cu_2S: $C_p = 20\cdot9$ cal (weighted average for three crystalline modifications)
O_2: $C_p = 7\cdot16 + 10^{-3}\,T$ cal
SO_2: $C_p = 10\cdot38 + 2\cdot54 \times 10^{-3}\,T$ cal

The sulphur content of liquid nickel will be reduced to one atomic percent at the temperature which makes the standard free energy change for the reaction

$$[S] + O_2 \rightarrow SO_2$$

and that for

$$2Ni + O_2 \rightarrow 2NiO$$

equal. The metal will also be saturated with oxygen.

The free energy of formation for two moles of solid NiO is given by the equation

$$\Delta G° = -111\,900 + 40\cdot58\,T \text{ cal from solid nickel (m.p. } 1455°C)$$
$$-120\,100 + 45\cdot30\,T \text{ cal from liquid nickel}$$

and it will be recalled that the free energy equation for the reaction

$$[S] + O_2 \rightarrow SO_2$$

for nickel as solvent is

$$\Delta G° = -54\,220 + 13\cdot78\,T \text{ cal}$$

These two equations are simultaneously satisfied at 2090 K.

If these equations are combined, we can obtain the free energy change for the oxidation of sulphur,

$$[S] + 2NiO \rightarrow 2Ni + SO_2$$
$$\Delta G° = 65\,880 - 31\cdot52\,T \text{ cal}$$

for the reaction involving liquid nickel and solid NiO (m.p. 1984°C). Assuming unit activity for the liquid metal, we can now calculate the pressure of SO_2 which would be exerted by the system at Ni/NiO equilibrium and with a sulphur content of one atom percent. The results show that this pressure reaches 0·175 atm at 1600°C and 0·20 atm at 1625°C. This latter pressure is, of course, the pressure of SO_2 which would be achieved under the conditions of oxidation by air. It is thus apparent that pure oxygen should be used to achieve a high temperature in this reaction, but finishing off with air would yield an acceptable sulphur content in the final product at a more modest temperature. A balance between these two requirements can clearly be met by the use of oxygen-enriched air during the early stages, and providing that the thermal ballast of the liquid metal was enough to sustain the temperature, reduction of the sulphur content to a quite low level (< 1 atom percent) could finally be achieved by further *dilution* of air with an inert gas. The precise balance between the needs for heat generation in order to sustain the temperature, and the need to reduce p_{SO_2} in order to reduce the final sulphur content of the metal is determined by the physical characteristics of the reactor.

The solubility of oxygen in liquid nickel at 1600°C in equilibrium with solid NiO is about 3 atom percent. Such a high oxygen content must therefore be considered as possibly affecting the activity coefficient of sulphur significantly and vice versa. The interaction coefficient of sulphur and oxygen in liquid nickel has been measured by Fischer and Ackermann (1966) by the use of a solid oxide electrolyte cell. They find the value $\varepsilon_S^O = -10$ at 1600°C with slightly higher values in liquid iron and cobalt as solvents. It therefore follows that an oxygen content of 3 atom percent in each of these metals would lower the sulphur activity coefficient to a value of about 0·75. For the present purposes and considering the experimental errors which are present in the number of results which have been brought together to make this calculation, it may be concluded that a sulphur content quite close to one atom percent as originally arrived at would be found in equilibrium with liquid NiO and SO_2 at 0·2 atm pressure in solution in liquid nickel at 1600°C.

REFERENCES

Byerley, J. J. and Takebe, N., (1972). *Met. Trans.* **3**, 559, Thermodynamics of the Fe–Ni–S ternary.

Colin, R. and Drowart, J., (1962). *J. Chem. Phys.* **37**, 1120, The vapour pressures of PbS and SnS.

Fischer, W. A. and Ackermann, W., (1966). *Archiv. Eisenhutten.* **37**, 779. Influence of sulphur on the activity of oxygen in Fe, Co and Ni at 1600°C.

Gerlach, J., Kantzer, K. P. and Pawlek, F., (1963). *Metall.* **17**, 1096. Solubility of SO_2 in liquid copper.

Jacob, K. T. and Jeffes, J. H. E., (1971). *Trans. Inst. Min. and Met.*, Section C80, 32, Thermodynamics of oxygen in liquid Cu, Pb and Cu + Pb alloys.—

Kellogg, H. H. and Basu, S. K. (1960). *Trans. AIME* **218**, 70.

Kameda, M. and Yazawa, A., (1961). *Phys. Chem. of Process Metallurgy*, Part 2, p. 963, Interscience Publishers. The oxygen contents of copper mattes.

Krivsky, W. A. and Schuhmann, R., (1957). *Trans. AIME* **209**, 981 (1957), Thermodynamics of the Cu–Fe–S ternary.

Matousek, J. W. and Samis, C. S., (1963). *Trans. AIME* **227**, 980. Thermodynamics of the Cu–Ni–S ternary.

Maurer, E., Hammer, G. and Mobius, H., (1942). *Arch. Eisenhutten.* **16**, 159. Thermodynamics of the Fe–S binary.

Rosenqvist, T. and Hartvig, T., (1958). Report No. 12, Royal Norwegian Council for Scientific and Industrial Research. Thermodynamics of iron–copper mattes and their equilibrium with magnetite.

Schuhmann, R. and Moles, O. W., (1951). *Trans. AIME* **181**, 235. Thermodynamics of the Cu–S binary.

Schuhmann, R., (1955). *Acta Met.* **3**, 219. Solution of the ternary Gibbs–Duhem equation.

Tuffley, J. R. and Russell, B., (1964). *Trans. AIME* **230**, 950. The roasting of lead sulphide in air to produce lead.

Wagner, C., (1952). "Thermodynamics of Alloys", Addison Wesley. Solutions of the ternary Gibbs–Duhem equation.

RECOMMENDED READING

Non-Ferrous Production Metallurgy. J. L. Bray. J. Wiley & Sons, N.Y. (1941).

11

REDUCTION OF METAL OXIDES BY CARBON

INTRODUCTION

The least stable metal oxides, those having free energies of formation less than 100 kcal mol^{-1} oxygen at around 1000°C, include the compounds of the metals iron, copper, lead and tin. The feature of the area of an Ellingham diagram containing these oxides which is important to note, is that the lines representing the standard free energies of formation for these oxides are all intersected at low temperatures by that for the reaction

$$2C + O_2 \rightarrow 2CO$$

This means that the standard free energy changes for the reduction of these oxides by carbon will all be negative at significant metal-making temperatures, i.e. above about 700°C. The intersections come about because the heats of formation of the oxides of the metals and of carbon are of the same order, but the entropy of formation of carbon monoxide is positive whereas for the formation of solid metal oxides from metal and oxygen gas the entropy change is always negative.

The entropy change for a reaction is largely determined by the balance between the number of gas molecules which are consumed on the left-hand side of the chemical equation, and the number of gaseous molecules which are produced on the right-hand side. The change for the reaction

$$\text{solid} + \text{solid} \rightarrow \text{solid}$$

is practically zero, as is that for the reaction

$$C\,(\text{solid}) + O_2\,(\text{gas}) \rightarrow CO_2\,(\text{gas})$$

where the number of gaseous molecules on both sides of the equation is the same.

Kubaschewski (1967) has shown that the entropies of diatomic gases at 298 K fall fairly closely to the curve

$$S^0_{298} = 53.8 + 0.043\,M - \frac{240}{M}\ \text{cal deg}^{-1}\,\text{mol}^{-1}$$

and of polyatomic gases the curve represented by the equation

$$S^0_{298} = 39.0 + 0.34\,M - 6.2 \times 10^{-4}\,M^2$$

where M is the molecular weight of the gas, with values of M between 20 and 300. Since the entropies of elements and simple solid compounds lie in the range $0-15$ cal deg^{-1} mol^{-1}, then it is clear that the change in the number of gas molecules will dominate the entropy change for a reaction because of the larger entropy contents of gases.

It follows from these considerations that metals forming the relatively unstable oxides can all be produced by direct reduction of the oxides with carbon provided that there is no subsequent metal + carbon reaction which leads to the formation of a stable carbide. The earliest metal-making processes involved these reduction reactions with carbon in the form of charcoal which could be produced very simply from wood. The industrial process in which carbon is used as the reducing agent on the large modern scale is the blast furnace. In this vertical shaft reactor, the solid charge of metal oxides, limestone and carbon, in the form of metallurgical coke, is introduced at the top of the furnace and moves downwards towards the hearth which is at the bottom. Air, usually preheated, is blown through gas ports, the tuyères, near the bottom of the furnace and the gaseous phase moves upwards through the furnace. The air is introduced just above the hearth where the reaction products, liquid metal and the non-metallic phase containing the impurities, the slag, are collected. The main source of heat for the chemical reactions which occurs in the reactor, comes from the burning of coke in the air injected at the tuyères. This furnace has the very valuable feature that the heat produced by this reaction is generated *within* the furnace, and thus the calorific value of the coke is used to very good advantage. The furnace operates with counter-current flow of the solid and the gaseous phases, and hence brings about good thermal exchange between these phases. One disadvantage, which will be immediately apparent, is that the volume of oxygen, which can be used very effectively in this manner, is accompanied by a fourfold greater volume of nitrogen, if air is used as the source of oxygen. The inert gas is to some extent a "passenger" in the reacting system, and can be considered largely as an inert diluent from the chemical point of view. There are some reactions of minor importance which involve nitrogen, and these will be discusesd later (p. 204).

THE COMBUSTION OF COKE IN THE BLAST FURNACE

The products of the oxidation of coke in air at high temperatures are carbon monoxide and dioxide. The proportion of each gas which is produced depends on the temperature and is the result of the kinetics and thermodynamics of the carbon–oxygen system. The thermodynamic properties of this system can be obtained from the two free energy equations for the formation of carbon monoxide and the dioxide

$$2C + O_2 \rightarrow 2CO \qquad \Delta G° = -53\,400 - 42\cdot0\,T \text{ cal mol}^{-1}$$

and

$$C + O_2 \rightarrow CO_2 \qquad \Delta G° = -94\,200 - 0\cdot2\,T \text{ cal mol}^{-1}$$

The difference in entropy change between these two reactions reflects the fact that one extra mole of gas is produced in the formation of CO, but the number of moles of gas is constant in the formation of CO_2.

By combination of these equations we find for

$$2CO + O_2 \rightarrow 2CO_2 \qquad \Delta G° = -135\,000 + 41\cdot6\,T \text{ cal mol}^{-1}$$

and for

$$CO_2 + C \rightarrow 2CO \qquad \Delta G° = +40\,800 - 41\cdot8\,T \text{ cal mol}^{-1}$$

These two equations are very important in blast furnace science. The first equation can be used to calculate the oxygen partial pressure for any given CO/CO_2 ratio and temperature. Thus from a knowledge of this partial pressure ratio, the oxidizing or reducing power of the gas for a given condensed oxide system may be obtained.

The second equation concerns the "solution loss" reaction. This reaction is the one whereby the reducing power of the gas is regenerated, and accounts for part of the rate of direct coke utilization in the furnace. The direct reduction of iron oxide by carbon through a solid–solid contact is now thought to be extremely unlikely and the principal reaction for the reduction of wustite, for example, is the reaction

$$FeO + CO \rightarrow Fe + CO_2.$$

Direct consumption of coke also occurs during the dissolution of carbon in the liquid metal and the oxidation of coke to almost pure CO near the tuyères of the furnace. From the temperature of the charge at the top of the furnace up to about 1300°C, where liquid metal is formed, the rate of consumption of coke in the blast furnace is determined by the kinetics and thermodynamics of the "solution loss" reaction. The broad features of the thermodynamic aspect of this reaction can be readily assessed from a consideration of the free energies of CO and CO_2 formation from carbon. Clearly, at low temperatures CO_2

will be more stable than CO, but at high temperatures greater than about 1000°C CO will predominate, increasingly so as the temperature is raised. Thus, in equilibrium with carbon, the major species changes from CO_2 at low temperature to CO at high temperature. The CO/CO_2 ratio at which FeO reduction occurs can be obtained if we combine the CO_2 formation reaction from CO with the free energy of formation of wustite

$$2CO + O_2 \rightarrow 2CO_2 \qquad \Delta G° = -135\,000 + 41.6\,T \text{ cal mol}^{-1}$$
$$2Fe + O_2 \rightarrow 2FeO \qquad \Delta G° = -124\,100 + 29.8\,T \text{ cal mol}^{-1}$$
$$FeO + CO \rightarrow FeO + CO_2 \quad \Delta G° = -5450 + 5.9\,T \text{ cal mol}^{-1}.$$

The CO/CO_2 ratio in equilibrium with Fe/FeO lies between 1/1 and 10/1 since the standard free energy changes for these two reactions are within ± 6 kcal of one another from 500–1500 K.

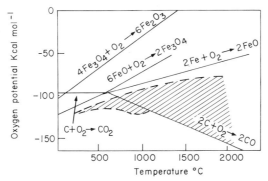

Fig. 64. The hatched area represents the range of oxygen potentials which have been found by measurement in live blast furnaces.

These considerations can be brought together in a manner first suggested by Goodeve (Figs 64 and 65) in which the fate of a gram mol of oxygen which is injected into a blast furnace at the tuyère level is followed. Firstly, the gas is rapidly heated to about 1900°C and reacts with coke to form CO almost entirely. The CO/CO_2 ratio is very close to that which is in equilibrium with carbon at this temperature. The CO/CO_2 ratio changes as the gas ascends the furnace and is cooled by heat transfer and chemical reaction. Finally, the gas leaves the furnace at around 200°C and should consist, mainly of CO_2 under equilibrium conditions. The extent to which the equilibrium with carbon is reached in the lower temperature region (less than 1000°C) depends on the kinetics of the carbon precipitation reaction

$$2CO \rightarrow C + CO_2$$

This process is known to occur slowly at temperatures below 700°C in the absence of a catalyst, and hence the final gas composition of the gas leaving the furnace is richer in CO than would be anticipated from the thermodynamic information.

This semi-quantitative picture has been corroborated by measurement on live blast furnaces, and the limits of CO/CO_2 ratios which were obtained throughout these furnaces are shown in the upper and lower B.F. lines in Fig. 64.

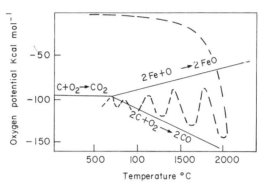

Fig. 65. The change in oxygen potential of a quantity of oxygen as a result of chemical reactions during transit through the iron blast furnace. (After Goodeve, loc. cit.)

It must be stressed that the real picture deviates from the equilibrium predictions due to kinetic factors which involve the gaseous reactions of the carbon–oxygen system. In laboratory studies of this system, most work has been carried out with graphite so that the material can be readily characterized with respect to its physical properties and surface area. The effects on the results of changing from graphite to various grades of coke are not such as to change the conclusions in a qualitative sense, but some quantitative variations are found depending on the source of the coke. The studies on graphite show that the rate of the reactions between air or CO_2 with graphite depend on a chemical step at the gas–solid surface up to about 1200°C, and above this temperature region the rates depend upon gaseous diffusion of reaction products to and away from the gas–solid interface.

The suggested mechanisms for the reactions of oxygen and CO_2 with graphite are represented by the following equations:

$$2C_s + O_2(g) \rightarrow 2C\text{—}[O] \rightarrow 2CO(g)$$

C_s is an active surface site on the graphite surface, and between one and five percent of the surface of most carbon samples plays an active part in providing these surface sites. C–[O] represents an oxygen atom which is adsorbed on an active site.

The complete reaction path for the reduction of CO_2 by carbon involves two steps thus:

$$C_s + CO_2(g) \rightarrow C-[O] + CO(g)$$
$$\downarrow$$
$$CO(g)$$

The experimental results for the reaction rate between carbon and CO_2 can be represented by the equation

$$\text{Rate} = \frac{k_1 p_{CO_2}}{1 + k_2 p_{CO} + k_3 p_{CO_2}}$$

where k_1, k_2 and k_3 are velocity constants, and this result can be accounted for in terms of the reaction mechanism given above in the following way.

If σ_1 and σ_2 are the fractions of active sites covered by adsorbed oxygen atoms and by CO molecules respectively, then $(1-\sigma_1-\sigma_2)$ is the fraction of surface sites which is available for further reaction. We may now write the rate equation for each step of the mechanism which was proposed above separately thus:

$$C + CO_2 \xrightarrow{i_1} CO(g) + C-[O]$$
$$C-[O] \xrightarrow{j_3} CO(g)$$
$$CO(g) + C_s \underset{j_2}{\overset{i_2}{\rightleftharpoons}} C-[CO]$$

where i_1 and j_2, etc., are the respective velocity constants and C-[CO] represents a CO molecule which is adsorbed on the surface. At the steady state, the rate of adsorption of CO_2, which is equal to the rate of formation of C-[O] complexes, is equal to the rate of destruction of these complexes by desorption of CO gas.

$$i_1 \cdot p_{CO_2} \cdot (1-\sigma_1-\sigma_2) = j_3 \sigma_1$$

Similarly, a balance is maintained between the number of C-[CO] complexes on the surface according to the equation

$$i_2 \cdot p_{CO} \cdot (1-\sigma_1-\sigma_2) = j_2 \sigma_2$$

By elimination of σ_1 and σ_2 through the use of these equations, it can be concluded that the rate of reaction of CO_2 with carbon, which is equal to $j_3 \sigma_1$, is given by the equation

$$\text{Rate} = \frac{i_1 p_{CO_2}}{1 + (i_2/j_2) p_{CO} + (i_1/j_3) p_{CO_2}}.$$

This is identical with the general equation which was given above setting

$$i_1 = k_1; \qquad i_2/j_2 = k_2 \quad \text{and} \quad i_1/j_3 = k_3.$$

An alternative mechanism which has been proposed applies if the adsorption of CO on to the active sites is negligible and then the third rate equation above is omitted. The effect of the CO pressure is now included through the reverse reaction

$$CO(g) + C-[O] \rightarrow CO_2(g) + C_s$$

Now the expression for the reaction rate is given by

$$j_3\sigma_1 = \frac{i_1 p_{CO_2}}{1 + (j_1/j_3) p_{CO} + (i_1/j_3) p_{CO_2}}$$

which is again identical with the general equation with

$$i_1 = k_1; \quad j_1/j_3 = k_2 \quad \text{and} \quad i_1/j_3 = k_3$$

On the basis of experimental information which is at present available, there seems to be some support for both mechanisms, but setting aside the differences, the general conclusion that CO reduces the rate of CO_2 reduction by decreasing the number of active sites which are covered by adsorbed oxygen atoms, is found in both mechanisms.

The fact that the rate of the carbon solution loss reaction is controlled by a chemical process, which occurs at the gas–solid interface up to temperatures of at least 1000°C, makes this reaction almost unique amongst high temperature reactions. It is generally concluded that diffusion across the gaseous boundary layer is the rate determining step in most gas–solid surface reactions at temperatures above about 600°C, but in this very important case the chemical control remains important up to metal-making temperatures.

It is not to be expected that the equilibrium CO/CO_2 ratio will be achieved in a live furnace by reacting carbon with air or CO_2 until the temperature region is reached where gaseous diffusion control takes over from chemical control. This means that equilibrium can only be rapidly achieved above 1000°C, which is *above* the temperature of the metal reduction in most non-ferrous blast furnaces such as those for lead, but well below the maximum temperature which is achieved in the iron blast furnace.

BLAST FURNACE PRODUCTION OF IRON

The reduction of iron ore to produce liquid metal saturated with carbon in the blast furnace accounts for the largest tonnage of metal which is obtained. The blast furnace is a counter-current device in which air is blown into the bottom of the furnace and rises against a descending burden of iron ore, coke and limestone. At the base of the furnace, the oxidation of coke to CO heats the ascending gas to approximately 2000°C. This is more than adequate to bring about the formation of liquid metal and a liquid slag phase by heat transfer to the burden. As well as heating the burden by simple heat conduction and radiation during its ascent of the furnace, the gas reacts with the burden in two ways. Firstly, metallic iron sponge is produced with CO_2 as the gaseous product. Secondly, carbon monoxide is regenerated by the solution loss reaction

$$CO_2 + C \rightarrow 2CO$$

The limestone is present to bring about slag formation only.

The mechanism and kinetics of iron ore reduction have been discussed previously, so in this section our principal interest will be focussed on the production of the liquid phases and the subsequent metal–slag reactions. The saturation of iron with carbon lowers the melting point from 1535°C to 1150°C. It is probable that all of the iron oxide has been reduced before the burden reaches the region of the furnace where this temperature prevails. The slag-forming temperature is thought to be somewhat higher than the temperature of fusion of carbon-saturated iron in the blast furnace. This is because the eutectics in the $CaO–SiO_2$, and $CaO–Al_2O_3$ systems are in excess of 1400°C, and only the more complex slag system $CaO–Al_2O_3–SiO_2$ has a lower eutectic which is at 1300°C. There will, therefore, be a zone between 1100°C and up to about 1400°C where liquids will begin to form in the burden, and run through to the hearth at the bottom of the furnace. The density of the metal phase is higher than that of the slag, and thus metal drops will sink through the liquid slag and metal-slag equilibrium can be approached under the optimum conditions of contact. The gas phase surrounding the liquids will be in equilibrium with carbon and will thus consist substantially of CO together with nitrogen.

At the bottom of the furnace, a fairly stagnant metal layer forms below the slag layer, and there is very little further chemical reaction. The hearth is below the point of injection of the air into the furnace and so can be largely ignored in the subsequent discussion. Following Michard *et al.* (1961), it is fruitful to consider the blast furnace as a counter-current process in which there are four principal zones. In the first zone, coke is combusted in pre-heated air to form CO at a temperature of 1900–2000°C. In the second zone, the melting and reduction reactions which involve the major impurities occur between 1200°C and 1600°C. The third zone is where oxide reduction to solid iron, together with replenishment of CO by the reaction of CO_2 and coke, occur, these reactions involving solids and gases only. The fourth zone occupies the rest of the furnace up to the stock line, and is one in which iron ore reduction occurs without solution loss, limestone is decomposed to CaO and CO_2 and the burden is dried by the rising hot gases (Fig. 66).

The temperature of demarcation between the third and fourth zones is set

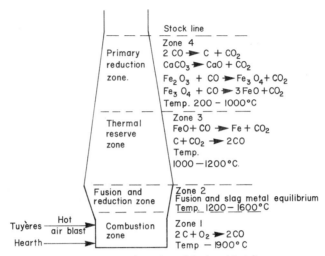

Fig. 66. Diagrammatic section of the iron blast furnace.

at 1000°C by Michard *et al.* (1961), but it is difficult to justify this choice on any theoretical basis. Analyses of blast furnace gas at several points along the furnace indicate that the gas is always reducing to wustite (Fig. 65). It has been shown by several workers on the basis of studies with operating blast furnaces that the solution loss reaction does not occur to any marked extent below a burden temperature of 1000°C, and this is the basis of Michard's choice of the temperature separating the third and fourth zones. We have shown earlier that this choice is consistent with the results of laboratory studies of the kinetics of this reaction. It must be remembered, however, that under the conditions of rapid transit of the gas phase in the blast furnace, the average oxygen atom spends only about ten seconds in passing through the furnace and therefore that gas–solid equilibria are not likely to be completely established even when the kinetics are rapid.

A most significant factor in the operation of the blast furnace is the exchange of heat between the solids and gases within the three upper zones of the furnace. Calculation and experimental tests have shown that the second zone, where reduction of the non-metallic elements and fusion occur, is relatively short, and the third zone is the longest. The heat transfer from gas to solids and liquids is large in the second zone, and in the fourth zone where the cool burden first enters the stack. The third zone, which is where the reduction of ore to solid iron occurs, is a region of very small heat exchange, and extends over the long central region of the furnace. Since the length of this zone could be altered without in any way changing the efficiency of the furnace with respect to heat transfer, it would seem important to attempt to minimize the length in order to optimize the operation of the furnace.

The top gases leaving the furnace are still at a temperature of about 200°C, and thus the heat exchange in the fourth zone of the furnace does not cool the gases down as much as could be permitted. This suggests that some reduction in gas throughput could be introduced to the normal practice. Increasing the blast temperature in the first zone and lowering the coke content of the charge reduces the volume of gas passing through the furnace without depleting the source of heat which is needed in the zone of fusion. This may also be achieved by enriching the oxygen content of the air supply before it is pre-heated and then passed through the tuyères. The residence time of the gas in the furnace could be increased by raising the pressure at the top of the furnace, thus allowing more time for gas–solid heat transfer. Each of these possibilities is under active industrial study at the present time.

A heat balance for an industrial furnace was presented by Michard et al. (1961) which shows the following distribution of the thermal load of the furnace (Table XXVI).

TABLE XXVI

Heat generation		Heat consumption	
Air blast heat content	32·7%	Decomposition, calcination of oxides and drying of coke	5·4%
Heat from coke combustion	67·3%	Solution loss and iron oxide reduction	29·9%
		Heat content of liquid iron	22·1%
		Heat content of slag	25·8%
		Heat losses	16·8%

These results confirm the point that in blast furnace production most of the heat which is generated by coke combustion is used to further the fusion and the iron-making reactions. Only 22·2% of the heat consumption is shown as being used otherwise in this heat balance. Much of the heat content of the air blast in the tuyères is obtained by burning blast furnace gas in stoves through which the air is driven before arriving at the tuyères. The failure of the carbon deposition reaction to reach equilibrium

$$2CO \rightarrow C + CO_2$$

in the cooler parts of the furnace is thus turned to advantage through the burning of the residual CO in these heat-exchanging stoves.

THE THERMODYNAMICS OF DILUTE SOLUTES IN LIQUID IRON

The reduction of iron ores by means of coke in the blast furnace leads to the formation of liquid metal which is saturated with carbon and frequently contains some manganese. There are a number of other non-metallic impurities in blast furnace metal, sulphur, silicon and phosphorus being the most important elements. These substances are present in dilute solution, sometimes even less than one atomic percent, and they must be removed in the subsequent refining or "Steelmaking" process. The distributions at metal–slag equilibrium and the extent of the elimination of these impurities during the steelmaking process can only be accurately predicted from a knowledge of the thermodynamic behaviour of these dilute solutes in liquid iron. A great deal of experimental work has gone into establishing these properties both in the simple binary solutions and the more complex solutions which more nearly approximate to the industrial phases.

Iron-oxygen and iron-sulphur

The elimination of impurities in the steelmaking process requires the presence of oxygen in solution in the liquid metal. The summary of results

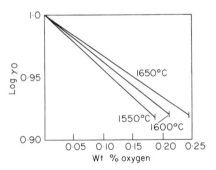

Fig. 67. The dependence of the activity coefficient of oxygen in liquid iron at 1600°C on concentration.

shown in Fig. 67 demonstrates that thermodynamic knowledge of this system is now reasonably complete. By measurement of the equilibria

$$H_2O \rightarrow H_2 + [O]$$

where $[O]$ represents oxygen dissolved in the iron, it has been shown that the saturation solubility of oxygen in iron which is in equilibrium with liquid FeO at steelmaking temperatures, around 1600°C, is about 0·70 atomic percent. Even is such dilute solutions, the oxygen atoms influence one another, and the activity coefficient of oxygen decreases with increasing oxygen content (Fig. 67). The value of the equilibrium constant is usually given with γ_O equal to unity in the infinitely dilute solution.

$$K = \frac{p_{H_2}}{p_{H_2O}} \cdot \gamma_O X_O$$

Similarly, studies in which liquid iron was brought to equilibrium with H_2/H_2S mixtures show that the activity coefficient of sulphur dissolved in liquid iron also decreases with increasing sulphur content.

These studies show that in the Taylor expansion of the activity coefficient as a power series of the concentration of the dilute solute

$$\ln \gamma_i = \ln \gamma_i^0 + X_i \frac{\partial \ln \gamma_i}{\partial X_i} + \frac{X_i^2}{2!} \cdot \frac{\partial^2 \ln \gamma_i}{\partial X_i^2}$$

the first two terms on the right-hand side are of measurable significance. The first term which is zero when the reference state is the infinitely dilute solution and finite for deviations from Raoult's law, reflects the interaction of the solute atoms with the solvent. The second term involves the differential $\partial \log \gamma_i/\partial X_i$ which is called the "interaction coefficient" since it can be interpreted as arising from the mutual effect of solute atoms on the activity coefficient of the solute. The symbol "ε_i^i" is generally used in the literature to designate the interaction coefficient of the atoms of the ith species with one another. (See p. 173).

When alloying elements are added to the dilute solutions of oxygen or sulphur in liquid iron, the activity coefficients of the dilute solutes usually change. A further linear term can be added to the expansion for $\ln \gamma_i$ which is given above, to represent the effect of an alloying element j on the activity coefficient of the ith species, thus

$$\ln \gamma_i = \ln \gamma_i^0 + X_i \frac{\partial \ln \gamma_i}{\partial X_i} + X_j \frac{\partial \ln \gamma_i}{\partial X_j}$$

$$= \ln \gamma_i^0 + X_i \varepsilon_i^i + X_j \varepsilon_i^j$$

Providing only a small addition of alloying element, up to a few atomic percent, is made, the value of ε_i^i remains practically constant. Finally, in a complex alloy containing many dilute solutes the expansion can be written in the form

$$\ln \gamma_i = \ln \gamma_i^0 + \sum_j \cdot X_j \varepsilon_i^j \qquad (j = 1, 2, 3 \ldots i \ldots)$$

The free energy of formation of a mixture is a homogeneous function, which means that the order of differentiation with respect to the compositional variables is reversible. It follows that

$$\frac{\partial^2 \Delta G^M}{\partial X_i \partial X_j} = \frac{\partial^2 \Delta G^M}{\partial X_j \partial X_i} \quad \text{and hence} \quad \varepsilon_i^j = \varepsilon_j^i$$

For practical applications, the interaction coefficients have been worked out using the weight percentage composition of the alloy system, and for logarithms to the base 10 of the activity coefficient, and the symbol "e" is used for the coefficients

$$e_0^A = \frac{\log_{10} \gamma_0}{\partial [\text{wt} \% A]}$$

Figures 68 and 69 show the effects of a number of the common alloying elements on the activity coefficients of oxygen and sulphur in liquid iron. It can be seen that the general trend is that those elements which have a smaller

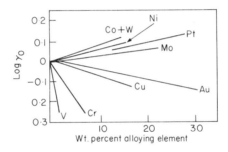

Fig. 68. Results for the effects of alloying elements on $\log \gamma_0$ in liquid iron at 1600°C.

affinity for non-metals than iron raise the activity coefficients, whereas those which have a greater affinity than iron, lower, the activity coefficients. We shall deal more thoroughly with this problem in a subsequent chapter.

Experimental methods for the investigation of steel-making equilibria have been brought to a very satisfactory state of development largely by Chipman and his co-workers (1949, 1952). The technique which was most successful in the study of the solutions of oxygen and sulphur in liquid iron at temperatures between 1550 and 1700°C involved induction heating of a metal sample which was equilibrated with a pre-heated gas mixture. The sample, usually about 50–100 g of pure iron or iron alloys, was held in an alumina crucible. The crucible was mounted inside a water-cooled quartz tube around which the induction heater work coil was wound (Fig. 70).

The mixture of hydrogen and steam or H_2S was passed down a resistance-heated pre-heater tube so that thermal segregation of the gas mixture at the point of contact with the liquid metal was eliminated. The stirring effect of

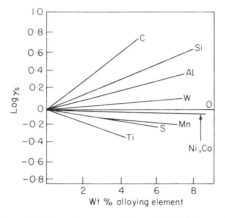

Fig. 69. Effects of alloying elements on log γ_S in liquid iron at 1600°C.

the induction field on the metal sample ensured both thermal and compositional homogeneity, and equilibrium could be achieved rapidly. Because the gas phase was enclosed in the quartz tube, leakage of air into the system was negligible. After equilibration, the sample was rapidly quenched and analysed for oxygen by vacuum fusion, and for sulphur by dissolution in mineral acid followed by determination of the H_2S which was evolved. The temperature of the experiment was obtained by optical pyrometry, the pyrometer being focussed on the surface of the molten steel. The accuracy of such a measurement at steelmaking temperatures is about ± 5°C.

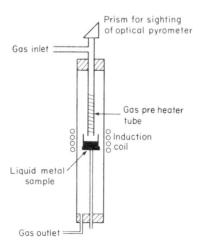

Fig. 70. Induction furnace for gas–liquid equilibration at steelmaking temperatures.

H

The thermodynamic properties of the binary dilute solutions can be represented by a simple two-term equation for the free energy of solution of the gaseous elements in liquid iron. Both gases obey Sievert's Law at low concentrations, that is to say the solubility is proportional to the square root of p_{O_2} or p_{S_2}. This demonstrates that the solutes dissolve in liquid iron as atomic species, which are fully coordinated with metal atoms.

We can assemble the data using as the reference state either the infinitely dilute solution, that is with γ_O° and γ_S° equal to unity, or with the real or hypothetical one atom percent solution.

The latter standard state is more frequently used in practice, and the relevant free energy equations are then as follows:

$$O_2 \rightarrow 2[O]$$

$$K = \frac{\text{atom } \% \, [O]^2}{p_{O_2}} \qquad \Delta G^\circ = -55\,800 - 6 \cdot 0 \, T \text{ cal}$$

$$S_2 \rightarrow 2[S]$$

$$K = \frac{\text{atom } \% \, [S]^2}{p_{S_2}} \qquad \Delta G^\circ = -60\,000 + 7 \cdot 18 \, T \text{ cal}$$

Carbon and silicon

The thermodynamic properties of these elements in dilute solution in liquid iron have been studied by Richardson and co-workers (1953 and 1964) using gas–liquid equilibration. The reactions used were

$$2CO \rightarrow [C] + CO_2 \qquad K = a[C] \frac{p_{CO_2}}{p_{CO}^2}$$

and

$$SiO(g) + H_2 \rightarrow [Si] + H_2O \qquad K = \frac{a[Si] \, p_{H_2O}}{p_{SiO} \cdot p_{H_2}}$$

The pressure of SiO was fixed in the latter study by passing H_2O/H_2 gas mixtures through a bed of pure silica at the reaction temperature.

In both of these studies, there is a possible complication that the results might have been affected by the presence of oxygen in the melts. However, the oxygen potentials which were defined by p_{CO_2}/p_{CO} and p_{H_2O}/p_{H_2} in the two experiments were probably too low for this effect to be significant. In both cases, these were much lower than the potentials at which oxygen can saturate liquid iron at the experimental temperatures. Both carbon and silicon show an activity coefficient which is small at low concentrations, but which increases with increasing carbon or silicon contents. The self-interaction coefficients ε_c^c and ε_{Si}^{Si} are therefore positive, and average values of 9·5 and 4·4 are obtained from these results at 1600°C. The composition ranges over which these constant values were obtained were 0·005–0·025 for the carbon atom fraction and 0·02–0·12 for the silicon atom fraction.

The iron–carbon–silicon system

Chipman and co-workers (1965) have used the liquid–liquid partition technique for silicon between liquid silver and iron-carbon alloys to determine the activity of silicon in this system at 1420°C. The activity of silicon in solution in liquid silver was obtained by calculation from the Ag–Si phase diagram. The solutions in equilibrium with solid silicon in this system extend from $X_{Si} = 1·0$ at the melting point of silicon (1410°C) down to $X_{Si} = 0·154$ at the eutectic (830°C). The activity coefficients were calculated at the liquidus compositions in the standard procedure for a simple eutectic system, and these gave values of the regular solution function $\ln \gamma_{Si}/(1 - X_{Si})^2$ which were a linear function of composition. By extrapolation a value of $\log \gamma_{Si}^\circ = -0·3 \pm 0·1$ was obtained for the dilute solution of silicon in silver at 1420°C. This value was used for the partition study in which the atom fraction of silicon in the silver phase varied from 0·001 to 0·0236 at 1420°C.

The partition of silicon between silver and iron–carbon alloys was next studied as a function of both the silicon and carbon contents of the iron phase. It was found in the concentration range of silicon, $X_{Si} = 0·17–0·36$, that the effect of carbon on the activity coefficient of silicon can be expressed by the equation

$$\log \gamma_{Si}^C = \log \gamma_{Si} + 5·5\, X_C$$

for carbon atom fractions from 0 to 0·1.

In the absence of carbon, it was found that the activity coefficient of silicon varied in a simple manner with silicon content in the composition range $X_{Si} = 0·15–0·40$.

$$\log \gamma_{Si} = \log \gamma_{Si}^\circ + \frac{\varepsilon_{Si}^{Si}}{2·303}\, X_{Si}$$

$$= -3·2 + 5·37\, X_{Si}$$

Hence $\varepsilon_{Si}^{Si} = 12·37$, which is a much larger value than that found for the dilute solutions by Richardson et al. Chipman (1965) used calorimetric data for liquid Fe + Si alloys (Fig. 71) to correct the activity coefficients at 1420°C to 1600°C with the standard thermodynamic equation

$$\frac{\partial \ln \gamma}{\partial (1/T)} = \frac{\Delta \bar{H}}{R}$$

This yields the equation for $\log \gamma_{Si}$ in the carbon-free solutions at 1600°C

$$\log \gamma_{Si} = -2.9 + 5.25 \, X_{Si}$$

giving a value of ε_{Si}^{Si} at this temperature of 12·10.

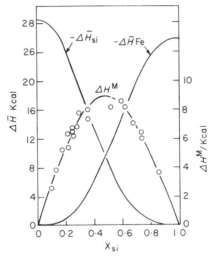

Fig. 71. Calorimetric results for the heat of mixing of liquid Fe + Si at 1600°C.

Saturation with silicon carbide or carbon

These results may be combined with the data for the free energy of formation of silicon carbide to obtain the conditions under which this substance would separate from an iron–silicon–carbon alloy.

For the reaction

$$Si(l) + C \rightarrow SiC$$

the standard free energy of formation is given by

$$\Delta G° = -25\,100 + 8.53 \, T \text{ cal mol}^{-1}$$

$$K_{SiC} = \frac{a_{SiC}}{a_{Si} \cdot a_C}$$

For the solutions in iron of the elements we have obtained the equations

$$\ln \gamma_{Si} = -6.68 + 12.1 \, X_{Si} + 12.67 \, X_C \text{ at } 1600°C$$

and

$$\ln \gamma_C = -0.47 + 12.67 \, X_{Si} + 9.5 \, X_C$$

The standard state for silicon is liquid silicon, and for carbon is pure graphite. The value of $\ln \gamma_C^\circ$ (-0.47) was obtained from the results of Richardson and Dennis (1953). The equations for the chemical potentials of these elements at 1600°C are very simply obtained

$$\Delta\mu_i = RT \ln \gamma_i + RT \ln X_i$$

and hence

$$\Delta\mu_{Si} = \overbrace{(-24\,850 + 45\,000\,X_{Si} + 47\,150\,X_C)}^{RT \ln \gamma_{Si}}$$
$$+ RT \ln X_{Si}$$
$$+ (8570 \log X_{Si})\,\text{cal}$$

and

$$\Delta\mu_C = \overbrace{(-1750 + 47\,150\,X_{Si} + 35\,350\,X_C)}^{RT \ln \gamma_C}$$
$$+ RT \ln X_C$$
$$+ (8570 \log X_C)\,\text{cal}$$

when the calculation is rounded out to the nearest 10 cal. We could now plot the results for these functions at 1600°C to show the compositions where

$$\Delta\mu_{Si} + \Delta\mu_C = -RT \ln K_{SiC}\,(a_{SiC} = 1)$$
$$= \Delta G^\circ_{SiC}$$

However, in the context of studies of the blast furnace, the most interesting region of composition is near the saturation limit for carbon where $\Delta\mu_C$ has the value zero. The solubility of carbon in iron-silicon alloys would then give the appropriate value of X_C which should be used in the equation above for $\Delta\mu_{Si}$. The solubility study has been made by Chipman and Fuwa (1959) with the results which are shown in Fig. 72. They also determined the compositions

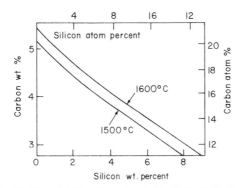

Fig. 72. The solubility of graphite in liquid Fe + Si alloys.

of the three phase equilibrium [Si]–C–SiC and found at 1600°C that the atom fraction of carbon at this point was 0·015 and that of silicon was 0·371. Substitution of these values in the equations for $\Delta\mu_{Si}$ and $\Delta\mu_C$ which are given above yield the values $-10\,600$ and $+640$ cal. respectively. Of course, $\Delta\mu_C$ should be equal to zero at saturation, but we will use the calculated value to obtain

$$\Delta\mu_{Si} + \Delta\mu_C = -9960 \text{ cal.}$$

The value for ΔG°_{SiC} which is obtained from the equation given earlier is -9120 cal at 1600°C, which is in very good agreement with the value calculated only from the chemical potentials of silicon and carbon and assuming that the interaction coefficients have a very simple linear form in the Fe–Si–C system. This is quite remarkable when it is remembered that the saturation concentrations of silicon and carbon which are in equilibrium with SiC are relatively large, and how much inter-relating of separate studies must be made to arrive at the calculated values.

The chemical potential of silicon in carbon-saturated Fe + Si + C alloys which is at equilibrium with SiC can be used together with the oxygen potential of the C–CO equilibrium at 1600°C to calculate the silica activity of a slag which would be in equilibrium with this alloy under the conditions which prevail in the hearth of the blast furnace. The oxygen potential is found from the Ellingham diagram to be -132 kcal for p_{CO} equal to one atmosphere, and the free energy of formation of SiO_2 at this temperature is -126 kcal, which is almost the same as $\Delta\mu_{O_2}$ for the C–CO equilibrium. Since

$$-\Delta G^{\circ}_{SiO_2} = \Delta\mu_{SiO_2} - (\Delta\mu_{Si} + \Delta\mu_{O_2})$$

it follows that

$$\Delta\mu_{SiO_2} \cong \Delta\mu_{Si}$$

The activity of silica in the slag phase is therefore about the same as that of silicon in the metal when saturation of the metal phase with SiC occurs: that is, about 9×10^{-2}. From the data for the activities of silica in liquid $CaO–Al_2O_3–SiO_2$ slags which were collected by Chipman, it is clear that such a slag must have a mole fraction of silica less than 0·45. The contour for this activity in the $CaO–Al_2O_3–SiO_2$ system at 1600°C runs roughly parallel to the lime-alumina base of the composition triangle, and therefore the CaO/Al_2O_3 ratio is relatively unimportant from the point of view of chemical equilibrium in this system (Fig. 73). However, the lime content of the blast furnace burden is easily controlled through limestone addition, and the viscosity of the slag phase would be higher the lower the CaO/Al_2O_3 ratio. The optimum Al_2O_3 content for a blast furnace slag is around 15 wt per cent and the CaO/SiO_2 ratio about 1·4 according to practical experience, and this fits in well with the simple physico-chemical picture. Fulton and Chipman (1954) in a laboratory study of this equilibrium at 1600°C found that SiC separated from the system Fe + Si + C and CaO–SiO$_2$ slags under 1 atmos. pressure of CO when the mole fraction of silica was 0·43. At this composition the activity of silica in the slag is close to 0·1.

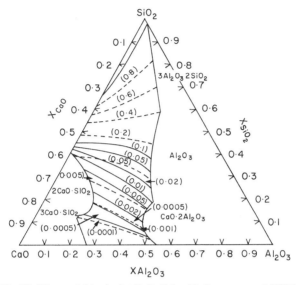

Fig. 73. Silica activities in the CaO–SiO$_2$–Al$_2$O$_3$ system at 1600°C.

Manganese

The interaction coefficients for ε_{Mn}^{Mn} and ε_{Mn}^{C} are quite small in solution in liquid iron at 1600°C (0·0 and −0·5 respectively) and so the partition of this element between liquid metal and slag in the blast furnace can be handled very simply, as far as the alloy phase is concerned, with the approximation that the activity of manganese is equal to the mole fraction. The results for MnO activities in liquid slags show that the activity coefficient is probably about 0·1 on the SiO$_2$-rich side of the CaO/SiO$_2$ equimolar composition, and it increases almost linearly from this value as the lime content is increased, up to a value of 2·0 when the slag is saturated with CaO. In a typical blast furnace slag the MnO activity coefficient is close to unity and therefore the slag-metal partition of manganese can be calculated approximately from the standard free energy data

$$Mn + \tfrac{1}{2}O_2 \rightarrow MnO; \qquad \Delta G_{1873}^{\circ} = -58\,000 \text{ cal}$$

$$\frac{X_{\{MnO\}}}{X_{[Mn]}} = K_{MnO} \cdot p^{\frac{1}{2}}_{O_2}; \qquad \log K_{1873} = 6\cdot77$$

If we insert the oxygen pressure of the C–CO equilibrium into this equation ($p_{O_2} = 10^{-15}$ atm) it is found that

$$\frac{X_{\{MnO\}}}{X_{[Mn]}} = 0\cdot186 \text{ at blast furnace equilibrium}$$

The alloy ferromanganese is made by blast furnace reduction and the manganese atom fraction of the product can be adjusted within certain limits by the control of the materials charged to the furnace.

Because of the high chemical stability of MnO when compared with FeO, the CO_2/CO ratio in equilibrium with a manganese oxide–magnanese mixture of equal activities is very much lower than for a ferrous oxide–iron mixture and even at 1600°C this ratio only reached 4×10^{-4}

$$MnO + CO \rightarrow Mn + CO_2; \qquad \Delta G^{\circ}_{1873} = +29\,000 \text{ cal}$$

$$K = \frac{p_{CO_2}}{p_{CO}} \cdot \frac{a_{Mn}}{a_{MnO}} = 4 \times 10^{-4}$$

The CO_2/CO ratio for the C–CO equilibrium at the same temperature is 3×10^{-5} and hence the capacity of the gas phase which is equilibrated with carbon to reduce MnO to the metal at the same activity is 0·037% by volume. At temperatures below 1400°C where the standard free energy change for the C–CO reaction is more positive than that for MnO formation, carbon monoxide could not be used to reduce MnO. It thus emerges from these considerations that the indirect reduction of MnO to manganese metal by the reaction

$$MnO + CO \rightarrow Mn + CO_2$$

cannot contribute significantly to the production of manganese in the blast furnace. Since slag-making temperatures in the furnace are probably between 1200–1400°C, it seems most likely that the main reaction path is the direct reduction of MnO from a liquid slag through contact with solid coke particles or carbon-saturated liquid iron droplets. It is therefore necessary in this process to maintain as large a volume of the furnace as possible at temperatures in excess of 1400°C in order to optimize the possibility of carbon-slag reaction which will produce the metal. This is achieved by the combustion of coke at a much higher rate in ferromanganese production per ton of metal produced than for iron.

It would seem desirable to run the furnace to as high a temperature as possible in order to optimize the chemical reaction. Unfortunately, manganese has a high vapour pressure (approximately 250 times that of iron at 1600°C), and if too high a temperature is reached around the tuyères, manganese evaporation becomes significant. The metal will be recycled in the furnace due to oxidation at the lower temperatures which prevail in the higher sections of the furnace stack, or blown out of the furnace as fine particles in the discharge gases.

Phosphorus

Many of the low-grade iron ores contain phosphate minerals such as apatite, $Ca_3(PO_4)_2$, and in blast furnace production the phosphates are reduced to phosphorus in the metal phase. The thermodynamics of the dilute solutions of phosphorus in liquid iron has been studied by Urbain (1959) who used liquid–liquid partition between iron and silver. The atom fraction of phosphorus dissolved in the iron phase varied between 0·1 and 0·35 whilst that in the silver phase was about 100 times smaller. The free energy of transfer

$$[P]_{Ag} \rightarrow [P]_{Fe}$$

could be represented by the equation

$$\Delta G^\circ = -35\,300 + 5\cdot3\,T \text{ cal} (X_P \rightarrow 0)$$

and the partition coefficient $X_{[P]_{Ag}}/X_{[P]_{Fe}}$ was related to the mole fraction of phosphorus in the iron phase by the equation

$$\log X_{[P]_{Ag}}/X_{[P]_{Fe}} = \log R = \log R^\circ + 17\,X^2_{[P]_{Fe}}$$

$$(X_{[P]_{Fe}} = 0 - 0\cdot30)$$

where

$$R^\circ = X_{[P]_{Ag}}/X_{[P]_{Fe}} (X_P \rightarrow 0).$$

and $\log R^\circ = -3$ at 1660°C
$\qquad\qquad -3\cdot75$ at 1300°C

Since the silver phase was such a dilute solution, it can be assumed that the activity coefficient of phosphorus dissolved in silver was constant in the study and therefore all of the variation of R above reflects the variation of $\gamma_{[P]_{Fe}}$ with composition. Hence,

$$\log \gamma_{[P]_{Fe}} = \log \gamma^\circ_{[P]_{Fe}} + 17\,X^2_{[P]_{Fe}} (X_{[P]} = 0\cdot11{-}0\cdot32)$$

in the stated composition range.

The effect of carbon on the activity coefficient of phosphorus can be obtained through the reciprocal relationship

$$\varepsilon^P_C = \varepsilon^C_P$$

and from measurements of the effect of phosphorus on the solubility of carbon in liquid iron. The value 6 for this interaction parameter can be used over the whole range of composition. Hence in the blast furnace metal, if we combine Urbain's results with this information, the variation of the activity coefficient of phosphorus can be represented by the equation

$$\log \gamma_P = \log \gamma^\circ + 17\, X_P^2 + 2 \cdot 6\, X_C$$

ignoring the effects of other solutes. The value of $\log \gamma^\circ$ in this equation has not been satisfactorily determined through experiment.

Schenck and co-workers (1966) used an isopiestic method, in which phosphorus in a vessel at about room temperature was equilibrated via the gas phase with a phosphorus solution in liquid iron at 1530°C, to obtain results for the equilibrium constant of the reaction

$$P_2 \rightarrow 2[P]; \qquad K = \frac{[a_P]^2}{p_{P_2}} = \frac{\gamma_{[P]}^2 X_{[P]}^2}{p_{P_2}}$$

in the composition range $X_P = 0 \cdot 15 - 0 \cdot 4$ at 1530°C. The results for the logarithm of the mass action constant of this reaction, K', were represented as a linear function of $X_{[P]}$, where

$$-\tfrac{1}{2} \log K' = \log (p_{P_2}^{\frac{1}{2}}/X_{[P]}) = \log \gamma_{[P]} - \tfrac{1}{2} \log K$$
$$= \log \gamma_{[P]} - \log \gamma_{[P]}^\circ$$

and it is found experimentally that

$$\log \gamma_{[P]} = 6 \cdot 95\, X_{[P]} - 3 \cdot 0\, [X_P = 0 \cdot 15 - 0 \cdot 4]$$

The results can also be accommodated to Urbain's equation in good approximation and we obtain $\log \gamma_{[P]}^\circ$ equal to $-2 \cdot 25$ for the infinitely dilute solution. The combined results are shown plotted together in Fig. 74. The vapour phase over phosphorus at low temperatures is predominantly composed of P_4 molecules, but at liquid iron temperatures, dissociation of these molecules to the P_2 species is virtually complete. No direct measurement of the partial pressure of the monatomic species has been made at this high temperature, but it may be concluded from a combination of spectroscopic data and of thermal data for the P_2 molecule that the dissociation of P_2 is not significant at the pressures which were involved in the Fe–P liquid system studies. The free energy of dissociation of P_2 can be represented, within the accuracy of the extant information by the equation

$$P_2 \rightarrow 2P(g)$$
$$\Delta G^\circ = 118\,750 - 28 \cdot 39\, T \text{ cal mol}^{-1}$$

This equation demonstrates that at 1500°C the partial pressure of the monatomic species only becomes important when p_{P_2} is around 10^{-9} atm or less.

The two studies which have been discussed here only cover the composition range above $X_{[P]} = 0.1$ and in the binary Fe–P system there is no further information. However, there appears to be little signifiant error in accepting the combined results and using Urbain's extrapolation to $X_{[P]} = 0$. It is now possible to calculate the iron activity from $X_{Fe} = 1.0$ down to the composition Fe_2P for which other thermochemical data are available.

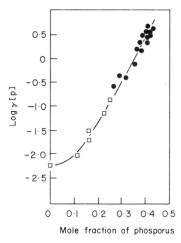

Fig. 74. The variation of the activity coefficient of phosphorus with composition at temperatures around 1530°C.

–☐—☐– = Distribution results of Urbain.
–●—●– = Gas equilibration results of Schenck *et al.*

The Gibbs–Duhem equation, with the substitution for the phosphorus activity coefficient from Urbain's equation is as follows:

$$\log \gamma_{Fe} = - \int \frac{X_{[P]}}{X_{Fe}} \, d \log \gamma_{[P]}$$

$$\log \gamma_{[P]} = 17 \, X_{[P]}^2 + \log \gamma_{[P]}^{\circ}; \qquad d \log \gamma_{[P]} = 34 \, X_{[P]} . d \, X_{[P]}$$

Hence

$$\log \gamma_{Fe} = -34 \int \frac{X_{[P]}^2}{1 - X_{[P]}} . d X_{[P]}$$

$$= 34[\tfrac{1}{2}(1 - X_P)^2 - 2(1 - X_P) + \ln(1 - X_P)]\dagger \, {}^{X_{Fe} = X_{Fe}}_{X_{Fe} = 1}$$

$$= [17 \, X_{Fe}^2 - 68 \, X_{Fe} + 34 \ln X_{Fe}]^{X_{Fe} = X_{Fe}}_{X_{Fe} = 1}$$

$$\dagger \left\{ \int \int \frac{x^2 dx}{(a + bx)} = \frac{1}{b^3} \left[\tfrac{1}{2}(a + bx)^2 - 2a(a + bx) + a^2 \ln(a + bx) \right] \right.$$

At $X_{Fe} = 0.66$, $\log \gamma_{Fe} = -0.56$

$$a_{Fe} = 0.182$$

From the results of Schenck et al. (1966), the phosphorus activity in liquid Fe_2P can be obtained with respect to P_2 gas at one atmosphere as standard state. Thus from these data

At $X_{[P]} = 0.33$; $\log(p_{P_2}^{\frac{1}{2}}/X_P) = -0.5$. Hence

$$p_{P_2} = 0.0108 \text{ atm}$$

and

$$\Delta G^{\circ}_{Fe_2P} = -RT \ln \frac{a_{Fe_2P}}{a_{Fe}^2 \cdot p_{P_2}^{\frac{1}{2}}} = 2RT \ln a_{Fe} + \tfrac{1}{2}RT \ln p_{P_2}$$
$$(a_{Fe_2P} = 1)$$
$$= -(-12\,190) + (-8100)$$
$$= -20\,290 \text{ cal mol}^{-1} \text{ at 1800 K.}$$

Now Weibke and Schrag (1941) measured the heat of formation of Fe_2P by reaction calorimetry at about 600°C, and Kubaschewski and Catterall (1956) have combined this result with other thermal data to obtain the value

$$2Fe + P(s) \rightarrow Fe_2P; \qquad \Delta H^{\circ}_{298} = -38\,100 \pm 3000 \text{ cal mol}^{-1}$$

Using the value for the heat of sublimation of white phosphorus to P_2 gas at 298 K, we then obtain for the reaction

$$2Fe + \tfrac{1}{2}P_2(g) \rightarrow Fe_2P; \qquad \Delta H^{\circ}_{298} = -54\,900 \pm 3500 \text{ cal mol}^{-1}$$

If we ignore the free energy of fusion of Fe_2P above the congruent melting point at 1365°C and up to the temperature of 1527°C (1800 K) we can now obtain the entropy of formation of the phase at this latter temperature.

From the Gibbs–Duhem calculation and the calorimetric studies for the reaction

$$2Fe + \tfrac{1}{2}P_2 \rightarrow Fe_2P \qquad \Delta G^{\circ}_{1800} = -20\,290 \text{ cal}$$
$$\Delta H^{\circ} = -54\,900 \text{ cal}$$
and hence $\qquad \Delta S^{\circ} = -19.2 \text{ cal mol}^{-1} \text{ deg}^{-1}$

Nitrogen

The droplets of liquid iron which are formed in the liquid reaction zone of the blast furnace are exposed to carbon monoxide/dioxide gas mixed with nitrogen. The nitrogen content of liquid iron which is in equilibrium with nitrogen gas at one atmosphere pressure is extremely small; at 1600°C the solubility of nitrogen is only 0·178 atom percent in pure liquid iron. The free energy equation for the solution reaction

$$\tfrac{1}{2}N_2 \rightarrow [N] \qquad K = \frac{\text{atom percent } [N]}{p_{\frac{1}{2}N_2}}$$

$$\Delta G° = 861 + 2·96\, T \text{ cal}$$

shows that the heat of solution of nitrogen is only 860 cal g atom^{-1} in liquid iron. The effect of carbon is to lower the solubility of nitrogen in iron, and at carbon saturation and 1600°C the solubility of nitrogen is only one tenth of that in pure iron.

The other reactions of nitrogen in the blast furnace which need to be considered here are those involving the formation of cyanogen, by reaction with carbon, and the possible formation of metal cyanides and carbo-nitrides. Richardson and Jeffes (1949) have shown that the formation of cyanogen gas can be represented by the free energy equation

$$2C + N_2 \rightarrow C_2N_2 \qquad \Delta G° = 73\,000 - 12·0\, T \text{ kcal}$$

and from this equation it can be concluded that this gas occurs in negligible proportions in the blast furnace. The possible formation of alkali metal cyanides was also discussed by these authors who concluded that sodium and potassium cyanide could be formed between sodium and potassium metal vapours, which were cooling during their ascent of the furnace, and coke particles in the nitrogen-laden atmosphere of the furnace. Such compounds are however unstable in the presence of silica particles since the alkali metal silicates would be preferentially formed, and in the presence of high enough CO_2 partial pressures to form the metal carbonates. The possibilities for such reactions taking place depend on the mode of operation of the furnace, the temperature and burden distribution, and the kinetics of the change of oxygen potential in the gas phase in the cooler parts of the furnace, but under equilibrium conditions they are not important.

The other way in which nitrogen can have an effect on reactions in the blast furnace is through the formation of the refractory carbo-nitrides. The elements of Groups IV and V which form stable refractory carbides usually also form relatively stable nitrides. These phases are usually miscible in all proportions in the solid state and the solid solutions such as TiC + TiN and VC + VN probably conform quite closely to Raoult's law. It follows that the presence of nitrogen in the gas phase of the blast furnace will favour the presence of such refractory solids, especially at the lower reaches of the furnace where the metal activities are raised by the process of carbon reduction.

An approximate calculation of the effect of nitrogen in stabilizing the carbide phase which might be expected can be made as follows. The pressure of nitrogen is close to one atmosphere and the activity of carbon is approximately unity throughout the furnace. If we use the free energy of formation equations for TiC and TiN simultaneously, and assume ideality for TiC and TiN solid solution, we can find the respective mole fractions of these compounds in a carbo-nitride of titanium which is formed in the furnace.

$$\text{Ti} + \text{C} \rightarrow \text{TiC} \qquad K_{\text{TiC}} = \frac{a_{\text{TiC}}}{a_{\text{Ti}}} \simeq \frac{X_{\text{TiC}}}{a_{\text{Ti}}}$$

$$\Delta G^\circ = -44\,600 + 3\cdot16\,T \text{ cal}\,(a_{\text{C}} = 1)$$

$$\text{Ti} + \tfrac{1}{2}\text{N}_2 \rightarrow \text{TiN} \qquad K_{\text{TiC}} = \frac{a_{\text{TiN}}}{a_{\text{Ti}}} \simeq \frac{X_{\text{TiN}}}{a_{\text{Ti}}}$$

$$\Delta G^\circ = -80\,850 + 22\cdot77\,T \text{ cal}\,(p_{\text{N}_2} \simeq 1)$$

since a_{Ti} is common to both sub-systems

$$\frac{X_{\text{TiN}}}{X_{\text{TiC}}} = \frac{1 - X_{\text{TiC}}}{X_{\text{TiC}}} = \frac{K_{\text{TiN}}}{K_{\text{TiC}}}$$

This expression is also obtained by consideration of the free energy of the exchange reaction

$$\text{TiC} + \tfrac{1}{2}\text{N}_2 \rightarrow \text{TiN} + \text{C} \qquad K = \frac{a_{\text{TiN}}}{a_{\text{TiC}}}\,\frac{a_{\text{C}}}{p_{\text{N}_2}^{\frac{1}{2}}}$$

$$\Delta G^\circ = -36\,250 + 19\cdot61\,T \text{ cal}$$

At 1600°C this equation and the substitution of the approximation for the activity ratio of nitride/carbide yields a mole fraction of TiN in the carbo-nitride of approximately one-half. Using this value for aTiN in the formation of the solid from nitrogen at one atmosphere and titanium at 1600°C, we obtain further information

$$\text{Ti} + \tfrac{1}{2}\text{N}_2 \rightarrow \text{TiN} \qquad \Delta G^\circ_{1873} = -38\,200 \text{ cal}$$

and hence it is concluded that under equilibrium conditions, the activity of titanium above which TiN will form at this temperature at an activity of 0·5 with a nitrogen pressure of one atmosphere, is therefore $1\cdot74 \times 10^{-5}$, which is extremely small. Such a small activity of titanium is found at concentrations of about 0·3 wt percent in carbon-saturated liquid iron at 1600°C according to the solubility studies of Delve, Meyer and Lander (1961). These workers confirmed the effects of nitrogen in the formation of titanium carbo-nitride and found an activity coefficient of titanium of 10^{-2} in the carbon-saturated liquid iron solvent.

Similar considerations should also apply for vanadium, niobium and zirconium to those presented here for titanium, and it must therefore be concluded that nitrogen plays an important part in promoting the formation of refractory metal carbonitride solid phases in the blast furnace.

Approximate free energies of formation for these refractory carbides and nitrides can be obtained from data for 298 K to yield the following equations:

$$V + C \rightarrow VC \qquad \Delta G^\circ = -24\,100 + 1\cdot50\,T \text{ cal}$$
$$V + \tfrac{1}{2}N_2 \rightarrow VN \qquad \Delta G^\circ = =51\,900 + 20\cdot97\,T \text{ cal}$$
$$Nb + C \rightarrow NbC \qquad \Delta G^\circ = -31\,100 + 0\cdot40\,T \text{ cal}$$
$$Nb + \tfrac{1}{2}N_2 \rightarrow NbN \qquad \Delta G^\circ = -56\,800 + 23\cdot36\,T \text{ cal}$$
$$Zr + C \rightarrow ZrC \qquad \Delta G^\circ = -44\,100 + 2\cdot20\,T \text{ cal}$$
$$Zr + \tfrac{1}{2}N_2 \rightarrow ZrN \qquad \Delta G^\circ = -87\,900 + 23\cdot11\,T \text{ cal}$$

It must be pointed out that each of these phases shows some departure from non-stoichiometry. The monocarbide phases range from $VC_{0\cdot75}$–$VC_{0\cdot88}$, $NbC_{0\cdot70}$–$NbC_{1\cdot0}$ and $ZrC_{0\cdot60}$–$ZrC_{0\cdot98}$, but in the context of blast furnace operation the upper limit of composition, which is in equilibrium with carbon, will probably be a good approximation. Similar remarks must also be applied to the nitrides and the carbonitrides, but this point will not be pursued further here.

A comparison of the nitrides and carbides in terms of their free energies of formation shows that the heat of formation terms for the nitrides are approximately double those for the carbides, and the entropy terms are dominated by the change in entropy for the condensation of nitrogen. The carbide–nitride exchange reactions therefore favour the carbide for vanadium and niobium more than is the case for titanium and zirconium. Nitrogen should therefore play a less important role in the blast furnace chemistry of vanadium and niobium. Making the assumptions, as before, of unit activity for carbon and nitrogen and Raoultian behaviour of the MC + MN solid solutions, the mole fractions of VN and NbN in VC and NbC respectively, are found to be one tenth and one hundredth respectively.

DESULPHURIZATION IN THE BLAST FURNACE

Richardson and Fincham (1954) studied the sulphide capacities of a number of simple binary slag systems to hold sulphur, as well as the more complex blast furnace slags. The studies consisted of equilibrating the slag samples with a gas of fixed oxygen and sulphur potential at a known temperature. The gas mixture which provided these two potentials was of H_2, CO_2 and SO_2. The sulphide capacity was then expressed as

$$C_S = \{S\}. \frac{p^{\frac{1}{2}}_{O_2}}{p^{\frac{1}{2}}_{S_2}}$$

It was found that the sulphide capacities of the synthetic slags paralleled the metal oxide activity; thus the rapid increase of the activity of CaO in the binary CaO–SiO_2 system which occurs on the CaO-rich side of the compositions is accompanied by a rapid increase in C_S. The sulphide capacities of FeO–SiO_2 slags, which are practically Raoultian solutions, were much higher than those of corresponding composition in the CaO–SiO_2, but the MgO–SiO_2 solutions had lower sulphide capacities in the same range of compositions.

The relationship between slag sulphide capacity and slag composition can be brought out in the following manner. The exchange reactions

$$FeO(l) + \tfrac{1}{2}S_2 \rightarrow FeS(l) + \tfrac{1}{2}O_2$$
$$CaO(s) + \tfrac{1}{2}S_2 \rightarrow CaS(s) + \tfrac{1}{2}O_2$$

have about the same value of equilibrium constant, approximately 2×10^{-3} at 1500°C, whereas the value for the reaction

$$MgO(s) + \tfrac{1}{2}S_2 \rightarrow MgS(s) + \tfrac{1}{2}O_2$$

has the value of $1 \cdot 5 \times 10^{-5}$ at the same temperature. The sulphur contents of the silicates are quite small so that the sulphide probably obeys Henry's law, and the activity of the oxide is relatively unchanged by the small exchange of sulphur for oxygen during the sulphidation equilibration. Hence,

$$C_S = \{\% S\} \cdot \frac{p^{\frac{1}{2}}_{O_2}}{p^{\frac{1}{2}}_{S_2}} = \{X_{MS}\} \cdot \frac{p^{\frac{1}{2}}_{O_2}}{p^{\frac{1}{2}}_{S_2}} \qquad (M = Ca, Fe, Mg)$$

whereas the equilibrium constant for the exchange reaction

$$K_{exch} = \frac{\{a_{MS}\}}{\{a_{MO}\}} \cdot \frac{p^{\frac{1}{2}}_{O_2}}{p^{\frac{1}{2}}_{S_2}} = \frac{\gamma_{MS}}{\gamma_{MO}} \cdot \frac{\{X_{MS}\}}{\{X_{MO}\}} \cdot \frac{p^{\frac{1}{2}}_{O_2}}{p^{\frac{1}{2}}_{S_2}} = \frac{C_s \cdot \gamma_{MS}}{\{a_{MO}\}}$$

for a roughly constant MO/SiO_2 ratio in the silicate phase. Therefore, the lower value of the exchange constant for magnesia and the higher value of a_{FeO} in the FeO–SiO_2 system accounts for the sequence of sulphide capacities in the silicates.

In the more complex CaO–Al_2O_3–SiO_2 slags, the sulphide capacities follow the pattern of CaO activities, thus showing that γ_{CaS} is not particularly sensitive to composition.

The summary of results shown in Fig. 75 also includes points for typical blast furnace and open hearth slags. The major effect of the CaO activity in making the sulphide capacity highest in the basic open hearth furnace is clearly in line with the results on the simple binary laboratory silicate mixtures.

The slag-metal equilibrium sulphur distribution which is found under blast furnace conditions

$$[S] + \{O\} \rightarrow \{S\} + CO(g) \qquad K = \frac{\{S\}\, p_{CO}}{[S]\{O\}}$$

is clearly the result of a number of inter-related effects. Firstly, the activity coefficient of sulphur in the metal is increased above the value in the pure iron-sulphur system by the saturation of the metal with carbon ($\varepsilon_S^C = 13 \cdot 3$). Secondly, the presence of CO at one atmosphere pressure with a carbon activity of unity ensures that the oxygen pressure of the system is very low. The sulphur content of the slag is raised, for the given sulphide capacity of the appropriate slag composition, by these two effects which produce a low value of the ratio $p^{\frac{1}{2}}_{O_2}/p^{\frac{1}{2}}_{S_2}$. It therefore follows that conditions in the hearth of the

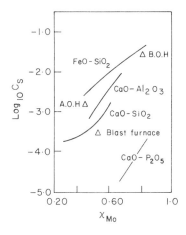

Fig. 75. Sulphide capacities of liquid slag systems at 1600°C as a function of composition.

△ B.O.H. = Basic open hearth slag.
△ A.O.H. = Acid open hearth slag.
△ Blast furnace = B.F. slag.

blast furnace favour the transfer of sulphur from the metal to the slag more than would be so under highly oxidizing conditions. The next stage in the process of steelmaking occurs under conditions where carbon must be removed from the metal and the simplest means for achieving this refining step is by oxidation. The removal of carbon from iron also leads to a lowering of the activity coefficient of sulphur in solutions in iron. It can obviously be concluded that, under equilibrium conditions, desulphurization of iron is much more readily achieved in the blast furnace than in the later stages of steelmaking.

THE CALCULATION OF THERMODYNAMIC PROPERTIES OF COMPLEX SLAGS

The structural concepts which have been formulated for the understanding of binary silicate solid systems have been applied to complex liquid systems in an attempt to predict the thermodynamic and transport behaviour of slag systems. The basic structural unit is the oxygen ion tetrahedron containing silicon, aluminium or phosphorus atoms. The charges on these atoms at the centre of the tetrahedra of oxygen ions in silicates, aluminates and phosphates are probably less than would be indicated by Si^{4+}, P^{5+} and Al^{3+} in these compounds, and the binding with the oxygen ions is probably partially covalent, but there is no precise information on this aspect. It is possible to make a satisfactory model of slag systems if it is assumed, however, that the bonds are purely ionic and that the structures come about as a result of the packing of oxide ions around Si^{4+}, P^{5+} and Al^{3+} ions. We thus obtain anionic units SiO_4^{4-}, PO_4^{3-} and AlO_4^{5-}, and would expect the substitution of PO_4^{3-} or AlO_4^{5-} for SiO_4^{4-} to be made electrically neutral by the removal or addition respectively of a cationic charge. The further addition of metal oxide MO beyond the composition M_2SiO_4 provides free oxide ions in the melt according to this model. In the light of our general knowledge about the statistical nature of structural arrangements in liquids, it is very unlikely that a few simple structures would exist in the liquid silicates. It would seem more likely that, at a given metal oxide/silica ratio, the melt would contain a number of structural units, of which one may predominate. Thus, in a composition which is slightly richer in SiO_2 than the ortho silicate, it is probable that the species

$$SiO_4^{4-}, Si_2O_7^{6-}, Si_3O_9^{6-} \ldots \ldots \text{etc. and } O^{2-}$$

exist to a greater or less extent, depending on the temperature as well as the molar ratio of the components, but SiO_4^{4-} would predominate. It is thus a "quasi-solid" approximation to assume that the predominating species shall be counted as the sole structural unit for a given composition. If this approximation is accepted, however, it is then possible to calculate the concentrations of free oxide ions in metal oxide-rich silicate systems or slags.

The reactions in which the slag takes part in blast furnace production involve oxygen ions in sulphur and phosphorus removal from liquid metal. The desulphurizing and dephosphorizing powers of a slag depend also on the metal cation nature and content. Thus, a high-lime slag behaves very much differently from a high silica slag, and at a given metal oxide/silica ratio, a high CaO slag content is more effective than a high FeO content. To account for the effect of the metal oxide, Flood and co-workers (1953) combine the thermodynamic properties of the reactions involving single metal cations to obtain the results for complex slags. To show how this is done, we will consider the exchange of one anion for another in a solution containing a mixture of two cationic species, the exchange being brought about by means of a gas-liquid reaction.

As an example

$$(\text{Na}, \text{K})\text{Cl} + \tfrac{1}{2}\text{Br}_2 \rightarrow (\text{Na}, \text{K})\text{Br} + \tfrac{1}{2}\text{Cl}_2$$

The free energy change of this reaction may be obtained, according to Flood's approximation, from consideration of the following cycle

$$(\text{Na}, \text{K})\text{Cl} + \tfrac{1}{2}\text{Br}_2 \xrightarrow{\text{I}} (\text{Na}, \text{K})\text{Br} + \tfrac{1}{2}\text{Cl}_2$$

$$\left.\begin{array}{c} \downarrow \text{II} \\ \\ \text{NaCl} + \text{KCl} \underset{\text{IIIb}}{\overset{\text{IIIa}}{\rightleftarrows}} \left.\begin{array}{l} \text{NaBr} \\ \text{KBr} \end{array}\right\} \end{array}\right. \nearrow \text{IV}$$

The free energy changes for II–IV are

$$\Delta G_{\text{II}} = -\Delta G^{\text{M}}_{\text{NaCl–KCl}} = -(\Delta G^{\text{ideal}}_{\text{NaCl–KCl}} + \Delta G^{\text{XS}}_{\text{NaCl–KCl}})$$

$$\Delta G_{\text{IIIa}} = X_{\text{NaCl}} \cdot \Delta G^\circ; \text{ for the reaction NaCl} + \tfrac{1}{2}\text{Br}_2 \rightarrow \text{NaBr} + \tfrac{1}{2}\text{Cl}_2$$

$$\Delta G_{\text{IIIb}} = X_{\text{KCl}} \cdot \Delta G^\circ; \text{ for the reaction KCl} + \tfrac{1}{2}\text{Br}_2 \rightarrow \text{KBr} + \tfrac{1}{2}\text{Cl}_2$$

$$\Delta G_{\text{IV}} = \Delta G^{\text{M}}_{\text{NaBr–KBr}} = (\Delta G^{\text{ideal}}_{\text{NaBr–KBr}} + \Delta G^{\text{XS}}_{\text{NaBr–KBr}}).$$

Now in the summation of the steps to arrive at the cyclic relationship

$$\text{I} = \text{II} + \text{IIIa} + \text{IIIb} + \text{IV}$$

it should be observed that

$$\Delta G^{\text{ideal}}_{\text{NaCl–KCl}} = \Delta G^{\text{ideal}}_{\text{NaBr–KBr}} = RT(X_{\text{Na}} \ln X_{\text{Na}} + X_{\text{K}} \ln X_{\text{K}})$$

the atom fractions of the sodium and potassium compounds being identical in the two solutions. Flood also assumes that the difference between the two excess free energy functions is zero

$$\Delta G^{\text{XS}}_{\text{NaCl–KCl}} - \Delta G^{\text{XS}}_{\text{NaBr–KBr}} = 0$$

Hence

$$\Delta G_{\text{I}} = \Delta G_{\text{IIIa}} + \Delta G_{\text{IIIb}}$$

The free energy change for the anion exchange reaction with the mixture of cations is therefore partitioned between that for IIIa and IIIb according to the mole fractions of NaCl and KCl. This is also in the proportion of the number of equivalents of Na^+ and K^+ in 1 g.mol of the mixture.

For another exchange reaction such as

$$(\text{Na}, \text{Ca})\text{Cl} + \tfrac{1}{2}\text{Br}_2 \rightarrow (\text{Na}, \text{Ca})\text{Br} + \tfrac{1}{2}\text{Cl}_2$$

it is clear that the free energy change for the reaction would be obtained once again by apportioning the anions amongst the cations according to the number of equivalents of the cations. Each g atom of Na^+ only takes up 1 g atom of Cl^- whereas each g atom of Ca^{2+} takes up 2 g atom of Cl^-. Hence the free energy change for the mixed cation solution can be obtained by using *the equivalent fractions* x_{Na^+} and $x_{Ca^{2+}}$ as weighting factors, and

$$\Delta G^\circ_{(Na. Ca)Cl} = x_{Na^+} \Delta G^\circ_{NaCl} + x_{Ca^{2+}} \Delta G^\circ_{CaCl_2}$$

with

$$x_{Na^+} = \frac{n_{Na^+}}{n_{Na^+} + 2n_{Ca^{2+}}}; x_{Ca^{2+}} = \frac{2n_{Ca^{2+}}}{n_{Na^+} + 2n_{Ca^{2+}}}$$

where n_{Na^+} is the number of sodium ions in the melt, and $n_{Ca^{2+}}$ the number of calcium ions.

The approximation makes use of the fact that the difference between the excess free energies of mixing of molten slag components with the same sign of ionic charge, e.g. NaCl + KCl which are involved in the cycle is negligible in comparison with the free energies of the exchange reactions of simple compounds. The comparison is approximately the same as that for the heats of mixing of the molten salts with the heat changes for the exchange reactions. Usually the latter are considerably larger than the former. The heats of mixing of molten chlorides or bromides are typically 0–5 kcal mol^{-1} whereas the chloride-bromide exchange reactions have heat changes which are typically in the range 10–15 kcal mol^{-1}.

This procedure can obviously be extended to include exchange reactions between metal and slag phases. Flood *et al.* (1953) demonstrated this in the case of the exchange of iron and manganese between metal and slag.

$$\{Fe^{2+}\} + [Mn] \rightarrow \{Mn^{2+}\} + [Fe]$$

where the slag might contain Fe^{2+}, Mn^{2+}, Ca^{2+} and Mg^{2+} cations together with F^-, O^{2-}, PO_4^{3-} and SiO_4^{4-} anions. The exchange reaction now involves the component reactions

$$FeF_2 + Mn \rightarrow Fe + MnF_2 \tag{i}$$

$$FeO + Mn \rightarrow Fe + MnO \tag{ii}$$

$$Fe(PO_4)_{\frac{2}{3}} + Mn \rightarrow Fe + Mn(PO_4)_{\frac{2}{3}} \tag{iii}$$

and

$$Fe(SiO_4)_{\frac{1}{2}} + Mn \rightarrow Fe + Mn(SiO_4)_{\frac{1}{2}} \tag{iv}$$

for the anions. Ignoring effects due to mixing of the salts, the slag may be decomposed into its components and the free energy changes of each of the above reactions should be summed to get the free energy change for iron and manganese, using equivalent fractions as the weighting factors. Thus

$$\Delta G° \text{ for } Fe^{2+} + Mn \rightarrow Mn^{2+} + Fe$$

is equal to

$$x_{F^-}\Delta G°_{(i)} + x_{O^2}\Delta G°_{(ii)} + x_{PO_4^{3-}}\Delta G°_{(iii)} + x_{SiO_4^-}\Delta G°_{(iv)}$$

with

$$x_{F^-} = \frac{n_{F^-}}{n_{F^-} + 2n_{O^{2-}} + 3n_{PO_4^-} + 4n_{SiO_4^-}}$$

The equilibrium constant for the exchange of sulphur and oxygen between metal and the slag containing Fe^{2+}, Mn^{2+}, Ca^{2+} and Mg^{2+} would similarly be calculated from the data for the component reactions

$$[O] + FeS \rightarrow FeO + [S] \tag{a}$$

$$[O] + MnS \rightarrow MnO + [S] \tag{b}$$

$$[O] + CaS \rightarrow CaO + [S] \tag{c}$$

and

$$[O] + MgS \rightarrow MgO + [S] \tag{d}$$

and

$$\Delta G° = x_{Fe^{2+}}\Delta G°_{(a)} + x_{Mn^{2+}}\Delta G°_{(b)} + x_{Ca^{2+}}\Delta G°_{(c)} + x_{Mg^{2+}}\Delta G°_{(d)}.$$

Of these four terms, those involving Fe^{2+} and Ca^{2+} are the dominant terms in the practical situation of ironmaking. This is in agreement with the measured effects for sulphide capacities of complex slags and their main dependence on the iron and calcium contents which was referred to in the previous section.

Having obtained the free energy changes for such exchange reactions involving complex slag phases, Flood and co-workers then assume that the mass action law can be used in relating the free energy change to concentrations of the slag components. In the two cases which have been outlined above, for iron–manganese exchange and for sulphur–oxygen exchange, the free energy changes are used thus

$$\Delta G°_{Fe-Mn} = -RT \ln K = -RT \ln \frac{\{Mn^{2+}\}}{\{Fe^{2+}\}} \cdot \frac{a_{[Fe]}}{a_{[Mn]}}$$

and

$$\Delta G°_{S-O} = -RT \ln \frac{\{O^{2-}\}}{\{S^{2-}\}} \cdot \frac{a_{[S]}}{a_{[O]}}$$

where { } represent concentrations in the slag phase. This procedure says nothing about the metal phase which is involved in exchange reactions with the slag, and hence it is appropriate to use activities only in this phase.

The difficulty, which was referred to earlier, of deciding on the make-up of the anionic structure of an oxide slag clearly affects the value of this procedure to practical application in metal-making reactions. Thus, for oxide slags, some decision must be made concerning the concentrations of the anionic species such as O^{2-} and SiO_4^{4-} in order that the exchange reactions which involve the cations can be apportioned to the appropriate anions. In the iron–manganese exchange between metal and a simple $FeO + MnO + SiO_2$ slag, Flood and Toguri (1963) successfully used the equation

$$\Delta G_{Fe-Mn} = x_{O^{2-}} \Delta G°(O^{2-}) + x_{SiO_4^-} \Delta G°(SiO_4^{4-})$$

with $\Delta G°(O^{2-})$ for the reaction

$$FeO + Mn \rightarrow MnO + Fe$$

and $\Delta G°(SiO_4^{4-})$ for the reaction

$$\tfrac{1}{2}Fe_2SiO_4 + Mn \rightarrow \tfrac{1}{2}Mn_2SiO_4 + Fe$$

The values of the equivalent fractions $x_{O^{2-}}$ and $x_{SiO_4^-}$ can only be obtained in this simple expression if the possibility of the concentrations of other complex silicate anions are ignored or can be accounted for.

THE BLAST FURNACE PRODUCTION OF LEAD BULLION

The reduction of lead and zinc oxides with carbon

The principal source of lead is the mineral galena, PbS, which usually arrives at the smelter in the form of a finely-divided flotation concentrate, containing Fe, Cu, Zn and As, Sb, Bi as the main impurities. The stages which are necessary before such a material can be used in the burden of a blast furnace are consolidation of the fine powder to more coarse-grained material, and oxidation of the sulphide to oxide. These two steps are accomplished by *reaction sintering* in a Dwight–Lloyd sinter machine. (Fig. 76).

The concentrate is mixed with silica and a small quantity of water to enable some agglomeration to be achieved in a rotating drum before the material is transferred to the sinter machine. The charge is ignited by a flame from above the moving sinter bed, and air is drawn through the agglomerate to bring about oxidation roasting. A considerable, although not complete, conversion of sulphide to oxide occurs, and lead silicates are formed by reaction between

the PbO which is produced, and the silica of the charge. Too high a temperature during this roating process, i.e., about 1000°C, must be avoided to
minimize the loss of lead by PbO vaporization. If the charge has a large
sulphur content, the excess heat which is evolved during the oxidation is
usually absorbed by admixing limestone to the charge. This addition does not
function completely as chemically inactive thermal ballast, but forms calcium
sulphate to a small extent. This exothermic process is more than compen-
sated for by the decomposition of the carbonate to CaO and CO_2 which is
an endothermic process.

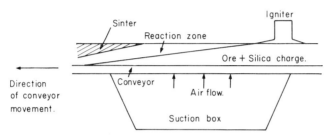

Fig. 76. Schematic diagram of a Dwight–Lloyd sinter machine.

In the blast furnace process, lead oxide is readily reduced by carbon, and
at these oxygen potentials the iron goes into the slag or combines with sulphur
to form a matte phase. The metals lead and iron are quite immiscible, but the
bullion can retain some copper. There are therefore at least three phases in
the lower part of the furnace, quite pure metal, a matte phase and a silicate
slag. Sometimes another phase, a speiss, which contains arsenides and anti-
monides, also appears as a separate layer. A glance at the oxide and sulphide
free energy diagrams shows that the lead metallic phase is relatively free of
iron, which forms a much more stable oxide and would thus go to the slag,
but PbS and FeS have roughly the same free energy of formation, and so the
matte phase can contain a significant amount of lead. For the same reason,
that of having a low affinity for oxygen but a high affinity for sulphur, copper
also appears in the matte phase. Typical analyses of the phases are given in
Table XXVII where the composition is given in weights percent.

One of the difficulties of lead production is connected with the presence
of zinc in the ingoing concentrate. At the low oxygen pressures and high
temperatures, this element is vaporized from the descending burden and
gathers as oxide accretions at the cooler parts of the furnace, creating an
obstruction.

The thermodynamic properties of dilute solutions of ZnO in CaO–FeO–
SiO_2 slags were established by Richards and Thorne (1961). These workers
equilibrated slags at 1200°C with controlled zinc pressures and CO/CO_2

ratios in the gas phase and determined the ZnO content of the resulting slags. These were contained in iron crucibles during equilibration; the gas phase was made by passing a controlled CO/CO_2 mixture through a bed of ZnO pellets at 900°C. The ratio $\gamma\{ZnO\}/\gamma\{FeO\}$ was found to have a value around unity in these studies, and it increased with increasing CaO/SiO_2 ratio in the slag. FeO behaves almost ideally in silicate systems, the activity coefficient having values between 0·5 and 2·0. It follows that the assumption of ideality

TABLE XXVII

Phase						
Metal	Pb	Bi	Sb	As	Fe	Ca
	98·6	0·14	0·84	0·01	0·002	0·009
Slag	Pb	S	SiO_2	FeO	CaO	
	0·94	1·1	35·0	28·7	22·2	
Matte	Pb	Fe	Cu	S		
	10·7	44·7	12·6	23·4		
Speiss	Pb	As	Fe	Cu	S	
	4·8	19·4	55·6	4·8	11·6	

for dilute solutions of ZnO in silicate slags with a Raoultian activity coefficient of unity is close to correct at blast furnace temperatures. Since the standard free energy change for the process

$$ZnO + C \rightarrow Zn(g) + CO$$

is -17 kcal g atom^{-1} at 1200°C and becomes increasingly more negative with increasing temperature, it follows that the zinc content of lead minerals which are supplied to the lead-making blast furnace would be eliminated from the slag to quite low levels, certainly less than 1 %, if the furnace were to be operated near C–CO equilibrium at 1200°C. The equilibrium constant has the value

$$K_{ZnO} = \frac{p_{Zn}}{a_{ZnO}} \cdot \frac{p_{CO}}{a_C} = 3 \times 10^2.$$

The oxygen potential corresponding to the C–CO_2 equilibrium has been shown to be an upper limit for the iron blast furnace operation, and ZnO could be retained in the slag phase of a furnace operated with such a high oxygen potential, and at a relatively low temperature. When the blast furnace is used for lead making, high temperatures and low oxygen potentials must be avoided if the objective is to retain ZnO in the slag. This does not affect the reduction of lead adversely since PbO is readily reduced not only by carbon but also by carbon monoxide.

The zinc metal can be recovered from the ZnO-rich slag by a treatment at high temperatures and low oxygen potential. This is the "zinc fuming" process in which powdered coal is blown through the slag to produce zinc vapour.

TABLE XXVIII. Analysis of gases in the Trail smelter for lead

	Volume	%CO	%CO$_2$	%O$_2$	%N$_2$
Top gas		5·1	17·2	4·0	73·7
Tuyère gas		13·8	10·5	3·9	71·8

Measurements have been made of the gas composition at tuyère level and at the top of some operating lead blast furnaces. It is always found that incomplete combustion of the ingoing air is shown by the presence of about 1% of oxygen at the tuyère level. It can be concluded that the non-ferrous blast furnaces, which operate at lower tuyère temperatures than the iron blast furnace, i.e. 1300°C rather than 1600°C, are also considerably more oxidizing due to lack of gas–solid equilibrium in the coke–air reaction.

THE REDUCTION OF ZINC OXIDE BY CARBON

The reduction of zinc oxide to the metal is carried out under conditions where the metal is produced in the gaseous phase. The region of temperature in which the metal is solid or liquid gives a positive value for the free energy change

$$ZnO + C \rightarrow Zn(s, l) + CO.$$

However, above the boiling point of zinc, the entropy of vaporization of zinc adds significantly to the upward slope of the standard free energy line. The standard free energy change of the reaction

$$ZnO + C \rightarrow Zn(g) + CO$$

becomes zero at around 950°C, and this is the ideal situation for a gaseous transport process. It will be remembered that in such a process, material is transported from one part of a system to another by means of vapour molecules which may or may not then revert to the original components. In order for the process to be efficient, there must be a significant change in the pressures of the transported species between the initial point of reaction and the final point of deposition. Because of the large entropy change of the reaction which produces zinc vapour such a large change in pressure will occur over a relatively small temperature range, and zinc will condense at the lower temperature end of a reaction system. This is the basis of the Belgian retort method for the extraction of zinc. It is also obviously necessary that the partial pressure of the material to be transported must be high in order that the *amount* of material which is transported shall be high. Table XXIX

TABLE XXIX. Partial pressure of zinc produced by zinc oxide reduction

T° C	700	800	900	1000	1100
p_{Zn} (atms)	10^{-2}	7×10^{-2}	0·37	1·48	4·95

shows how p_{Zn} varies from 700 to 1100°C if the equilibrium constant is simplified

$$K = \frac{p_{Zn}p_{CO}}{a_{ZnO} \cdot a_C} = p_{Zn}^2 \text{ when } \left\{ \begin{matrix} p_{Zn} = p_{CO} \\ a_{ZnO} = a_C = 1 \end{matrix} \right\}$$

The economic shortcomings of the Belgian retort process mainly centre around the batch nature of the process and the method for the supply of heat to the charge. Since the reactions of zinc formation and condensation are carried out in the sealed retort, the heat for the endothermic reaction must be supplied through the sides of the retort. This disadvantage was overcome by direct electrical heating of the charge in the St. Joseph Lead Company's process.

The lead-zinc blast furnace

There is one further difficulty associated with both of these processes, and that is that there is no provision for the removal of gangue material in the form of a liquid slag phase. It follows that these processes are only viable when a rich source of ore is available. The low-grade ores of zinc contain a high proportion of iron, and in many ore bodies there is also a considerable lead content. The Imperial Smelting Process in which the iron collects in a slag phase, lead as liquid metal at the bottom, and zinc vaporizes from the top of a blast furnace combines all the desirable properties of direct heating within the furnace, and the separation of the two principal metals from one another and from the gangue material. (Fig. 77).

The need to make a slag with a high FeO content which will remove iron from the charge clearly means that at the slag level in this process the oxygen potential must be relatively high. The gas, therefore, has an unusually high

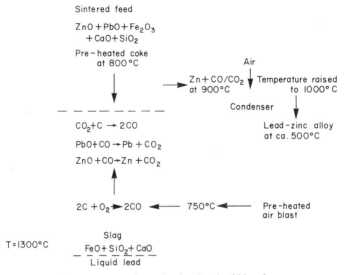

Fig. 77. Process scheme for the zinc–lead blast furnace.

CO_2 partial pressure when compared with that in typical iron blast-furnace operation. Zinc oxide is reduced by gas-solid interaction

$$ZnO + CO \rightarrow Zn(g) + CO_2$$

rather than through a solid-solid reaction

$$ZnO + C \rightarrow Zn(g) + CO_2$$

The CO/CO_2 ratio at the top of the furnace where the off-take temperature is 1000°C is quite low compared with that which is achieved in zinc production by the retort process where the oxygen potential is near the C/CO equation.

The re-oxidation of the metal vapour which would occur if this CO/CO_2 mixture and zinc vapour were condensed together slowly, is avoided by entraining the liquid zinc in liquid lead droplets which fall into a moving pool of liquid metal. This molten alloy is then cooled when zinc separates as a relatively pure phase from the lead pool at about 550°C.

Figure 78 shows the logarithm of p_{CO_2}/p_{CO} plotted against log p_{Zn}. The lines for various temperatures show the relationship between p_{CO_2}/p_{CO} and p_{Zn} at equilibrium with pure ZnO, using the free energy equation.

$$ZnO + CO \rightarrow Zn(g) + CO_2 \quad \rightarrow \quad \Delta G° = 42\,600 - 26.6\,T \text{ cal}$$

$$K_{ZnO} = p_{Zn}\frac{p_{CO_2}}{p_{CO}}$$

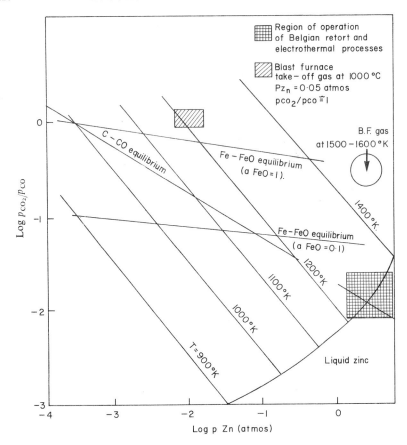

Fig. 78. CO_2/CO ratios and zinc pressures as a function of temperature at equilibrium with solid zinc oxide.

Also shown for comparison are lines giving the p_{CO_2}/p_{CO} ratios for Fe–FeO equilibrium and the C–CO equilibrium. The Fe–FeO lines are for $a_{Fe} = a_{FeO}$ on the upper line, and $a_{Fe} = 10a_{FeO}$ on the lower line. The vapour pressure curve of liquid zinc is shown together with the areas of p_{CO_2}/p_{CO} operation of the older processes for zinc-making, and the gas compositions at the slag level and gas off-takes of the blast furnace.

It can be seen at once that oxygen potential in the older processes is much lower than the blast furnace gas, and the zinc pressure is more than two orders of magnitude higher than in blast furnace operation. By measuring the displacement of the hatched-out area for the blast furnace off-take to the right of the point for the vapour pressure of zinc at 1000°C (1273 K) it can be shown that the activity of zinc is about 5×10^{-3} under these conditions, since

$$a_{Zn} = \frac{p_{Zn}}{p_{Zn}^{o}}$$

It can also be deduced that the activity of zinc in the off-take gas would be approaching unity at temperatures slightly below 900 K (627°C). The liquid lead shower in which the zinc vapour is then shock-cooled and trapped, enters the condenser at 440°C and leaves at about 550°C according to the published literature.

Since the furnace operates at a relatively high oxygen potential, when compared with the iron blast furnace, and at a low temperature, because of the need to make liquid lead, the thermal requirements are not entirely met by the burning of coke within the furnace followed by gas-solid heat transfer. It is necessary to preheat the coke in the furnace charge to about 800°C in order to meet the thermal requirements of the furnace.

THE REDUCTION OF TIN OXIDE

The production of tin is mainly carried out in the reverberatory furnace at 1000–1200°C, although some blast furnace production in the same temperature range is used under favourable circumstances. The main chemical problem in treating cassiterite, SnO_2, is connected with the presence of iron in the metallic product. Iron and tin form intermetallic compounds in the solid state, and show only limited miscibility in the liquid state (Fig. 79). The solid compounds are extremely hard and thus have a deleterious effect on the mechanical properties of tin. The standard free energy change for $Sn–SnO_2$

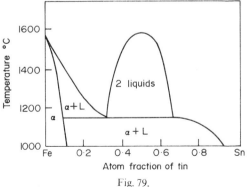

Fig. 79.

is close to that for $FeO-Fe_3O_4$, and just about 20 kcal above the line for reduction of FeO to iron at around 1000–1200°C. Hence, it is understandable that care must be exercised in the reduction of cassiterite in the presence of iron oxides to maintain a sufficiently high oxygen pressure so that too much iron reduction does not occur. In order to get relatively pure tin, the reduction is made in two steps. First under fairly oxidizing conditions, when there is

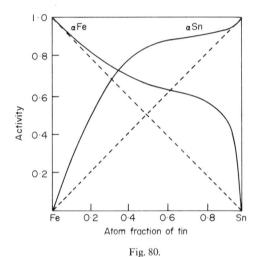

Fig. 80.

also a large tin loss to the slag, the purest tin is produced. Then the slag is remelted with more coke, lime and scrap iron to recover tin.

Davey and Floyd (1966) have suggested on the basis of metal–slag equilibration studies that the metal–slag distribution of iron and tin in the blast furnace process can be accounted for by using the thermodynamic data for the iron–tin system, and assuming that SnO and FeO have unit activity coefficients in the slag phase. The thermodynamic properties of iron-tin liquid alloys at blast furnace temperatures, around 1200°C, can be calculated from the results of Onillon and Olette (1969) Fig. 80 which were obtained at higher temperatures (1547°C) and using the regular solution model. These show that in the tin-rich range of liquids, tin may be taken as approximately ideal ($\gamma_{Sn} \simeq 1 \cdot 12$) whereas the activity coefficient of iron in dilute solution in tin may be described by the equation

$$RT \ln \gamma_{Fe} = 6500 \, X_{Sn}^2$$

For the exchange reaction with all liquid reactants and products in their standard states

$$SnO + Fe \rightarrow Sn + FeO \qquad \Delta G^\circ = 13{,}000 - 13 \cdot 9T \text{ cal mol}^{-1}$$

There are no accurate data for the reaction

$$Sn(l) + \tfrac{1}{2}O_2 \rightarrow SnO(l)$$

since SnO is always metastable with respect to the disproportionation reaction

$$2SnO \rightarrow Sn + SnO_2$$

There are experimental indications that the free energy change for this disproportionation reaction is very small, and so the equation for the reaction

$$SnO + Fe \rightarrow Sn + FeO$$

was obtained by combining the following two equations

$$\tfrac{1}{2}Sn(l) + \tfrac{1}{2}O_2 \rightarrow \tfrac{1}{2}SnO_2 \qquad \Delta G° = -69\,500 + 25\cdot7\,T$$

and

$$Fe(l) + \tfrac{1}{2}O_2 \rightarrow FeO(l) \qquad \Delta G° = -56,500 + 11\cdot8\,T$$

At 1200°C under conditions of slag–metal equilibrium

$$RT \ln \frac{X_{[Sn]}}{\gamma_{[Fe]}X_{[Fe]}} = RT \ln \frac{X_{\{SnO\}}}{X_{\{FeO\}}} + \Delta G° \,(7500 \text{ cal})$$

$$RT \ln \frac{X_{[Sn]}}{X_{[Fe]}} = RT \ln \frac{X_{\{SnO\}}}{X_{\{FeO\}}} + 7500 + 6500 X_{[Sn]}^2$$

If we approximate

$$X_{[Sn]} \cong 1$$

Then the iron content of the metal phase at 1200°C can be readily obtained for any degree of SnO content in the slag from the equation above in the simplified form

$$\log X_{[Fe]} = \log \frac{X_{\{FeO\}}}{X_{\{SnO\}}} - 2\cdot1.$$

It is clear from this equation that, if the ingoing raw material contains a significant amount of iron, relatively pure tin can only be obtained if the slag phase contains a large sacrifice of tin oxide. This, in turn, requires that the reduction process is carried out at any oxygen potential close to that of the Sn–SnO$_2$ line so that a$\{SnO\}$ is high. From the previous consideration of the iron blast furnace, it can be concluded that the oxygen potential in such a system at temperatures around 1200°C would tend to be closer to the C–CO line, and the oxygen potential which would be required in the tin blast furnace when there is a lot of iron to be contended with, is clearly very

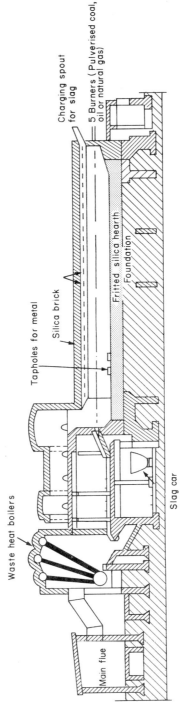

Fig. 81. Typical reverberatory furnace structure. The flames from the burners (right) heat the roof from which energy "reverberates" on to the slag–metal charge below.

much higher. It seems therefore desirable to use the reverberatory furnace (Fig. 81), where higher oxygen potentials than the C–CO equilibrium can be readily maintained for iron-containing materials, and to keep blast furnace operation only for the situation where the raw material has a low iron content, and thus the resultant tin, even if produced at a low oxygen potential, would contain only small amounts of iron.

REFERENCES

Abraham, K. P., Davies, M. W. and Richardson, F. D., (1960). *J. Iron Steel Inst.* **196**, 82. MnO activities in silicate melts.

Chipman, J., (1955). *J. Iron Steel Inst.* **180**, 97. Interaction coefficients of solutes in solution in liquid iron.

Chipman, J. (1965). "Steelmaking" p. 3. The M.I.T. Press. A general discussion of the Fe–Si–C–O system.

Chipman, J. and Dastur, M. N. (1949). *J. Metals* **1**, 441.

Chipman, J. and Gokcen, N. A. (1952). *J. Metals* **4**, 171.

Davey, T. R. A. and Floyd, J. M., (1966). *Proc. Aust. I.M.M.* **219**, 1. Slag-metal equilibria in tin smelting.

Delve, F., Meyer, H. W. and Lander, H. N., (1961). *Phys. Chem. of Process Met.* **2**, p. 1111. Interscience, N.Y., Titanium in the blast furnace.

Flood, H., Førland, T. and Grjotheim, K., (1953). *Phys. Chem. of Melts,* p 46, I.M.M., London. Exchange reaction thermodynamics in complex salt phases.

Flood, H. and Toguri, J. M., (1963). *Trans. Met. Soc. AIME* **227**, 525.

Fulton, J. C. and Chipman, J., (1954). *Trans. AIME* **200**, 1136. Slag-metal graphite equilibria involving silicon.

Fuwa, T. and Chipman, J., (1959). *Trans. AIME* **215**, 708. Carbon solubility in iron alloys.

Goodeve, C. F., (1948). *Discussions Faraday Soc.* **4**, 19.

Kubaschewski, O. and Catterall, J. A., (1956). "Thermochemical Data of Alloys", Pergamon Press, London.

Kubaschewski, O., Evans, E. H. and Alcock, C. B. "Metallurgical Thermodynamics", Pergamon (1967).

Michard, J., Dancoisne, P. and Chanty, G., (1967). "Blast Furnace, Coke Oven and Raw Materials Conference", *AIME* **20**, 329. Analysis of a blast furnace with 100% sinter burden.

Morgan, S. W. K., (1956). *Trans. I.M.M.* **66**, 553. The Imperial Smelting Co., Zinc–Lead Blast Furnace.

Onillon, M. and Olette, M., (1969). *Rev. Hautes Temp. et Ref.* **6**, 245. Activities in the Fe + Sn system.

Rao, Y. K. and Jalan, B. P., (1972). *Metallurgical Trans.* **3**, 2465. The kinetics of the C–CO$_2$ reaction between 839–1050°C and a short general survey.

Richards, A. W. and Thorne, D. F. J., (1961). *Phys. Chem. of Process Met.* **1**, p. 277. Interscience, N.Y., Activities of ZnO and FeO in slags.

Richardson, F. D. and Dennis, W. E., (1953). *Trans. Far. Soc.* **49**, 171. Thermodynamics of liquid Fe–C alloys.

Richardson, F. D., Bowles, P. J. and Ramstad, H. F. (1964). *J. Iron Steel Inst.* **202**, 113. Dilute solution thermodynamics of silicon in liquid iron.

I

Richardson, F. D. and Fincham, J. B., (1954). *Proc. Roy. Soc.* **A233**, 40. Sulphide capacities of slags.

Richardson, F. D. and Jeffes, J. H. E., (1949). *J. Iron and Steel Inst.* **163**, 397. Thermodynamics of the blast furnace.

Schenck, H., Steinmetz, E. and Gohlke, R., (1966). *Arch. Eisenhutten* **37**, 775. Phosphorus in liquid iron.

Schurmann, E., Zischkale, W., Ischebeck, P. and Geynert, G., *Stahl und Eisen* **80**, 854. Blast furnace study showing lower temperature limit of the "solution loss" reaction.

Urbain, G., (1959). *Revue de Métal* **56**, 529. "Thermodynamics of Phosphorus Dissolved in Liquid Iron".

Walker, P. L., Ruskino, F. and Austin, L. G., (1959). *Advances in Catalysis* **11**, 134. Academic Press, N.Y.

Weiblee, F., and Schrag, G. (1941). *Z. Elektrochem.* **4 7**, 222.

RECOMMENDED READING

"Physical Chemistry of Iron and Steel Manufacture", C. Bodsworth and H. B. Bell, Longmans (1972).

12

THE REDUCTION OF OXIDES AND HALIDES BY REACTIVE METALS

INTRODUCTION

The metal oxides in the intermediate range of stability include some oxides which are less stable than CO at temperatures above 1000°C, but the metals form stable carbides, e.g. Si, V, Cr, Nb and, hence, carbon reduction is not preferred as a means of winning the metal. For these metal oxides, other reducing agents must be found, and the reactive metals such as aluminium, magnesium and calcium are suitable for this purpose.

Examples of metals which are prepared in the pure state by aluminium reduction in the so-called "aluminothermic" process are manganese and chromium. The appropriate oxide is mixed with aluminium powder in a brick-lined crucible and the mixture is ignited. The heat evolved, as a result of the chemical reaction, raises the reaction mixture to a temperature above the melting point of alumina, and two liquid phases, metal and slag, are formed. The heat of the reaction also supplies the heat loss of the reaction mixture to the surroundings.

For chemical reasons which were discussed in Part I, a number of metals which form very stable solid solutions of oxygen cannot be readily obtained through the reduction of oxides. These are therefore chlorinated before the metal extraction process. In the Kroll process, which is used in preparing the refractory metals, zirconium, titanium and tantalum, which have very high melting points and high affinity for oxygen, the chlorides are reduced with calcium or magnesium. After reaction in a sealed crucible, the solid metal product is separated from the salt product by vaporization of the latter. The thermal requirement for the heat of reaction to be sufficient to

227

sustain the reduction is only that the final reaction temperature be high enough to melt the salt phase; $CaCl_2$ and $MgCl_2$ melt at 772°C and 714°C respectively. An advantage in using magnesium as the reducing agent is found in the vaporization separation of product salt from the metal phase. $MgCl_2$ boils at 1418°C and is much more easily vaporized than $CaCl_2$ which boils at about 2000°C.

Sodium is also used as a reducing agent for $TiCl_4$ and the product NaCl melts at 801°C and boils at 1465°C. There is little to choose between sodium and magnesium as reducing agents for the Kroll process, since the chemical stabilities and physical properties of their chlorides are very similar. The lower boiling point of sodium than magnesium means that care must be exercised in bringing about the reaction when sodium is used in order to avoid excessive pressure in the reactor. Due to the fact that the boiling point of $TiCl_4$ is only 137°C, it is relatively simple to introduce the chloride as a gas to the reactor and thus control the rate of heat generation through the input flow of $TiCl_4$.

THE THERMAL BALANCE IN REACTIVE METAL REDUCTIONS

The reduction reactions in which there is an exchange of combination with oxygen or chlorine from one metal to another usually only involve condensed phases. It follows that the entropy changes for such reactions will be small and thus that the difference in heats of formation of the compounds determines the feasibility of a given reaction.

Many of the metals which are prepared by reduction of their compounds with reactive metals show a number of valencies, e.g. VCl_2, VCl_3, VCl_4, VCl_5, whereas the elements which are used as reducing agent form only one compound, e.g. $CaCl_2$, $MgCl_2$. The heat change for the reduction reaction depends on the valency state of the metal being reduced, becoming larger the higher the valency. This heat is used to raise the reaction products to a temperature at which a physical separation can be achieved between the metal which is produced and the reactive metal compound. However, the higher the valency, the larger the amount of reactive metal which must be used to bring about the reduction. We must therefore enquire whether it is better to work with a high valency starting compound and to sacrifice reactive metal, or to keep the valency down and to lower the final temperature which can be achieved.

The most appropriate basis for comparison would seem to be the heat of formation of each compound per equivalent of halogen. This value gives the average heat which is available per atom of halogen, and this after subtraction of the heat of formation of the reducing metal compound per equivalent, gives the heat available to raise the temperature of the reactants.

Thus, if we wish to compare the reduction of the trichloride, tetrachloride, and pentachloride of uranium by calcium, we should consider the reactions

$$\tfrac{1}{3}UCl_3 + \tfrac{1}{2}Ca \rightarrow \tfrac{1}{3}U + \tfrac{1}{2}CaCl_2$$

$$\tfrac{1}{4}UCl_4 + \tfrac{1}{2}Ca \rightarrow \tfrac{1}{4}U + \tfrac{1}{2}CaCl_2$$

and $\tfrac{1}{5}UCl_5 + \tfrac{1}{2}Ca \rightarrow \tfrac{1}{5}U + \tfrac{1}{2}CaCl_2$

These reactions produce 1·83, 1·75 and 1·70 g atom of product respectively, and the heat which is generated by the change of one average uranium–chlorine bond to one average calcium–chlorine bond is therefore required to heat less reaction product as the valency of the starting compound increases. Generally, it may be concluded that the final temperature which is achieved in a crucible reduction reaction must increase with increasing valency unless the heats of formation of the chlorides per equivalent decrease with increasing valency drastically.

Heat available in the Reaction $1/x\ UCl_x + \tfrac{1}{2}Ca \rightarrow 1/x\ U + \tfrac{1}{2}CaCl_2$ at 298 K.

Compound	ΔH°/valency of uranium	Heat available per g* atom of product
UCl_3	$- 71{\cdot}2$ kcal equiv^{-1}	13·4 kcal
UCl_4	$- 62{\cdot}3$	19·0
UCl_5	$- 52{\cdot}3$	25·5

* The heat of formation of $CaCl_2$ at 298 K has been taken as $-191{\cdot}4$ kcal mol^{-1}.

The corresponding results for titanium and tantalum show the same pattern when calcium is used as the reducing medium.

Compound	ΔH°_{298}/equivalent of chlorine	ΔH°_{298}/g atom of product
$TiCl_2$	$- 61{\cdot}4$ kcal	17·2 kcal
$TiCl_3$	$- 57{\cdot}2$	21·0
$TiCl_4$	$- 48{\cdot}9$	26·7
$TaCl_3$	$- 44{\cdot}1$	28·2
$TaCl_4$	$- 42{\cdot}6$	30·3
$TaCl_5$	$- 41{\cdot}1$	32·1

THE "ADIABATIC-SYSTEM" APPROXIMATION

The temperature to which the reaction products will be raised as a result of the exothermic heat of reaction is determined mainly by the heat capacities and heats of transformation of the substances involved. We may make use of the adiabatic approximation as a starting point and neglect heat losses to the surroundings of the reactor. The final temperature can now be calculated by applying the heat of reaction to the products only, since Hess's Law states that the heat change in a closed cycle of chemical reactions is independent of the path which is followed. Therefore, the heat required to raise the reactants to a given temperature and react them at the high temperature is the same as that required to react the system at room temperature and heat the products up to the final temperature. In order to calculate this temperature, we must know the heat capacities of the reaction products, and for most condensed substances a value of $7\,\mathrm{cal\,g\,atom^{-1}\,degK^{-1}}$ can be used as a good approximation. Hence, if we calculate the heat change accompanying the reaction in calories per g atom of product, the final temperature can be assessed by division by seven. This does not allow for fusion or possible vaporization of the reaction products, but as this will vary from one system to another it is difficult to write a general rule. The salt which is produced should always be liquid in the final state in order to provide good contact with and hence protection for the metallic sponge which is co-produced. This narrows our interest to a few salts, e.g. those of the reducing metals such as $CaCl_2$, $MgCl_2$. These two salts have heats of fusion of 2·3 and 3·4 $\mathrm{kcal\,g\,atom^{-1}}$ respectively, and the appropriate value must be subtracted from the heat available for the reaction before the calculation of the final temperature.

THE CHOICE OF REDUCING AGENT

The corresponding heat changes for the reduction of halides by magnesium are significantly smaller than for calcium reduction. This is because the heat of formation of, for example, $MgCl_2$ ($-153\cdot4\,\mathrm{kcal\,mol^{-1}}$) is 38 $\mathrm{kcal\,mol^{-1}}$ smaller than that of $CaCl_2$. The reduction of UCl_4 by Mg only has a thermal balance of 8·0 $\mathrm{kcal\,g\,atom}$ of product compared with 19 kcal for calcium. It follows that a high final temperature is much more readily reached by calcium reduction and the use of magnesium whilst normally more economic represents a thermal sacrifice. The use of magnesium in production can always be made more thermally effective by the use of fluorides e.g. UF_4 rather than UCl_4 as starting material. This can be seen from consideration of the respective heats of reaction

$$UF_4 + 2Mg \rightarrow U + 2MgF_2 \qquad \Delta H^{\circ}_{298} = -79 \text{ kcal}$$

$$UCl_4 + 2Mg \rightarrow U + 2MgCl_2 \qquad \Delta H^{\circ}_{298} = -55 \text{ kcal}$$

The heats of fusion of the magnesium compounds, which must be considered in the heat balance, are quite similar MgF_2, ΔH_{fusion} 13·9 kcal, and for $MgCl_2$, 10·3 kcal mol^{-1}. The extra heat which is made available when the fluoride is reduced, 16·8 kcal for the production of 1 g atom of uranium, can thus be applied usefully.

The adiabatic approximation suggests that the reduction of UF_4 by Mg would not have a sufficient heat change to raise the MgF_2 product above its melting point 1263°C. After subtraction of the heat of fusion of MgF_2, 13·9 kcal mol^{-1}, the heat balance is only 51 kcal for the formation of one g atom of uranium. Since the reaction involves seven g atoms for the formation of one g atom of uranium, it is clear that even if we ignore heat losses, the expected temperature to which the system would rise as a result of reaction would only be about 1000°C. Hence, some preheating would be required, and the extent of this is found in practice to vary with the physical properties of the reductant. Thus, with pure magnesium, the temperature range where the reaction begins to occur rapidly is 600–700°C. This range is around the melting point of pure magnesium (650°C). When an alloy of magnesium which has a lower temperature eutectic is used as the reductant, e.g. Mb–Pb at 460°C, the temperature of reaction initiation is substantially lower. This suggests that even when the heat balance for fusion of MgF_2 is made up by external preheating of the reacting mixture of UF_4 and Mg, the initiation of the reaction at a useful velocity only occurs when sufficient heat is available and when the reactants can come into good physical contact. In this respect magnesium has an advantage over calcium (melting point 843°C) for use as a halide reducing agent.

THE THERMAL BALANCE IN THE ALUMINOTHERMIC PROCESS

If MnO_2 is mixed with aluminium as the starting mixture, the reaction is explosive, whereas when MnO is used, insufficient heat is evolved. A balanced charge of predetermined Mn/O ratio is therefore made by mixing MnO_2 and MnO so that a reasonable heat balance is achieved.

In the case of Cr_2O_3, there is not sufficient heat available from the reaction

$$Cr_2O_3 + 2Al \rightarrow 2Cr(l) + Al_2O_3(l)$$

and the deficiency is usually met by preheating the charge to about 500°C or by adding $Na_2Cr_2O_7$ (sodium dichromate). The addition of CrO_3 is not practical on the industrial scale because the solid is hygroscopic and stores badly.

The standard heat change for the reduction of Cr_2O_3 by aluminium is -130 kcal. Since the reaction involves one mole of Cr_2O_3 for every 2 g atoms of aluminium, the heat capacity of this amount of material is about 49 cal deg^{-1}; there are seven g atoms involved. To raise the temperature of this mass by 2200°C, which is slightly above the melting point of Al_2O_3, requires 108·0 kcal. The heats of transformation and melting total 26 kcal for one mole of Al_2O_3 and approximately 10 kcal for two g atoms of chromium. The total heat requirement is, therefore, 144 kcal. Since this quantity is 14 kcal larger than the heat change for the reaction which produces chromium, it follows that the pre-heat to 500°C must be necessary to make up for this difference and the heat losses by convection and radiation from the reaction assembly. This amount is then 24 kcal which is the heat capacity of 49 cal deg^{-1} for the reaction formula multiplied by the pre-heat temperature. The heat loss due to radiation and convection therefore seems to be about 10 kcal g mole of reaction.

Referring back to the manganese production, we can now estimate what metal/oxygen ratio would be required in the original charge to raise the reacted mass to the melting point of Al_2O_3. For the reactions involving each oxide

$$\text{(A)} \quad 3MnO + 2Al \rightarrow 3Mn + Al_2O_3 \qquad \Delta H° = -124 \text{ kcal}$$

$$\text{(B)} \quad \tfrac{3}{4}Mn_3O_4 + 2Al \rightarrow \tfrac{9}{4}Mn + Al_2O_3 \qquad \Delta H° = -152 \text{ kcal}$$

$$\text{(C)} \quad \tfrac{3}{2}MnO_2 + 2Al \rightarrow \tfrac{3}{2}Mn + Al_2O_3 \qquad \Delta H = -214 \text{ kcal}$$

The approximate heat capacities, using the Kopp–Neumann law, are then (A) 56 cal deg^{-1}; (B) 50·8 cal deg^{-1}; (C) 45·5 cal deg^{-1}. The heats required to melt the charge are:

$$\text{(A)} \ 160 \text{ kcal}; \quad \text{(B)} \ 145 \text{ kcal}; \quad \text{(C)} \ 131 \text{ kcal}$$

Clearly from this point of view, the choice should be to operate with as high a manganese to oxygen ratio as is feasible in order to produce the necessary heat. On the other hand, the Mn/Al ratios are

$$\text{(A)} \ 1·5; \quad \text{(B)} \ 1·125; \quad \text{(C)} \ 0·75$$

and, hence, the necessity for the maximum economy in the use of the expensive reductant, aluminium metal, points to the lowest Mn/O ratio as being the most desirable. The operator must choose a compromise between these two opposing possibilities and seek, if possible, to lower the heat requirements of the production. This can either be done by pre-heating as much as is feasible, or by attempting to reduce the liquidus temperature of the slag (Al_2O_3) phase by suitable additions.

THE THERMODYNAMICS OF ALUMINOTHERMIC REDUCTION

The products of an aluminothermic reduction are not pure metal and oxide phases, and it is necessary to study the equilibria of these reactions in order to assess the completeness of reaction which is to be expected. No experimental information is available at present, but it is possible to use theoretical laws to obtain what is probably a close approximation to the truth. For the metallic phase, Raoult's law can be assumed, but we must now give consideration to the form of the ideal mixing law in molten salt phases.

Raoult's law requires that the activity of each component is equal to its atomic fraction in the alloy. This behaviour is shown by a system which is athermal ($\Delta H^M = 0$) and in which the partial entropy of each component is that appropriate for a random mixture. The following equations therefore apply to such a system:

$$a_i = X_i$$

$$\Delta \bar{G}_i = RT \ln a_i = \Delta \bar{H} - T \Delta \bar{S}$$

$$\Delta \bar{H}_i = 0 ; \Delta \bar{S}_i = - R \ln X_i$$

The thermodynamic properties thus arise entirely from the athermal random mixing of the components.

If we now turn to a binary salt mixture AX–BX, then the random mixing of A and B cations on the cationic sub-lattice of the mixture would again lead to Raoult's law for the thermodynamic behaviour of the components.

In the case of the mixture A_2Y–B_2Y the random mixing of cations would lead to the mixing of *two* g atoms of cations per mole of solution. Hence, the activities of the components in this randomly mixed athermal solution would be given by

$$a_{A_2Y} = X_{A_2Y}^2 ; \quad a_{B_2Y} = X_{B_2Y}^2$$

$$\Delta \bar{S}_{A_2Y} = - 2R \ln X_{A_2Y} = - R \ln X_{A_2Y}^2$$

Similarly for the A_3Z–B_3Z athermal random solution

$$a_{A_3Z} = X_{A_3Z}^3 ; \quad a_{B_3Z} = X_{B_3Z}^3$$

$$\Delta \bar{S}_{A_3Z} = - 3R \ln X_{A_3Z} = - R \ln X_{A_3Z}^3$$

Temkin (1945) suggested that the generalized equation for activities in solutions containing species A_pX_q, for example, should be obtained by assuming an athermal mixing with a random mixing of cations on one sub-lattice and of anions on the other sublattice.

Hence

$$a \, A_p X_q = \left(\frac{n_A}{\Sigma n_{\text{cations}}}\right)^p \left(\frac{n_X}{\Sigma n_{\text{anions}}}\right)^q$$

where n_i is the number of g atoms of the ith ionic species.

In solid solutions which contain a mixture of cationic valencies, e.g. Li_2O–NiO, it is necessary to consider the vacant site which is formed each time a molecule of NiO is substituted for one of Li_2O in calculating the entropy of mixing. The vacant site must be included as a "component" and considered in the $\Sigma n_{\text{cations}}$ calculation. This is tantamount to considering the thermodynamics of a system in which the number of cation sites is held constant, which seems reasonable for solids. In liquid systems, on the other hand, it is not necessary to consider the number of sites which will be occupied by the ions of each component, since this strict lattice model of the liquid does not seem applicable. Instead, it is merely the *number* of ions of each species which are present which must be added to give the summation denominators for cations and anions.

Returning now to the aluminothermic reaction, we shall assume that liquid Al–Cr alloys and Al_2O_3–Cr_2O_3 liquid slags are ideal, and that the equilibrium constant for the reduction reaction is

$$K = \left(\frac{a_{Cr}}{a_{Al}}\right)^2 \cdot \left(\frac{a_{Al_2O_3}}{a_{Cr_2O_3}}\right) \simeq \left(\frac{X_{Cr}}{X_{Al}}\right)^2 \cdot \left(\frac{X_{Al_2O_3}}{X_{Cr_2O_3}}\right)^2$$

(According to Temkin's rule, in the slag phase the activity of the molecular components must equal the *square* of the mole fractions.)

It thus follows that

$$K^{\frac{1}{2}} = \frac{X_{Cr}}{X_{Al}} \cdot \frac{X_{Al_2O_3}}{X_{Cr_2O_3}} = \exp - \left(\frac{\Delta G^\circ}{2RT}\right)$$

Obviously, this reaction will proceed very nearly to completion, and the equilibrium constant may be approximated by assuming that the mole fractions of Al_2O_3 and Cr are sufficiently close to unity so that

$$\frac{\Delta G^\circ}{2RT} = \ln X_{Al} \cdot X_{Cr_2O_3}$$

At 2400 K, the value of $K^{\frac{1}{2}}$ is roughly 2×10^{-4}; hence, there will be a hyperbolic relationship between the aluminium content of the metal phase and the unreduced Cr_2O_3 content of the slag phase such that

$$X_{Al} \cdot X_{Cr_2O_3} = 2 \cdot 0 \times 10^{-4}$$

and thus to reduce the Cr_2O_3 content of the slag much below 1 per cent would require the wastage of important amounts of aluminium which would only render the product chromium metal impure ($2\% \ Cr_2O_3$ in slag for $1\% \ Al$ in metal).

Using the $Al-Al_2O_3$ line on the diagram, if X_{Al} is 10^{-3}, and this is equal to a_{Al}, then the chemical potential of aluminium metal ($\Delta u_{Al} = RT \ln a_{Al}$) at $2400\,K$ is $-33\,000\,cal\,g\,atom^{-1}$. For the reaction

$$\tfrac{4}{3}Al + O_2 \rightarrow \tfrac{2}{3}Al_2O_3$$

$$\Delta G^\circ = RT \ln K = -\tfrac{2}{3}RT \ln a_{Al_2O_3} + \tfrac{4}{3}RT \ln a_{Al} + RT \ln p_{O_2}$$

$$= -144\ kcal$$

If $a_{Al_2O_3} = 1$, $\Delta\mu_{Al_2O_3} = 0$; hence, $RT \ln p_{O_2} = -144\,000 + \tfrac{4}{3}(33\,000)$ cals

Hence, the oxygen pressure of the system will be about 10^{-9} atm corresponding to an oxygen potential of $-100\,kcal\,mol^{-1}$.

Referring now to the literature data for the free energies of formation of V_2O_3 and TiO_2 and NbO, it is clear that these oxides will not be reduced so completely by aluminium as are manganese and chromium oxides, but useful reduction to make alloys of these metals with iron could be achieved with aluminium. The free energy for $Nb-NbO$ can be extrapolated to a value of $-95\,kcal$ at $2200°C$, and thus relatively pure solid niobium could be produced by this process. The metal would still be solid, however, and there are difficulties associated with separation. The reaction is used, industrially, to make alloys of this metal and others of the so-called refractory metals with iron in ferro-alloys. The point of fusion of the ferro-alloys is below $2200°C$ over most of the composition ranges, and the use of magnetite in the starting material to provide iron also has obvious thermal advantages.

SILICON REDUCTION OF MAGNESIA

The important industrial example of the use of the reduction of activities in producing metals is the Pidgeon process: The basic reaction

$$2MgO + Si \rightarrow 2Mg(g) + SiO_2 \qquad \Delta G^\circ = 146\,000 - 61\cdot8\ T\ cal$$

has a free energy change of $+55\,kcal$ at $1200°C$. If the solid reactants and products are at unit activity, the pressure of magnesium which is produced is less than 10^{-4} atm. However, if the magnesia is present in the reaction mixture as dolomite ($MgO.CaO$) the lime reacts with the silica which is produced, lowering a_{SiO_2} considerably. The reaction should therefore be represented by the equation

$$2(MgO.CaO) + Si \rightarrow 2Mg(g) + Ca_2SiO_4$$

and the pressure of magnesium, which is developed, is now found to be 2×10^{-2} atm at 1200°C. The large effect of the formation of the orthosilicate rather than the pure oxide, SiO_2, can be represented as a decrease in the silica activity in the basic reaction.

The free energy equation for the formation of the orthosilicate is

$$2CaO + SiO_2 \rightarrow Ca_2SiO_4 \qquad \Delta G° = -30\,200 - 1·2\,T \text{ cal.}$$

The low entropy change for this reaction is to be expected, since only solids are involved.

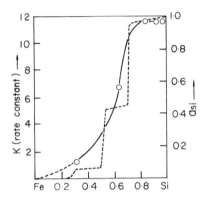

Fig. 82. Initial rate constant for the formation of magnesium gas, and activities of silicon as a function of X_{Si} in Fe–Si alloys at 1228°C.
O——O Initial rate constant
– – – – Activity of silicon.
(after Toguri and Pidgeon, loc. cit.).

At 1200°C, this free energy change has the value $-32\,\text{kcal}$ and thus the formation of the orthosilicate during the silicon reduction of magnesia lowers the standard free energy change of that process to $+23\,\text{kcal}$.

The metal silicon is not used in the pure state in the industrial process, but as ferrosilicon. Interactions in the iron-silicon system are very large, and the activity of silicon varies by a large amount across the composition range. The final equilibrium pressure of magnesium which can be achieved is thus dependent on the choice of ferrosilicon grade. From the equilibrium constant for the production of magnesium vapour, it is obvious that the magnesium pressure varies as the square-root of the silicon activity.

Toguri and Pidgeon (1962) have made a laboratory study of the rate of reaction of dolomite and ferrosilicon particles pressed into briquette form in the temperature range 1050–1560°C. They found that the initial rate constant for the reaction was directly proportional to the activity of silicon in the ferrosilicon phase. This conclusion was drawn by comparing their measured rates of production of magnesium vapour with the silicon activities for the solid Fe + Si system which were obtained by Rosenqvist and Ellingsaeter (1956) (Fig. 82). The authors then suggest that the rate-determining steps in magnesium production involve the distribution of the silicon potential in the system by means of vapour transport in the form of gaseous SiO. The reaction scheme which was proposed was

$$MgO + Si \rightarrow Mg(g) + SiO(g)$$

in which $p_{SiO} = 4 \cdot 29 \times 10^{-4}$ atm at 1277°C, followed by calcium silicate formation with the regeneration of silicon

$$2SiO(g) + 2CaO \rightarrow Ca_2SiO_4 + Si$$

It was also found that the rate of the reaction was increased in the presence of compounds which would produce low melting liquids with the $CaO + SiO_2$ system, e.g. CaF_2 to form the eutectic at 1125°C.

These studies suggest that the physical contacts between solid and liquid phases are of paramount importance in determining the rates at which reactive metal reduction reactions can be carried out, but that the presence of a volatile species should not be overlooked in elucidating the principal reaction mechanism.

REFERENCES

Rosenqvist, T. and B. Ellingsaeter, (1956). *J. Metals* **212**, 1111. Silicon activities in Fe + Si solid alloys.

Temkin, M. (1945). *Acta Physica Chimisa URSS* **20** 411.

Toguri, J. M. and Pidgeon, L. M., *Can. J. Chem.* **40**, 1769. Thermal reduction of dolomite with silicon.

13

THE ELECTROLYTIC REDUCTION OF MOLTEN SALTS

INTRODUCTION

In most metal extraction reactions, the metal compound is reduced to yield the relatively pure metal, and the non-metallic component is removed in a combined form with the reducing agent. This process may be expressed in its electrochemical analogue as the transfer of electrons to the metallic cation to produce the neutral species

$$M^{n+} + ne^- \rightarrow M^0$$

It is obvious that an alternative to the chemical reduction of a metallic cation is the supply of the necessary electrons through an electrical circuit. Such an electrolytic procedure is not limited by considerations of chemical stability, since the voltage which is applied to the sample of the compound to be reduced need only be slightly greater than the decomposition voltage to effect reduction of the compound to the metal. The decomposition voltage in the limit of zero current is related to the standard free energy of formation of the compound *per equivalent* by the equation

$$\Delta G^\circ = -FE^\circ$$

where F is Faraday's constant ($23\,060$ cal volt^{-1})

$$F = \frac{96\,500 \text{ C equiv.}^{-1}}{4 \cdot 185 \text{ J cal}^{-1}}$$

The free energies of formation of oxides, sulphides and halides seldom have values greater than 90 kcal per equivalent (360 kcal per mole of O_2 or

238

S_2 for oxides and sulphides. 180 kcal per mole Cl_2 for chlorides, etc.) and thus an applied voltage as small as 4 V should be more than adequate to dissociate most compounds. The rate at which decomposition can be carried out depends upon the current which is passed through the compound. If all of the electrical energy goes into the reduction system, then the ohmic voltage drop, I ($R_{elec.}$ + $R_{contact}$) must be added to the applied voltage for a required current (I amps). The resistance is mainly the electrolytic resistance of the compound ($R_{elec,}$ ohms), but some contact resistance of a variable nature is invariably present in practical electrolytic circuits. The I^2R energy dissipation due to the passage of a heavy current serves as a useful source of heat for the system.

The electrochemical reduction of a compound can evidently be carried out at quite low applied voltages and at high efficiency providing that the electric current which passes through the compound is almost completely drawn as a result of the migration of ions. In solid systems, there are very few compounds which are sufficiently electrolytic to behave in this way. This is because the mobilities of the ions are small when compared with those of electron defects. These arise from impurities or defects which can donate or remove electrons from the electronic structure of the compound, or from the presence of ions in more than one valency state.

In the liquid state, the mobilities of ions are much greater than in the corresponding solids. The mobility, μ, of an ionic component of a system is related to the diffusion coefficient of the component, D, by the Nernst–Einstein relationship

$$\frac{\mu}{e} \, kT = D.$$

The constant ratio k (Boltzmann's constant, $1{\cdot}3804 \times 10^{-16}$ erg deg^{-1}) divided by e (the electronic charge, $4{\cdot}803 \times 10^{-10}$ e.s.u. of charge) has the value $2{\cdot}874 \times 10^{-7}$, but since electrical mobilities are usually expressed with practical rather than absolute volts, this quantity must be multiplied by 300 which yields the value $8{\cdot}622 \times 10^{-5}$ eV deg^{-1}.

The equation now takes the form

$$8{\cdot}622 \times 10^{-5} \mu T = D$$

Since a typical diffusion coefficient in the liquid state is 10^{-5} cm^2 s^{-1}, it follows that μ for ions in a molten salt at 1000 K is about 10^{-4} cm^2 volt^{-1} s^{-1}.

Now a typical density for a condensed phase is 10^{-1} mol per cm^3; hence there are roughly 10^{22} ions in each unit volume of a molten salt. From measurements on a number of inorganic solids, it can be concluded that the mobilities of electron defects are between 10 and 100 cm^2 volt^{-1} s^{-1}. A concentration of electron defects (electrons or positive holes) of $10^{16} - 10^{17}$ particles per cm^3 will therefore yield an equal contribution to the electrical conductivity of the system as that due to ionic migration. This is because the conductivity is related to the concentration and mobility of species by the equation

$$\sigma = \sum_i n_i \, | \, z_i e \, | \, \mu_i$$

where $|z_i e|$ is the absolute value of the charge on the ith species.

A typical value for the specific conductances of molten electrolytic salts is about $1 \, ohm^{-1} \, cm^{-1}$, and the temperature variation follows to a good approximation the Arrhenius equation

$$\sigma = \sigma_0 \exp - (\Delta E / RT)$$

The experimentally measured activation energy for electrical conductance of a molten salt is usually quite small, which is in keeping with the relationship between the mobility and the diffusion coefficient referred to above. The thermally activated part of both processes is the same, and it is usually assumed that this reflects the activation energy for place exchange between the migrating ion and a vacancy in the liquid pseudo-lattice. As might be expected, this energy is only a few kilocalories and, as a typical example, the specific conductance of molten $CaCl_2$ can be represented by the equation

$$\sigma = 19 \cdot 63 \exp - \left(\frac{4750}{RT} \right).$$

There is very good reason for thinking, however, that transport processes in liquid systems should not be expressed in terms of an Arrhenius expression, and the data are in many cases fitted as well by a power series in the temperature range as by the exponential expression. Thus, the results for NaCl may be fitted by the equation

$$\sigma = -2 \cdot 497 + 8 \cdot 043 \times 10^{-3} \, T - 2 \cdot 223 \times 10^{-6} T^2$$

or in the Arrhenius form

$$\sigma = 2 \cdot 204 \exp - \left(\frac{1980}{RT} \right).$$

Molten sulphides are almost invariably semiconductors and the total conductivities have typically larger values than that quoted for the average molten electrolyte. Thus, molten Ag_2S has a specific conductance which is described by the equation

$$\sigma = 41 \cdot 51 \exp \left(\frac{2570}{RT} \right).$$

The conductance is some hundred times larger than that of $CaCl_2$. and it should be noted that the exponential term now has a positive exponent. which is the opposite from the electrolytic behaviour.

It is clear. then. that few substances will be electrolytes even in the liquid state unless a high degree of purity and a very low concentration of electron defects can be maintained. The requirement of purity usually means that a careful pre-treatment is required of a material which is to be electrolyzed for the production of a metal. The condition that the substance must have a low concentration of electron defects limits the molten salts which may be usefully employed mainly to the halides. since simple oxide and sulphide melts are mostly substantially semiconducting in nature.

As an example. the production of metal is relatively simple in the case of magnesium where the molten salt phase is the chloride. and it might be thought that electrochemical reduction would prove to be the ideal method for the production of a number of metals via their molten salts. At the present time. however. this potential has not been fully realized. and in a number of cases. such as magnesium. there is still some doubt concerning the value of electrochemical reduction *vis-à-vis* alternative procedures. such as the Pidgeon Process (see p. 235). In some other instances. such as the refractory metals. there still do not appear to be satisfactory electrolytes for the production process. Especially is this so when the product would be solid metal rather than the liquid products which are obtained in aluminium and magnesium production. An indication of the origins of the difficulties associated with this method of producing metals can be found by considering the mechanisms of electrical conduction in molten halides and the thermodynamics of metal–metal halide systems in the condensed state.

The products of electrolysis of molten salts containing the refractory metals are seldom the dense. coherent deposits which would be desirable. More often than not. there is considerable dendrite formation on the initial deposit or. worse still. a large fraction of the electric power is used up in producing metallic powder which settles very slowly out of the electrolyte. Either of these effects is undesirable for the smooth running of the cell and the routine production of tonnage metal. In inquiring into the fundamental reasons for the lack of success. it is intended to survey first of all the properties of metals in contact with their molten salts at high temperature.

METAL SOLUBILITY IN MOLTEN CHLORIDES

Bredig (1964) has reviewed the very considerable amount of quantitative data which is now available concerning the solutions of metals in their molten halides. In typical instances. a complete series of liquids is formed

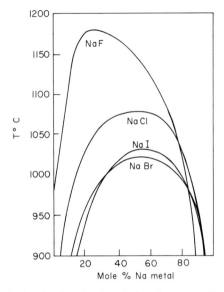

Fig. 83. Liquidus lines for the solutions between sodium and its halides.

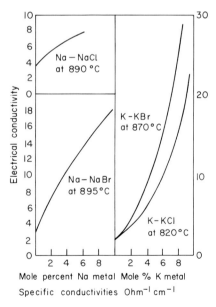

Fig. 84. Effects of metal solution on the electrical conductivities of alkali halides.

across the composition range at quite modest temperatures, and there is an upper consolute temperature below which two liquids separate, the metal containing halide and the halide with some metal in solution (Fig. 83). Typical examples of the consolute temperature are 1080°C in Na–NaCl, 790°C in K–KCl, and 706°C in Rb–RbCl. The temperatures are higher for the alkaline earth chlorides, 1340°C for Ca–CaCl$_2$, and 1010°C for Ba–BaCl$_2$, and higher still for rare earth systems. The important aspects of these studies to metal extraction are the existence of this significant solubility, and the effects which the dissolution of metal has on the electrical properties of the molten salt system. The introduction of metal atoms in solution nearly always brings about an increase in the specific conductivity of the corresponding metal chloride, which is typically 1 ohm^{-1} cm^{-1} (Fig. 84). In the case of K–KCl, it is possible to raise the specific conductivity by a factor of 10 by the intro-duction of 10% K metal into the chloride at 860°C. As might be anticipated, this enhanced conductivity is the result of the liberation of the electron from the dissolved atom so that the molten salt now takes on the properties of a metallic conductor. The mobilities of the free electrons which are introduced are high, about 20 cm^2 volt^{-1} s^{-1}, and thus the transport number of ions in the solutions containing metal atoms is less than unity. This process clearly reduces the current efficiency of any attempted electrolysis.

The electrons which are released do not always have high mobilities. Chemical explanations of this phenomenon involve the hypothetical forma-tion of electron traps in the molten salt, such as Na$_2^+$ covalently-bonded molecular species in sodium salts and Ca$_2^{2+}$ and Bi$_3^{3+}$ complex ions.

The facts that the dilute solutions of metals in molten salts can occur during electrolysis, and that this can bring about a drastic modification to the elec-trical properties of these electrolytes have further consequences for the metal-production attempts. This is because the metal chemical potential which is produced immediately in the vicinity of the cathode by the passage of current through a halide electrolyte is very rapidly transmitted through the electrolyte. The diffusion of the free electron through the liquid suffices to transmit this chemical potential, and therefore any reduction process involving this metal could occur significantly in parts of the system remote from the electrodes much more rapidly than if the speed depended on the diffusion of the ionic species, e.g. Na$^+$ in NaCl.

DEPOSITION OF METALS FROM SALT MIXTURES

When no complication arises from the solubility of metal in the molten salt phase, electrodeposition of metals can be carried out from molten salts usually without any appreciable over-potential. For example, it has been shown by Graves and Inman (1968) that the deposition of liquid lead from the molten salt mixture $PbCl_2$–LiCl–KCl at $450°C$ occurs at the potential calculated from thermodynamic data.

For an electrode to behave reversibly, the exchange current at the metal–melt interface must be large. This current is brought about by the exchange of metal ions between metal and melt under equilibrium conditions when no current is being drawn. After applying a potential, neutralization of a metal ion depends upon two factors assuming diffusion processes are rapid. The first of these involves the change (charging) in the double layer at the electrode–electrolyte interface where an excess of the metal ion in the electrolyte surface faces a balancing charge of electrons in the electrode surface. The second factor then depends on the speed of electron transfer reaction across the interface (Faradaic charging).

The complication of the uneven growth of electrodeposited metal from aqueous solutions is frequently eliminated by the use of complexes. Thus, the deposition of silver in aqueous solution is greatly improved if it is carried out from a solution with a high cyanide content rather than from a solution which contains a simple salt, e.g. $AgNO_3$. The explanation which is usually advanced to account for this smoother growth is that the complex ions which are formed, $Ag(CN)_2^{-1}$, are negatively charged and less easily decomposed at the cathode which leads to diffusion control. Accordingly, lattice energies of surface atoms become unimportant resulting in a more even growth of the electrodeposit.

This same idea has been applied in the high temperature field, and Mellors and Senderoff (1965) have had considerable success with the electrolysis of molten fluorides containing refractory metal complexes. The refractory metal fluoride is dissolved in alkali metal fluorides, and the deposition is carried out under an argon atmosphere. The refractory metals each have a number of valencies, and it was found that the deposition was most successful for a particular valency of each metal, e.g. Mo^{3+}, Nb^{4+}, Zr^{4+} and Ta^{5+}. The corresponding chlorides or even chloride–fluoride mixtures were found to yield dendritic products only.

The success of this application of molten fluorides suggests that it is the strong interactions in fluoride systems leading to complex anion formation which brings about the condition for satisfactory electrodeposition. A measure of the stability of the complexes which are formed in molten salt media is the activity of the anion-forming metal halide. This is found to vary with the chemical composition of the solvent as well as the choice of the anion.

For a given metal chloride dissolved in the alkali chlorides, the complexes may be postulated to have four or six-fold coordination. For example, $NiCl_2$ and $FeCl_2$ would form $NiCl_4^{2-}$ and $FeCl_4^{2-}$ in solution whereas $TiCl_4$ and $ZrCl_4$ would form $TiCl_6^{2-}$ complexes. An indication for the possible stoichiometries of such complexes can be obtained from the crystal structures of the inter-halide compounds which are formed in the solid state. It would further seem likely that the heats of formation of these solids would indicate the relative strengths of the interactions in the corresponding liquid systems.

In support of this suggestion, it has been found that the compounds of the $TiCl_6^{2-}$ and $ZrCl_6^{2-}$ anions with the alkali halides increase in heat of formation

Table XXX. Heat of formation of solid complex halides (nearest integral value)

Compound	$-\Delta H°(kcal\ mol^{-1})$
$2MCl(s) + ACl_2(s) \rightarrow M_2ACl_4(s)$	
Rb_2CoCl_4	6
CS_2CoCl_4	10
Rb_2ZnCl_4	12
Cs_2ZnCl_4	16
: $MCl(s) + RCl_3(s) \rightarrow MRCl_4(s)$	
$NaFeCl_4$	1
$KFeCl_4$	7
: $2MCl(s) + XCl_4(g) \rightarrow M_2XCl_6(s)$	
K_2TiCl_6	27
Rb_2TiCl_6	33
Cs_2TiCl_6	38
Na_2ZrCl_6	26
K_2ZrCl_6	52
Cs_2ZrCl_6	52
: $MCl(s) + TaCl_5(g) \rightarrow MTaCl_6(s)$	
$NaTaCl_6$	16
$KTaCl_6$	28
$CsTaCl_6$	34

from the separated chlorides with increasing ionic radius of the alkali metal ion (Table XXX); the liquids show increasing stability in the same manner. Thus, Flengas and Pint (1969) have demonstrated by vapour pressure measurements of the partial pressures of $TiCl_4$ and $ZrCl_4$ over the molten halide melts that the lowest vapour pressures are obtained from caesium chloride mixtures with the Group IV halides. An explanation which can be advanced for the greater stabilities of caesium chloride melts than those containing sodium and potassium chlorides is as follows. If the structures of the complex anions are entirely ionic, then the repulsive contribution due to interaction between the Ti^{4+} or Zr^{4+} ions will be reduced more by the insertion of the large Cs^+ ion than by smaller alkali metal ions. This requires that an extreme ionic model be adopted for the description of bonding in the complex anions. Although there are grounds for doubting that the Ti–Cl and

Zr–Cl bonds are purely ionic in the complexes, it can be shown that a satisfactory, quantitative, account of the heats of mixing of $TiCl_4$ and $ZrCl_4$ with alkali halides is obtained by calculating the heat of mixing of M^+, Cl^- and XCl_6^{2-} ions where M is an alkali metal and X the Group IV metal ions.

The improvement of the physical properties of the electro-deposit from molten salts in which stable complex anions can be formed over the situation when the solvent–solute molten salt interaction is weak has not been described in physico-chemical terms, but it should be pointed out that the decomposition voltage of a given halide in solution *increases* as the thermodynamic activity of the salt *decreases*. Therefore, the voltage at which zirconium, for example, would deposit from a molten salt solution would be greater with caesium chloride as solvent than with sodium or potassium chlorides. Since the free energies of formation of the alkali metal halides remain roughly constant in magnitude with increasing cationic radius, the activity of alkali metal, which would be produced in the solution by the application of the decomposition voltage of $ZrCl_4$ to the molten salt solutions would therefore be smaller for caesium than the other alkali metals, this effect being strengthened by the higher stability of $CsCl–ZrCl_4$ mixtures. The EMF for decomposition of $ZrCl_4$ in solution to form pure zirconium and chlorine gas at one atmosphere varies as the square root of the $ZrCl_4$ activity. The alkali metal activity which is produced in equilibrium with the melt and chlorine gas varies as the logarithm of the square of the chloride activity (Fig. 85). Hence, the lower value of the activity of CsCl than that of the other alkali halides has the most significant effect in electrolysis. It will be remembered that the alkali metals dissolve substantially in their molten chlorides to produce semi-conducting liquids; hence, it may be concluded that it is in the caesium solutions that the smallest alkali metal activity and conduction electron concentration will be found. These two facts may be related to the superior electrodeposition of zirconium which has been reported by Flengas and Pint from caesium chloride solutions.

It is possible by the method of chronopotentiometry to show that the deposition of metals from the molten fluorides does not occur in one step but two electron transfers occur before the metal is produced. In chronopotentiometric measurements, a short pulse of constant current is drawn through the electrolyte and the change in potential at the electrode of deposition is measured as a function of time. The curve of potential vs. time shows a plateau at each particular dischage process, and for example, the curve for the electrolysis of TaF_5 in the KF–LiF–NaF eutectic mixture, which is thought to contain the $(TaF_7)^{2-}$ ion has two steps corresponding to

$$(TaF_7)^{2-} + 3e^- \rightarrow TaF_2(s) + 5F^-$$

$$TaF_2 + 2e^- \rightarrow Ta(s) + 2F^-$$

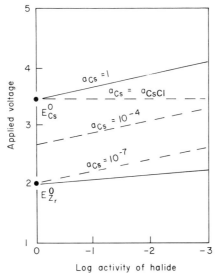

Log activity of halide

Fig. 85. Electrolytic decomposition at the applied potential E_I°

(I) MCl_4 to give pure metal $M + Cl_2$ at one atmosphere pressure

$$- \Delta G_{MCl_4}^\circ = \tfrac{1}{2} \Delta\mu_{MCl_4} \quad \text{when} \quad \Delta\mu_M = \Delta\mu_{Cl_2} = 0.$$
$$= 2FE_I^0$$

(II) RCl to give R in solution $+ Cl_2$ at one atmosphere pressure.

$$- \Delta G_{RCl}^\circ = 2\Delta\mu_{RCl} - 2\Delta\mu_R \quad \text{when} \quad \Delta\mu_{Cl_2} = 0$$

$$= 2FE_{II}^0$$

For simultaneous decomposition in $RCl-MCl_4$ salt melts to produce pure metal M and R in solution at a chemical potential $\Delta\mu_R'$

$$E = E_I^0$$

and

$$2\Delta\mu_R' = 2\Delta\mu_{RCl} - \tfrac{1}{2}\Delta\mu_{MCl_4}$$
$$\Delta\mu_R' - \Delta\mu_R = F(E_{II}^0 - E_I^0).$$

The figure shows the variation of the decomposition voltage for $ZrCl_4$ to give the pure metal and for $CsCl$ to give Cs in solution, together with one atmosphere pressure of chlorine.

THE DOWNES CELL FOR SODIUM PRODUCTION

Although the direct electrolysis of $NaCl$ to give metallic sodium cannot be exploited commercially, the fused salt mixture of $NaCl-CaCl_2$ can be used which avoids the problem arising from metal dissolution in the molten salt phase. The electrolysis of 40% $NaCl$–60% $CaCl_2$ at 600°C produces sodium metal, containing some calcium, which is separated on cooling.

The standard potentials for the pure molten salts $NaCl$ and $CaCl_2$ are given as a function of temperature by the equations

$$Ca(l) + Cl_2 \rightarrow CaCl_2(l)$$
$$E^{\circ} = 4{\cdot}24 + 5{\cdot}3 \times 10^{-4}T \log T - 2{\cdot}44 \times 10^{-3}T \text{ volt}$$
$$Na(l) + \tfrac{1}{2}Cl_2 \rightarrow NaCl(l)$$
$$E^{\circ} = 4{\cdot}00 - 7{\cdot}55 \times 10^{-4}T \text{ volt}$$

At 873 K these two equations yield the standard potentials for $CaCl_2$ and NaCl of 3·47 V and 3·34 V, respectively. It is clear that separation of pure sodium from this melt which is probably nearly ideal is difficult because these potentials are so close to one another. The solubility of calcium in liquid sodium is however only a few atomic percent at the temperature of the electrolysis (ca. 4%) and therefore a metallic alloy which can hold a relatively high activity of calcium at quite low concentration, can readily be produced by electrolysis. (Fig. 86).

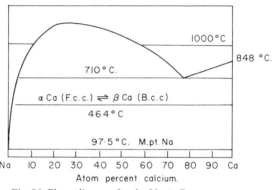

Fig. 86. Phase diagram for the Na + Ca system.

The range of composition of the salt phase is limited because the temperature must be kept below 600°C, the eutectic being at 520°C. This limitation on temperature results from the fact that sodium boils at 880°C and at 600°C has already reached a pressure of 1 millimetre. Operation at a higher temperature than 600°C would therefore raise the sodium pressure to a value which is too high for easy handling.

THE PRODUCTION OF ALUMINIUM

The solution of alumina in cryolite, Na_3AlF_6, is an electrolyte in which the current is almost entirely carried by aluminium and oxygen ions. In the industrial process, the electrolyte (20 wt % Al_2O_3) is electrolyzed between carbon electrodes at about 1000°C. The products of electrolysis are aluminium metal at the cathode and carbon dioxide at the anode.

The theoretical decomposition voltage is obtained from the oxide free energy diagram as follows:

$$\tfrac{4}{3}Al + O_2 \rightarrow Al_2O_3 \qquad \Delta G^0_{1273} = -201\,000\,cal$$
$$C + O_2 \rightarrow CO_2 \qquad \Delta G^0_{1273} = -95\,000\,cal$$

Therefore.

$$\tfrac{2}{3}Al_2O_3 + C \rightarrow \tfrac{4}{3}Al + CO_2 \qquad \Delta G^0_{1273} = 106\,000\,cl$$

Now, according to the standard electrochemical equation

$$-\Delta G^\circ = nFE^\circ$$

where $n = 4$, since the reactions involve one g molecule of oxygen, and F is Faraday's constant, $23\,060$ cal volt^{-1} equiv^{-1}. This yields a voltage of $1\cdot15$ V at $1000°C$. In practice, a voltage of between 7 and 8 volts D.C. at a current of about $10\,000$ amps per square foot of electrode is used. A number of cells are placed in series so that the total applied voltage to a group of cells is 250 volts. Since the atomic weight of aluminium is 27 g, the passage of 10^4 amps through one cell liberates

$$\frac{10^4}{96\,500} \times \frac{27}{3} = 0\cdot93 \text{ g of metal per second.}$$

The anode consumption (carbon) is $0\cdot31$ g during the same time if the anode reaction only involves CO_2 formation. The equilibrium composition of the gas would be a CO/CO_2 mixture in the ratio $142/1$ at $1000°C$ according to the reaction

$$CO_2 + C \rightarrow 2CO \qquad \Delta G^\circ = 40,800 - 41\cdot8T \text{ cal}$$

The consumption of carbon at the anode would, therefore, be double that which is observed if equilibrium were achieved, since the gas would be predominantly carbon monoxide. The fact that this is not so suggests that the kinetic barrier to the achievement of equilibrium in this reaction which was discussed earlier also operates in the electrolytic process. There is some evidence, of Welch and Richards (1962), that the slow step in the sequence of reactions

$$O_2 + C_s \rightarrow C - [O] + 2e \quad \text{(cf. p. 185–6)}$$
$$2C - [O] \rightarrow CO_2 + C_s$$

is the first; however, Antipin and Khudyakov (1956) showed that only CO is formed at very low current densities, and hence these kinetic effects are difficult to separate.

The gas which is evolved from industrial cells normally contains about 20–40% CO, but this is not due to a close approach to equilibrium at the point of discharge of the oxygen ions. It has already been stated that the rate of consumption of carbon indicates the almost exclusive formation of CO_2 at this location. The reduction of CO_2 after the formation of the bubbles of this gas is most probably the result of the presence of dissolved metal atoms in the molten electrolyte. These could either be aluminium or sodium atoms, since the close values of the deposition voltages of these metals from a cryolite melt saturated with Al_2O_3 have been established experimentally (Feinleib and Porter, 1960). The reduction reaction for CO_2 which involves aluminium atoms is obviously

$$2\{Al\} + 3CO_2 \rightarrow \{Al_2O_3\} + 3CO$$

Since this reaction causes a reversion of aluminium from the metallic state to that of a cation in the electrolyte melt, it represents a source of loss in current efficiency of the overall metal-winning reaction. It should be remembered, however, that CO evolution could partly come about as a result of the kinetics of the oxygen-carbon reaction at the electrode. The net effect can be summarized in Pearson and Waddington's (1947) equation

$$\text{Current efficiency} = \tfrac{1}{2}(\% \, CO_2 \text{ in anode gas}) + 50\%$$

Gjerstad and Welch (1964) studied the reduction of CO_2 bubbles which were passed through a column of the electrolyte placed in contact with aluminium. They found that the extent of the reduction to CO increased with the distance travelled by the bubble in traversing the electrolyte column. This would be consistent with mass transfer control of a reaction between the gas bubble and a species which is dissolved in the melt. Thonstad (1964) also showed that, in a similar experiment with bubbles of CO, carbon particles were produced in the melt. This is to be expected because of the low oxygen potential of the system Al/Al_2O_3.

The presence of dissolved metal in the electrolyte is expected to reduce the transport number of the ions in the melt, and in the case of aluminium, the metal atoms could release one or three electrons per atom. Univalent aluminium compounds are well known, e.g. AlCl, and either Al^+ or Al^{3+} ions could be produced together with virtually free electrons.

Sodium atoms are expected to be singly ionized releasing one electron per atom. The presence of either metal in solution in the electrolyte must, therefore, be another source of loss in current efficiency, not only by the chemical effects in reducing CO_2 but also by introducing a significantly finite electron transport number in the electrical conductivity of the electrolyte. This effect must be relatively small, however, when compared with the chemical reduction effect, since most experiments indicate a reasonably close agreement with the Pearson–Waddington equation. The stoichiometry of this equation is based on the reduction effect only, and does not take into account the deviation of the salt melt from truly electrolytic behaviour.

MIXED-ANION CONDUCTORS

The successful use of an oxide dissolved in a molten fluoride as the electrolyte for metal deposition suggests that other possibilities might exist for mixed-anion electrolytes. The aluminium cell has a mixture of oxide and fluoride anions, and the decomposition voltage of the oxide is reduced through the use of a chemically active electrode. Thus, alumina is reduced to metal and oxides of carbon rather than metal and oxygen gas. The formation of the mixed-anion electrolyte has "diluted" the oxygen sublattice of the oxide and, in the process of dissolution, the semiconduction properties of the oxide are eliminated in the electrolytic fluoride solvent.

TABLE XXXI. Heats of formation of sulphides, chlorides and fluorides

Compound	$-\Delta H^\circ_{298}$	kcal per mol S_2, Cl_2 and F_2			
2CdS	100·0	2PbS	76·0	2Cu₂S	70·2
CdF₂	167·4	PbF₂	158·5	2CuF	120
CdCl₂	93·0	PbCl₂	85·8	2CuCl	64·4
2ZnS	127·4	2FeS	76·6	$\frac{2}{3}$Bi₂S₃	59·0
ZnF₂	182·7	FeF₂	168·0	$\frac{2}{3}$BiF₃	212·0
ZnCl₂	99·5	FeCl₂	81·8	$\frac{2}{3}$BiCl₃	90·6

It should be possible to extend this principle to predict other possible systems and thus achieve new electrochemical processes for metal production. The main requirements for success appear to be a sufficient solubility of the semiconducting material in an electrolytic medium and a lower decomposition voltage for the semiconducting solute.

The sulphides offer some interesting possibilities in this respect, since the free energies of formation of sulphides are much less negative than those of the fluorides and similar to those of the chlorides in most circumstances where the naturally occurring mineral is a sulphide. Since the electrolysis of a sulphide solution in a halide solvent could produce a gaseous sulphur/halogen mixture as the anode product at the most probable operating temperatures, it is useful to compare the heats of formation of the corresponding sulphides and halides from metal and the diatomic gaseous species in order to appraise this possibility (Table XXXI). The entropies of formation of sulphides and halides from metal and one mole of diatomic gas will be approximately the same, and hence the heats of formation can be used to compare the relative stabilities.

The free energy change for the reaction

$$2MCl_2 + S_2 \rightarrow 2MS + 2Cl_2$$

can be approximated through the endothermic heat change to which must be added an entropy term for the vaporization of one mol of gas. Hence, the general reaction above can be represented by the equation

$$\Delta G^\circ = \Delta H^\circ_{298} - 26T \text{ cal}$$

and therefore the equilibrium constant

$$K = \frac{a^2\{MS\}}{a^2\{MCl_2\}} \cdot \frac{p^2_{Cl_2}}{p_{S_2}}$$

will increase with increasing temperature and the chlorine/sulphur ratio in the anode gas will correspondingly increase.

It can be seen at once that, excepting in the case of bismuth sulphide where almost pure sulphur always should be obtained, the electrolysis of sulphide–halide mixtures would most probably produce metal and sulphur gas in a fluoride solvent melt, or metal and sulphur + chlorine gas in a chloride solvent metal. Flengas and Garbee (1972) have shown that current efficiencies of about 80% for the production of copper could be obtained by the electrolysis of 10–20 mole per cent Cu_2S dissolved in CuCl at temperatures between 450 and 600°C. The sulphide–chloride mixtures for a given metal usually show only small positive deviations from Raoultian behaviour and so, under conditions of reversibility at the electrodes, the anode gas should contain both sulphur and chlorine. The possibility of using a fluoride solvent for a molten sulphide appears small because of solubility problems, but quantitative phase diagram information is lacking at present. It is probable that only oxides will mix significantly with fluorides, and sulphides with chlorides to any large extent because of the similarity of the anionic radii in each pair. These radii are as follows:

$$O^{2-} \text{ 1·40 Å, } F^- \text{ 1·36 Å, } S^{2-} \text{ 1·84 Å and } Cl^- \text{ 1·81 Å.}$$

REFERENCES

Antipin, L. N. and Khudyakov, A. N., (1956). *Russ. J. App. Chem.* **29**, 985. Anode reactions in aluminium production.

Bredig, M. A., (1964). "Molten Salt Chemistry" (M. Blander, Ed.) p. 367, Interscience, N.Y. Metal solutions in molten salts.

Feinleib, M. and Porter, F., (1960). *J. Electrochem. Soc.* **103**, 231. Sodium and aluminium deposition from cryolite–alumina melts.

Flengas, S. N. and Pint, P., (1969). *Can. Met. Quarterly* **8**, 151. Potential chloride electrolytes for Ti, Zr and Hf production.

Garbee, A. K. and Flengas, S. N., (1972). *J. Electrochem. Soc.* **119**, 631. Electrical properties of metal sulphide–chloride melts.

Gjerstad, S. and Welch, B. J., (1964). *J. Electrochem. Soc.* **111**, 976. Reaction of CO_2 with Al/cryolite-alumina melts.

Graves, A. D. and Inman, D., (1968). "Electromotive Force Measurements", p. 183, I.M.M., London. Electron transfer at metal–alkali halide melt interfaces.

Mellors, G. W. and Senderoff, S. (1965). *J. Electro Chem. Soc.* **112**, 266. *Electron deposition of refractory metals from molten fluorides.*

Pearson, T. G. and Waddington, I., (1947). *Disc. Faraday Soc.* **1**, 307. The electrolytic production of aluminium.

Thonstad, J., *J. Electrochem. Soc.* **111**, 955. The reduction of CO bubbles to carbon by A/cryolite-alumina melts.

Welch, B. J. and Richards, N. E., (1962). "Extractive Metallurgy of Aluminium" (G. Gerard, Ed.), Vol. 2, p. 15. Interscience, N.Y.

Part III

Metal Refining Processes

14

INTRODUCTION

The final stage of metal processing which will be considered now is the removal, largely by chemical means, of the impurities which remain in the metal phase after the initial extraction reaction. The impurities can range from non-metallic elements, such as sulphur and oxygen remaining from the original mineral, carbon which was introduced during reduction, or nitrogen from air which has passed through the system, down to other metals which were present in the reducing materials. The removal of metallic impurities is sometimes imperative because of the deleterious effects on the physical properties of product metal. Small amounts of an impurity metal having a higher affinity for traces of non-metals can finish up in the form of non-metallic oxide or sulphide inclusions. Some metals such as silver, gold and the platinum metals must be recovered from a major base metal because of their significantly greater economic value.

In all circumstances, new physical and chemical processes must be carried out with the crude metal phase, after separation from any attendant slag phase, in order to produce the refined metal. Due to the highly selective nature of the refining reactions which are normally employed, there are no general rules about refining, and the reactions are usually employed in specific instances and with specific separative objectives.

The main aspect of refining processes which is common to all refining processes is that the operator is always concerned with the removal of one or more minor constituents from a metal which contains a number of minor impurities. The thermodynamic and transport properties of dilute solutions are therefore of great interest in this context. Reactive gases, such as oxygen and chlorine, are frequently used as refining agents for the removal of impurities from liquid metals, and therefore gas bubble–liquid interactions are also important in affecting the rates of refining processes.

K

We will now begin the account of the refining of a number of specific elements with a general survey of dilute solution properties.

THE THERMODYNAMIC PROPERTIES OF DILUTE SOLUTIONS

Binary solutions

It appears to be reasonably well established that most solutes conform to Henry's law in the composition range 0–1 atom per cent within the limits of normal experimental error. Beyond this range of composition, the interactions between the solute atoms and the changes in the free electron concentration when species from different Groups of the Periodic Table are mixed, influence the variation of the activity coefficient of the dilute solute. It seems that elements which raise the electron/atom ratio of the alloy system show an activity coefficient which increases with increasing concentration. Thus, the values of the self-interaction coefficient, ε_i^i, tend to increase in solution in liquid copper and silver, as one passes from zinc to germanium or tin as solute. Table XXXII shows the experimental values for these solutes in the copper or silver solvent for the composition range 1–10 atom per cent.

TABLE XXXII. Self interaction coefficients for liquid alloys

Solute	Copper Solvent ε_i^i	Silver Solvent ε_i^i
Zinc	3·4	3·8
Indium	3·4	6·0
Gallium	6·7	3·4
Germanium	10·7	—
Tin	9·2	6·3

There is very little information concerning self-interaction coefficients for non-metallic solutes. but the indications are that these species exert an attractive force on one another in dilute solution. In many cases, this interaction leads to the separation of two liquids as the content of the non-metallic species increases, e.g. $Cu–Cu_2S$, whereas in others the effect is not so marked and two liquid phase separation does not occur, e.g. Fe–FeS. Very few accurate values of the self-interaction coefficient for non-metallic solutes have been obtained because the range of concentration in the dilute solution is normally very limited. The variation of the activity of sulphur in solution in liquid silver in which there is a reasonably wide range of metal-rich liquid does differ markedly from the behaviour of other metallic solutes (Fig.

87) in a manner which supports this suggestion. Further support comes from the experimental results which show that the self-interaction coefficient of sulphur in liquid iron has the value

$$\varepsilon_s^s = -3.7 \text{ at } 1600°C.$$

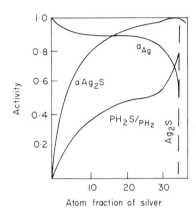

Fig. 87. Results for the liquid system Ag + S at 1125°C. The ratio pH_2S/pH_2 is a measure of the sulphur activity. (After T. Rosenqvist, *J. Metals* **1**, 451 (1949).)

Dilute ternary solutions

The very important role of the interaction between dilute solutes dissolved in a single major element which is to be refined has been the subject of considerable theoretical development. The model which has been most successful in accounting for experimental data for dilute element interactions is based on the constant pairwise bonding model. The most primitive form of this model is the random distribution, strictly, regular solution model, and the activity coefficient of a dilute solute, S, in solution in a binary alloy A + B varies according to the equation

$$\log \gamma_S^{A+B} = X_A \log \gamma_S^A + X_B \log \gamma_S^B - \frac{\Delta G_{A+B}^{XS}}{RT}$$

where γ_S^{A+B}, γ_S^A and γ_S^B are the activity coefficients of the infinitely dilute solution of S in A + B alloys, pure A and pure B respectively. ΔG_{A+B}^{XS} is the excess free energy of formation of the A + B alloy at the mole fractions X_A and X_B.

When the interaction between A and S is much greater than that between B and S, the distribution of the Z metal atoms in the coordination shell of each S atom is not directly in the ratio X_A/X_B as required by the random model. The appropriate expression, the quasi-chemical model, must now include a Boltzmann weighting factor which shows the relative probability of finding A and B atoms in the coordination shell of each S atom. This expression is

$$\frac{n_{A-S}}{n_{B-S}} = \frac{X_A}{X_B} \exp \frac{\{ - \Delta G° \text{ (exchange)}\}}{RT}$$

where n_{A-S} and n_{B-S} are the number of $A - S$ and $B - S$ contacts respectively and $\Delta G°$(exchange) is the free energy change for the exchange of contacts thus

$$B - S + A \rightarrow A - S + B$$

In calculating the free energy change for this reaction, it is necessary to take account of the fact that the A atom on the left-hand side must break a bond in the bulk of the alloy and the B atom on the right hand side must make a bond in the bulk of the alloy in order to make the $A - S$ and replace the $B - S$ contacts. The equation for the activity coefficient γ_S^{A+B} as a function of composition now takes the form

$$\left(\frac{1}{\gamma_S^{A+B}} \right)^{1/Z} = X_A \left(\frac{\gamma_A^{A+B}}{\gamma_S^A} \right)^{1/Z} + X_B \left(\frac{\gamma_B^{A+B}}{\gamma_S^B} \right)^{1/Z}$$

These two equations have found accurate predictive value when S was a metallic solute which was dissolved in a binary solvent. A good example is to be found in the author's work with Jacob (1973) on the effect of solvent composition on the dilute solution activity coefficient of indium dissolved in the solid α solutions (f.c.c. structure) of the $Cu + Au$ solvent. The comparison between the regular solution model and the quasi-chemical model is shown together with the experimental results in Fig. 88.

These equations fail for dilute solutions of oxygen and sulphur in metallic alloy solvents, and it has recently been shown that this fact can be accounted for by a consideration of the nature of the model in terms of the free electron theory of alloys. If these solutes dissolve in liquid metallic solvents in the form of negatively charged species, e.g. S^- or S^{2-}, then the charge on the ions will repel the conduction electrons from the solvent which is in immediate contact with the solute ions. Translating this thought into the chemical bond model, the equivalent statement is that the metal–metal interaction energies, or bonds, in the solvent coordination shell of the S ions will be weakened. This can be represented in the bonding theory either by a decrease in the number of bonds of constant pairwise energy or in a weakening of a constant number of bonds. All that the theory can deal with is the product of the number of bonds multiplied by the bond energies, and if this product is reduced in value due to electron repulsion, it is not operational to attempt to separate these two alternative possibilities. By fitting a general equation to experimental data for oxygen and sulphur in a number of alloy systems, it has been shown that most situations can be accounted for on the assumption that oxygen

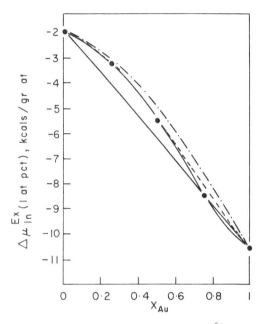

Fig. 88. Variation of the chemical potential of indium at 1 at.% concentration with Cu/Au ratio in Cu + Au + In alloys at 900 K.

—·—·— = Regular solution model.
– – – – – = Quasichemical model $Z = 12$.
———●——— = Experimental data.

and sulphur make four bonds only in solution in metallic alloys and that half of the metal–metal bonds which are normally made by the metal atoms in the bulk of the alloy are broken when these atoms are bonded to oxygen or sulphur.

The three models which have been described have all found application in metallic alloy systems containing two principal metallic species, but in the context of refining we are mainly interested in the situation where $X_A \cong 1 \gg X_B$ and the metallic species B must also be regarded as a dilute solute. In this case the predicted values for ε_S^B are

$$\varepsilon_S^B = \left(\frac{\partial \ln \gamma_S}{\partial X_B}\right)_{X_B \to 0} = \ln \gamma_S^B - \ln \gamma_S^A - \ln \gamma_B^A \quad \text{(Regular solution)}$$

$$= -Z(K-1) \quad \text{(Quasi-chemical solution)}$$

where

$$K = \left(\frac{\gamma_S^B}{\gamma_S^A} \cdot \frac{1}{\gamma_B^A}\right)^{1/Z}$$

$$= -n(K'-1) \quad \text{(Electron repulsion model)}$$

where

$$K' = \left(\frac{\gamma_S^A}{\gamma_S^B}\right)^{1/n} \cdot (\gamma_B)^\alpha$$

n is the number of bonds made by each S atom, and α is the fractional reduction of the metal–metal bonds in the S coordination shell.

$$\alpha = \frac{ZE - ZE^*}{ZE} = \frac{Z - Z^*}{Z}$$

where Z is the number of metal–metal bonds made by each atom in the absence of oxygen or sulphur, and Z^* is the number when O or S is present. E^* is the reduction in bond energy which must be assumed if the bond number, Z, remains constant.

Some examples of the measured interaction coefficients for oxygen in liquid iron solutions and those obtained from the electron repulsion model are shown in Table XXXIII.

The values for γ_O^B were estimated by the use of an approximation which was originally put forward by Richardson (1958)

$$\frac{\gamma_O^B}{\gamma_O^C} \cong \frac{\Delta H_{298}^\circ \text{ B oxide kcal g atom}^{-1}}{\Delta H_{298}^\circ \text{ C oxide kcal g atom}^{-1}}$$

TABLE XXXIII. The values for ε_O^B for oxygen from experimental data and the Electron Repulsion Model in liquid iron as solvent

Element B	$\gamma_B^{Fe}(X_B \to 0)$	ε_O^B measured	Theoretical
Aluminium	0·062	−105	−120
Vanadium	0·18	−23	−17
Manganese	1·33	−6	−29
Chromium	1·0	−9	−20

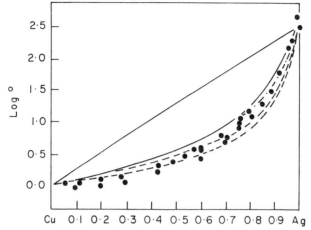

Fig. 89. The activity coefficient of oxygen in Cu + Ag + O alloys at 1200°C.
———●——— = Experimental results.
——————— = Theoretical equation.

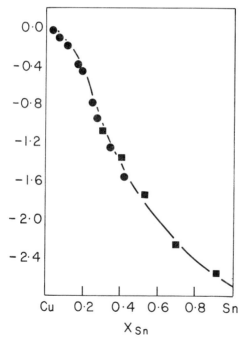

Fig. 90. Activity coefficient of oxygen in Cu + Sn + O alloys at 1100°C.
———●——— = Measured.
——————— = Theoretical.

Figure 89 shows the satisfactory agreement between theoretical and experimental results for the Cu + Ag–O system at 1200°C. The effects of the metal–metal interaction can be more clearly seen in the Cu + Sn + O system (Fig. 90) for which it has already been shown that the Cu + Sn binary system shows negative departures from Raoult's law (see Fig. 42).

DIFFUSION OF DILUTE IMPURITIES IN LIQUID METALS

The capillary-reservoir method (p. 140) for the measurement of the diffusion of constituents in liquid systems can be applied very simply to the case of a dilute solute. The capillary is initially filled with the dilute solution, and this is immersed in a large well-stirred volume of the pure solvent metal. Since the diffusion coefficient is independent of concentration for a dilute solute which obeys Henry's law, the efflux of the dilute constituent from the capillary may be analysed to determine the solute diffusion coefficient in the same way as that of a radioactive tracer.

A number of experimental studies have been made but the most systematic studies appear to be those of Ejima and co-workers (1966, 1968). The liquid solvents were silver and copper and the results demonstrate that for a number of solutes from the same Group of the Periodic Table, the diffusion coefficient decreases with increasing solute radius. The diffusion coefficients of S, Se and Te are lower in liquid silver than copper in the temperature range 1100–1300°C in which the study was made, and this reflects the lower self-diffusion coefficient of liquid silver when compared with liquid copper at a given temperature (see p. 149). According to these authors, the diffusion coefficients of sulphur in the solvents are approximately double those of selenium and about three times those of tellurium. According to the fluctuation model of liquid diffusion whereby a particle may move into a void which is created during the cooperative vibration of a number of its neighbours, it is to be expected, qualitatively, that the larger solute particle, which requires a larger volume of fluctuation, would find the opportunity to move less frequently than the smaller solute. Hence, it is to be expected, on volumetric grounds, that sulphur should diffuse more readily than selenium and tellurium. However, a quantitative theory is not available to describe this effect at present, and a first estimate for the diffusion coefficient of a solute, except for hydrogen which is a special case, should be the self-diffusion coefficient of the pure solvent. In comparing solutes, it may qualitatively be expected that the smaller solute will move more quickly than the larger solute. Some typical values for a few dilute solutions are shown in Table XXXIV.

TABLE XXXIV. Typical diffusion coefficients of dilute solutes in liquid metals

Solvent	Solute	$D(\text{cm}^2\,\text{s}^{-1} \times 10^4)$	Temperature (°C)
Iron	Self diffusion	1·7*	1600°C
	Hydrogen	13·2	
	Carbon	0·7	
	Oxygen	0·7	
	Sulphur	1·7	
Copper	Self diffusion	0·4	1100°C
	Hydrogen	9·9	
	Oxygen	0·8	
	Sulphur	0·9	
	Nickel	0·3	
	Silver	0·3	
	Gold	0·25	

* Extrapolated from data for Fe + C alloys in the temperature range 1300–1400°C.

DIFFUSION OF DILUTE CONSTITUENTS IN GASES

In a number of mass transfer operations involving gas bubbles, an impurity is transferred from a liquid metal solvent into a gas bubble within which there is a concentration gradient for the transferred species. The self-diffusion coefficients which have been given for gases in Part I again can be used as an approximation for the diffusion coefficient of the dilute, impurity gaseous species. The accurate calculation of interdiffusion coefficients for gases demands a knowledge of the collision cross section, which replaces d in the equation on p. 6, and of the forces through which atoms and molecules interact in the gaseous phase.

A more simple method for calculating the coefficients is to use empirical equations which have been obtained from a number of systems. The most successful of these is due to Andrussov (1950) and an experimental study by Schafer, Corte and Moesta (1957) demonstrated the satisfactory accuracy of the equation proposed by Andrussov

$$D_{A-B} = \frac{0{\cdot}06\ T^{1{\cdot}78}}{p(V_A^{\frac{1}{3}} + V_B^{\frac{1}{3}})^2}\left\{\frac{1 + \sqrt{M_A + M_B}}{\sqrt{M_A M_B}}\right\}$$

In this equation, for the interdiffusion of the gases A and B, the molar volumes are V_A and V_B, the molecular weights are M_A and M_B and the total pressure is p atmospheres.

Turkdogan (1965) in a valuable review of the subject considers the factors in the theoretical equation for D_{A-B} which can be expressed in the form

$$D_{A-B} = 1\cdot 86 \times 10^{-3} \frac{T^{\frac{3}{2}}}{p(\sigma_{AB})^2} \sqrt{\frac{1}{M_A} + \frac{1}{M_B}} \; (\Omega_{T^*_{AB}})^{-1}$$

where

$$\sigma_{AB} \simeq \frac{d_A + d_B}{2} \quad (d \text{ the molecular diameter})$$

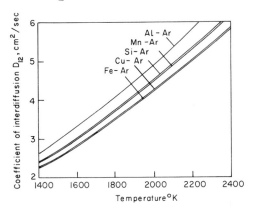

Fig. 91. Diffusion coefficients of monatomic metal vapours in argon. (After Turkdogan, loc. cit.)

and $\Omega_{T^*_{A-B}}$ is the collision integral, which has been tabulated by Hirschfelder et al., (1954) at the so-called reduced temperature T^*_{A-B}. This is related to the true temperature T by the expression

$$T^*_{A-B} = \frac{kT}{E_{A-B}}$$

k is Boltzmann's constant and

$$E_{A-B} = \sqrt{E_{A-A} \cdot E_{B-B}}$$

is the interaction energy between unlike atoms, here approximated to the geometric mean of the like atom interaction energies. Turkdogan shows how values for these energies may be arrived at empirically and derives calculated values for the diffusion of a number of dilute metallic species through argon at one atmosphere pressure (Fig. 91).

RECOMMENDED READING

"The Physical Chemistry of Melts in Metallurgy", F. D. Richardson, Academic Press, London and New York (1974).

15

ZONE REFINING

The elements germanium and silicon are needed in a degree of purity which is far beyond that normally required in industrial use for applications in the electronics industry. The high cost of the systems in which these very pure materials are applied makes it economically feasible to employ techniques of normally low productivity, measured in pounds of production of element per hour.

The method of zone refining which was first employed in the manufacture of very pure germanium depends for its success on the difference between the thermodynamic properties of a dilute constituent dissolved in a substantially pure solid solvent from the corresponding properties in the liquid system. The partition of a dilute constituent between co-existing solid and liquid phases under equilibrium conditions can be obtained from the phase diagram and, of course, the condition of equilibrium is that the chemical potential of the dilute solute, referred to a given standard state, is the same in both phases.

$$RT \ln a_S = RT \ln a_L$$

(The subscription S refers to solid alloys, and L to the corresponding liquids).

$$\Delta \bar{H}_S - T\Delta \bar{S}_S = \Delta \bar{H}_L - T\Delta \bar{S}_L$$

$$(\Delta \bar{H}_S - \Delta \bar{H}_L) - T(\Delta \bar{S}_S^{xs} - \Delta \bar{S}_S^{xs}) = RT \ln \frac{X_L}{X_S}$$

In the simple situation that the excess partial entropy of solution of the solute is the same in both liquid and solid solvents, the partition coefficient K_{S-L} is determined by the heat of transfer of the solute from the liquid to

267

the solid solvent $(\Delta \bar{H}_S - \Delta \bar{H}_L)$.

$$K_{S-L} = \frac{X_S}{X_L} = \exp - \frac{(\Delta \bar{H}_S - \Delta \bar{H}_L)}{RT}$$

The distribution of impurity after a small section of a cylindrical sample was melted would depend on the diffusion of impurity to and from the melted zone, and this, in turn would determine the distribution of the impurity within the sample on subsequent cooling. If sufficient time were

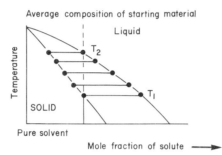

Phase diagram for a binary dilute system. Tie lines connect the possible solid–liquid distributions at a number of temperatures.

Solid–liquid distribution in a cylindrical sample held at T_1.

Distribution in the sample held at T_2.

Fig. 92.

always allowed for complete re-distribution of the solute between solid and liquid to take place, then the final compositions of the frozen section and the rest of the material would depend on the fraction of the total material which was melted, and hence on the temperature of equilibration (Fig. 92). Reference to Fig. 92 which shows the phase diagram for a binary dilute system in which the freezing point of the solvent is depressed, demonstrates that the solid–liquid distribution of the impurity depends upon the temperature of the operation, and this, in turn, depends upon the thermal characteristics of the system in terms of heat input and heat losses by convection and radiation. For a given thermal situation, the transfer of material from the solid to the liquid would depart more significantly from the equilibrium distribution, the shorter the time which is allowed for diffusion within the solid phase. It remains a good approximation to assume that the diffusion coefficients in the liquid state will be the same to within a factor of ten. In many systems of practical zone refining, heating is achieved by radiofrequency induction heating. The liquid phase is well stirred by the action of the r.f. field, and hence can be assumed to be uniform in composition except at the boundary layer which is adjacent to the solid surface.

If we make the simplifying assumption that diffusion to the liquid–solid interface through the solid is slow, then the liquid will appear with the composition shown at temperature T_2 in the figure. If the molten zone is then slowly moved through the length of the cylinder, the solid material which is left behind on freezing should ideally have the composition shown on the solidus at T_2. Hence, within the liquid, the composition would rise above that at the liquidus at temperature T_2, and dissolution of untreated material would occur, leading to an advance of the liquid front.

Since the cylindrical sample has a finite length, the chemical effect of passing a molten zone through the sample would be to move the impurity in the same direction, thus leaving purer material behind. As the impurity concentration in the remainder of the sample increased, the temperature of fusion would also decrease. When the liquid phase reaches the composition on the liquidus at T_1 which is in equilibrium with the solid of the original composition, no further refining continues. Finally, the liquid zone containing this amount of impurity is frozen at the end of the sample. The final distribution of the impurity after the passage of one liquid zone is shown in Fig. 93.

The zone refining system is clearly not easily analyzed because of the non-isothermal nature of the operation, as well as the balance between the thermal characteristics of the system, and the rate of advance of the molten zone which are parameters of choice for the operator. It is to be contrasted with the Parkes' process (p. 316) in which a small amount of solid phase is removed from a large body of melt, when equilibrium can be reasonably well approached. In practice, a compromise is drawn between having a number of rapid passes of the liquid phase, and hence having a small purification in each pass, and making fewer slow passes, each of high efficiency. A mathematical treatment of a very simplified but effective model of the zone refining process was given in the original paper by Pfann (1952) describing the process. Let the partition coefficient K_{S-L} of the solute between solid and liquid (hereafter written as K) remain constant and the length of the molten zone remain fixed at l, and C_0 be the initial concentration of the solute per unit volume, and let the molten zone advance dx cm.

The net change in the quantity of solute S is given by

$$\frac{\mathrm{d}s}{\mathrm{d}x} = \left(C_0 - \frac{Ks}{l} \right)$$

Here $Ks/l.dx$ is the amount of solute frozen out by movement of the zone, and $C_0\,dx$ is the amount taken into solution in the same step.

The differential equation has the solution

$$s = \frac{C_0 l}{K}\left\{ K + \left(\exp\frac{Kx}{l} - 1 \right) \right\} \exp - \frac{Kx}{k}$$

Writing $C = Ks/l$, we have

$$C = C_0\left\{ 1 - (1 - K).\exp - \left[\frac{Kx}{l}\right] \right\}$$

Some graphical solutions for this equation are shown in the accompanying figure with $C_0 = 1$ for all curves. The abscissa is drawn in units of one liquid zone length l.

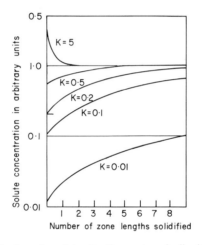

Fig. 93. Theoretical distribution of a solute after the passing of a liquid refining zone of length l according to the equation

$$C/C_0 = 1 - \left\{ (1 - K)\exp\left[-\frac{Kx}{l} \right] \right\}$$

The variation of the concentration of the solute along the cylindrical specimen will be changed by a number of refining steps from the simple expression which is given above for the single pass. Eventually, however, there will result a distribution of the solute which is not altered by further refining. Using the same simple model of the constant length of molten zone, l, then if the concentration is finally $C(x)$ as a function of the distance x from one end of the specimen, the concentration in the liquid zone, C'_{liquid}, must then be $C(x)/K$.

But

$$C_{\text{liquid}} = \frac{1}{l} \int_{x}^{x+l} C(x) \, dx = \frac{C(x)}{K}$$

The solution of this equation has the form

$$C(x) = A \exp(Bx)$$

where

$$A = \frac{C_0 BL}{\exp(BL) - 1}$$

and

$$K = \frac{Bl}{\exp(Bl) - 1}$$

Further experimental studies of zone refining have shown that the number of stages required to transform a system from the originally uniform, impure, state to that having the final concentration distribution which is given by this equation is approximately equal to the ratio of the specimen length to the length of the molten zone, L/l.

The values of the partition coefficients for solutes between solid and liquid germanium or silicon as solvents are particularly favourable for the application of zone refining. The solid solubilities increase with the valency of the solute, from about 10^{-5} atom percent for a Group IB (copper) in germanium up to about 1 atom percent for a Group IVB (tin) solute. The liquid phases with which these solid solutions are in equilibrium are approximately Raoultian at the germanium-rich end, showing small negative deviations for Cu + Ge and small positive deviations from Sn + Ge. We may therefore ignore, as a first approximation, the partial heat of solution of the solute in the liquid of composition along the germanium-rich liquidus of the phase diagram. It therefore follows that the partial heat of solution of the solute in solid germanium can be obtained in good approximation from the temperature dependence of the partition coefficient K_{S-L}. For the solute A

$$\frac{\partial \ln K_{S-L}^A}{\partial 1/T} = \frac{\Delta \bar{H}_L^A - \Delta \bar{H}_S^A}{R} \simeq \frac{-\Delta \bar{H}_S^A}{R}$$

Thurmond and Struthers (1953) have shown that the value of this heat of solution is $45 \cdot 2 \, \text{kcal g atom}^{-1}$ for Cu and $23 \cdot 1 \, \text{kcal g atom}^{-1}$ for Sb in solid solution in Ge. The highly endothermic nature of these heats of solution is probably related to the significant difference in structure between Ge and Si on one hand, and the usual solute species which are considered. The latter are usually metallic and have the typically high coordination number of close packing in the elementary state, whereas germanium and silicon have tetrahedral coordination which is brought about by essentially covalent bonding. The change in structure which a solute atom such as copper or zinc will undergo in the transfer from the metal to solid solution in Ge or Si is therefore extreme, and this could account for the overall decrease in binding energy which characterizes the transfer.

Another contributory factor to the relative success of zone refining for germanium and silicon also arises from the structure of these solid elements. The heats of fusion of germanium and silicon are 8·8 and 12·0 kcal g atom^{-1}, respectively. Using the melting points 1210 K and 1685 K and ignoring the differences in heat capacities between the solid and the corresponding liquid, we may obtain the free energy of fusion equations

$$\Delta G_f^\circ = 8800 - 7\cdot27T \text{ cal for germanium}$$

and

$$= 12000 - 7\cdot12T \text{ cal for silicon}$$

These equations should be compared with the equation for a typical metal such as copper (m.p. 1356 K) where

$$\Delta G_f^\circ = 3140 - 2\cdot32T \text{ cal}$$

Now the activity of the solid with respect to the liquid as standard state is given by the equation

$$RT \ln a_s = -\Delta G_f^\circ \text{ at the temperature } T.$$

Evaluation of the equations above show that the activities of solid germanium and silicon with respect to their liquid elements are lower for a given temperature below the melting point than is the case for copper. Thus, for a 300 degree depression

$$a_s = 0\cdot30 \text{ (Ge at 910 K)}$$

$$a_s = 0\cdot46 \text{ (Si at 1385 K)}$$

$$a_s = 0\cdot72 \text{ (Cu at 1056 K)}$$

Now it is because the activities of the Group IV solids are so abnormally low with respect to liquids as standard states and the liquids are nearly Raoultian that the phase diagrams, especially those of germanium, show an unusual convexity of the liquidus at the germanium- or silicon-rich end. Hence the value of the partition coefficient K_{S-L} is even more favourable for these elements than would be the case for a typical metallic solvent.

These two effects then, the endothermic heat of solution of a metallic element in the solids, and the convexity of the liquidus on the equilibrium diagrams, make the process of zone refining peculiarly useful for germanium and silicon.

16

LIQUID DISTILLATION

The production of high purity zinc with about 99·995% purity is achieved by distillation of the metal which is obtained from the primary extractive stage. This process also brings about the concentration of the cadmium impurity, which is always present in primary zinc, to yield high purity cadmium.

The vapour pressures of both zinc and cadmium are known with high precision as the result of a number of investigations. A range of techniques from direct boiling point determinations under reduced pressure at the highest temperatures, to Knudsen and gas transportation studies at the lower range of liquid temperatures have been employed with both metals, and the assessment of experimental values by Hultgren et al. (loc. cit.) suggests the following equations for the liquid vapour pressures.

$$\text{Zinc: log p (atmos)} = \frac{-6620}{T} - 1\cdot255 \log T + 9\cdot46$$

$$\Delta H_{evap.} = 27\cdot3 \pm 0\cdot4 \text{ kcal mol}^{-1} \text{ at boiling point (907 °C)}$$

$$\text{Cadmium: log p (atmos)} = \frac{-5819}{T} - 1\cdot257 \log T + 9\cdot41$$

$$\Delta H_{evap.} = 23\cdot9 \pm 0\cdot3 \text{ kcal mol}^{-1} \text{ at boiling point (765 °C)}$$

The thermodynamic properties of the liquid alloys between zinc and cadmium are also well known. The system behaves as a strictly regular solution with ideal entropy of mixing, and a heat of mixing which is given, as a function of composition, by the equation

$$\Delta H^{M} = 2000 \, X_{Zn} \cdot X_{Cd} \text{ cal gatom}^{-1}$$

At a temperature close to the boiling point of zinc, say 900°C, the vapour pressures of the two elements will be 0·921 and 3·90 atm for zinc and cadmium respectively, using the equations above. Then the equilibrium vapour phase will have equal amounts of zinc and cadmium at this temperature at a composition given by the expression

$$p_{Cd}^{\circ} \gamma_{Cd} X_{Cd} = p_{Zn}^{\circ} \gamma_{Zn} X_{Zn}$$

Using the regular solution equation for the respective activity coefficients, and substituting the values for the vapour pressures of the elements, which were obtained above, we find the following for the atom fraction of zinc under the conditions of equal vaporization:

$$p_{Zn} = p_{Cd}; \qquad \gamma_{Zn} X_{Zn} p_{Zn}^{\circ} = \gamma_{Cd} X_{Cd} p_{Cd}^{\circ}$$

$$0.92 \exp \left[\frac{2000 X_{Cd}^2}{RT} \right] \cdot X_{Zn} = 3.90 \exp \left[\frac{2000 X_{Zn}^2}{RT} \right] (1 - X_{Zn}^2)$$

$$\frac{2000}{RT} \{(1 - X_{Zn})^2 - X_{Zn}^2\} = \ln \left\{ 4.24 \frac{(1 - X_{Zn})}{X_{Zn}} \right\}$$

$$0.3726 (1 - 2X_{Zn}) = \log_{10} 4.24 + \log_{10} \left(\frac{1 - X_{Zn}}{X_{Zn}} \right)$$

From this equation, the value for X_{Zn} where the partial pressures of the two metals are the same is $X_{Zn} = 0.89$. When the alloy is richer in zinc than this atom fraction zinc concentrates in the vapour phase, but the converse happens and cadmium predominates in the vapour phase when the alloy contains a smaller amount of zinc.

If the composition of the vapour phase, when the total pressure is kept constant at one atmosphere, is plotted against the composition of the co-existing liquid as a function of temperature, it is found that the vapour is always richer in cadmium than the corresponding liquid (Fig. 94). The quantitative separation of zinc from a small amount of cadmium in such a way that only cadmium is removed may therefore be accomplished in a distillation column which is operated at one atmosphere total pressure. The less volatile zinc can be concentrated by refluxing, and cadmium escapes into the vapour phase.

A distillation column may be considered as being composed of two sets of stages. Those above the inlet of fresh material terminate in a condenser and are called "rectifying" stages. Those below the inlet terminate in a boiler and are called "stripping" stages. When a still is being run at a constant rate (steady state), a material balance at one stage applies simultaneously to all stages of each set. If the stages of the rectifying set are numbered successively from the top of the column downwards, then the material balance at the nth stage is given by

$$V_{n+1} = L_n + D$$

where V_{n+1} moles of vapour approach the stage from the $(n + 1)$th stage, L_n moles of liquid reflux, and D moles of product pass to the condenser. Calling

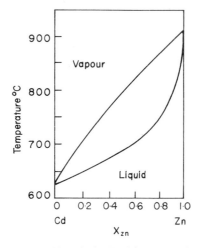

Fig. 94. Liquid and vapour compositions in the Zn–Cd system at 1 atmosphere pressure boiling points.

y^A the fraction of a component A in a binary A–B vapour mixture and x^A the fraction in the corresponding liquid,

$$V_{n+1}y^A_{n+1} = L_n x^A_n + D x^A_D$$

Hence

$$y^A_{n+1} = \frac{L_n}{L_n + D} x^A_n + \frac{D}{L_n + D} x^A_D$$

In this equation, all of the terms except y_{n+1} and x_n, but including x^A_D the atom fraction of component A in the product of the still, are constant. Hence, the relationship between y^A_{n+1} and x^A_n is linear of slope $L/(L_n + D)$ and a line representing the relationship must pass through $y_D = x_D$ when $x_n = x_D$ since the vapour and the liquid have the same composition in the product. McCabe and Thiele (1935) called this the "operating line" in their method for the graphical representation of distillation.

In an analogous manner, the material balance for the stripping stage is given by

$$V'_m = P_{m+1} - P$$

hence

$$V'_m y^A_m = P_{m+1} x^A_{m+1} - P x^A_p$$

and

$$y^A_m = \frac{V'_m + P}{V'_m} x^A_{m+1} - \frac{P}{V'_m} x^A_p$$

where V'_m is the flow rate of vapour from the mth stage and P is the flow rate of product at the bottom of the still. (Here the stripping stages are numbered

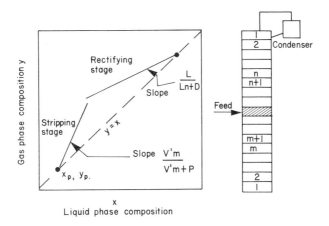

Fig. 95a. The operating lines for the rectifying and stripping stages of a distillation column.

from the *bottom* of the still.) The graphical representation of this function is then by a line of slope $V'_m/(V'_m + P)$ which passes through the point $y_p = x_p$. This is McCabe and Thiele's operating line for the stripping section (Fig. 95a).

Any point on each of these operating lines represents a liquid which is in equilibrium with a vapour phase. The composition of the corresponding gas phase is obtained by moving horizontally across the graph of y vs. x for the system at the given value of x to cut the curve of y.

Ideally, the composition of the ingoing material should be made equal to that at the point of intersection of the rectifying and stripping operating lines. If this coincides with a point on the y vs. x curve then the still operates with the minimum number of stages. The number of stages in each set can then be simply counted by the usual method of intercepts (Fig. 95b).

The slope of the rectifying stage operating line may be termed the "reflux ratio", since it defines the fraction of liquid which is returned to the stripping stage, the remainder passing on to the condenser. The extent of refluxing is partly determined by the capability of the descending liquid to extract heat from the ascending vapour phase. In liquid metal systems, this can be much higher than in the more conventional organic liquid systems which are normally separated by distillation, because of the high thermal conductivities

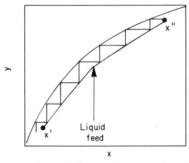

Fig. 95b. A still requiring seven theoretical stages for separation of a liquid feed into products x' and x''.

of liquid metals. The separation of metals by distillation can be expected to operate under practically the theoretically deduced conditions because of this property, and the slopes of the operating lines can be more readily controlled through stage design than would usually be the case. In this event, the compositions of the products of distillation can be determined from the reflux ratio, for a given liquid feed, in relationship to the theoretical line representing the gas composition which is used in the McCabe–Thiele diagram. For a given "gas-line" the number of stages increases with the reflux ratio, and for a given reflux ratio, the number of stages decreases the greater the difference between the gas-line and the line $y = x$. This difference is larger when the two metals have widely differing vapour pressures in the pure state, or show marked positive departures from Raoult's law.

THE NEW JERSEY ZINC PROCESS

The New Jersey method for zinc refining is carried out in one continuous process, but has two stages of distillation. In the first step, at the highest temperature, zinc and cadmium are volatilized together in a distillation column leaving behind a liquid phase which contains the lead impurity together with other minor impurities such as iron. The vapour pressure of pure lead at 900°C is 10^{-4} atm, and hence zinc is more volatile than lead by several orders of magnitude under conditions where distillation might be carried out.

In the second distillation column which is operated at a lower temperature, cadmium is removed at the top of the still in the vapour off-take, and zinc is collected at the bottom at the liquid metal of better than 99·99% purity. The melting point of zinc is 419·5°C, and although details of the temperature of operation of the second still do not appear to be in the literature, the range of temperature over which the preferential vaporization of cadmium occurs covers the whole liquid range of zinc as Table XXXV shows.

TABLE XXXV. Relative vapour pressures of zinc and cadmium

$T(^{\circ}C)$	$p^{\circ}_{Cd}/p^{\circ}_{Zn}$
450	12·69
550	9·30
650	7·30
750	6·06
850	5·09

The elimination of cadmium by volatilization obviously occurs more readily at the lower end of the temperature range not only because of the relative vapour pressures of the pure elements but also because of the fact that the activity coefficient of cadmium dissolved in liquid zinc becomes increasingly more positive as the temperature decreases. It will be remembered that the strictly regular solution model, which is appropriate here, requires an inverse dependence of log γ on temperature

$$\log \gamma_{Cd} = \frac{2000\, X^2_{Zn}}{4\cdot 575\, T}$$

It follows that, at any given liquid composition, the activity of cadmium is higher, the lower the temperature. It is also clear that the corresponding $p_{Zn}/p_{Zn} + p_{Cd}$ vs. X_{Zn} diagrams for temperatures lower than 900°C become increasingly more concave, favouring the enrichment of the vapour phase with cadmium.

The diagram which appears in the literature concerning this process shows that a large number of stages must be used in the still for the separation of zinc and cadmium, thus indicating that the reflux quantity at each stage must be very nearly equal to the amount arriving from the vapour phase. This conclusion could also be reached by considering a McCabe–Thiele diagram for a system which has an ingoing material of practically pure zinc, and from which it is required to draw purer zinc and a condensate which is rich in cadmium. Clearly, the stripping and rectifying operating lines are practically on the line $y = x$, and a large number of stages would be required to straddle the range of liquid compositions which would appear along the length of the distillation column.

17

STEELMAKING

The low oxygen potential and high carbon activity in the liquid metal-making region of the ironmaking blast furnace yield a product which is saturated with carbon and contains a number of other non-metallic impurities, namely silicon, phosphorus and sulphur. The next stage in the process which will yield liquid steel is largely a refining step in which these impurities are removed. The chemical reactions involve oxidation and absorption of the oxides of the impurities into the slag or gas phases.

The liquid steel which is finally achieved in this oxidation process is at a higher temperature than the material from the blast furnace and is, if anything, overoxidized. There must therefore be a final stage of deoxidation, and any required alloying additions, such as ferromolybdenum or ferroniobium, are made in the ladle which contains the product.

THE BESSEMER CONVERTER

Reference was made earlier to the bottom-blow converter invented by Bessemer for the decarburization of blast furnace iron (Fig. 96). Air was forced vertically through the molten metal pool, and the converter was lined with silica-brick in the original form. At the present time, dolomite ($MgO.CaO$) lining is preferred since an important reaction which is carried out simultaneously with carbon elimination is the oxidation of the phosphorus content of the metal. This is absorbed, together with SiO_2, MnO, and other metal oxides which are formed from the impurities in the blast furnace metal, in a slag with a high lime content, a basic slag. Typical blast furnace metal will contain about 20 atom per cent carbon, 4 atom per cent silicon and one atom per cent manganese as the principal impurities. Depending on the origin of the iron ore which was charged to the blast furnace, the phosphorus content varies considerably, but it may reach as high as 3 atom per cent in metal which is made from high-phosphate ore bodies.

One major disadvantage of the Bessemer process is that nitrogen is slightly soluble in liquid iron, and as the air contains principally nitrogen,

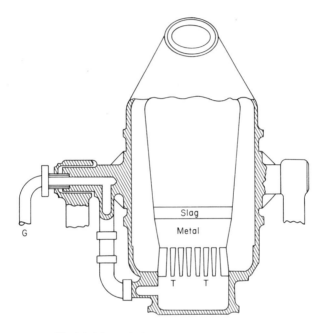

Fig. 96. Schematic drawing of the Bessemer converter.

G is gas port.
T are the gas tuyères.

the final iron product must contain this element. Nitrogen has a deleterious effect on the metal in the solid state because iron nitride Fe_4N which is precipitated on cooling embrittles the Bessemer iron giving a loss in ductility on ageing.

Carbon and silicon both raise the activity coefficient of nitrogen in solution in liquid iron and hence it is to be expected that the nitrogen dissolution will be at a minimum during the early stages of the blow when the carbon and silicon contents of the metal have their highest values. Towards the end of the blow, the nitrogen content can rise towards the equilibrium value for pure iron as solvent. This can be obtained from the two-term equation for the free energy of solution of nitrogen in pure liquid iron which was given earlier.

$$\tfrac{1}{2}N_2 \rightarrow [N] \qquad \Delta G^\circ = 860 + 3 \cdot 01\,T \text{ cal}$$

Here the standard state for nitrogen dissolved in liquid iron is one atomic per cent. The calculated solubility for one atmosphere of nitrogen is therefore 0·174 atom per cent at 1600°C.

The interaction parameters for carbon, silicon and manganese with nitrogen are approximately 6·5, 7·0 and −5·0 respectively, but the effect of carbon predominates over those of the other impurities by virtue of the higher concentration of carbon in the liquid metal at the start of the blow.

The sequence of removal of the impurities is that silicon and manganese are eliminated first, that is when the metal temperature is rising from the initial value of about 1300°C towards the final value of 1600°C. A glance at the Ellingham diagram for oxides shows that manganese oxide and silica have greater stability than CO at the lower temperature limit, but that the three oxides have roughly equal free energies of formation at 1600°C. The early oxidation of silicon and manganese is of the greatest significance to the thermal balance of the process. The heats of formation of SiO_2 and MnO are much greater than that of CO, but these effects must be considered together with the partial heat of solution of each element in liquid iron to yield the final thermal value during the oxidation process. Thus

$$Si + O_2 \rightarrow SiO_2 \qquad \Delta H^\circ = -217 \text{ kcal mol}^{-1}$$

$$Si \rightarrow [Si] \qquad \Delta \bar{H} = -28 \text{ kcal} \, [X_{Si} = 0 \cdot 01]$$

$$[Si] + O_2 \rightarrow SiO_2 \qquad \Delta H = -189 \text{ kcal mole}^{-1}$$

$$2[Mn] + O_2 \rightarrow 2MnO \qquad \Delta H^\circ = -184 \text{ kcal mol}^{-1}$$
(Mn forms a Raoultian solution in iron)

$$2C + O_2 \rightarrow 2CO \qquad \Delta H^\circ = -53 \text{ kcal}$$

$$2C \rightarrow 2[C] \qquad \Delta \bar{H} = +13 \text{ kcal} \, [X_C = 0 \cdot 01]$$

$$2[C] + O_2 \rightarrow 2CO \qquad \Delta H = -66 \text{ kcal}$$

It can be seen from these data that carbon has considerably less value in generating heat in the converter than do silicon and manganese. The thermal requirements to raise one mole of air from room temperature to 1600°C and one mole of iron from 1300°C to 1600°C are approximately 11 000 cal and 2 000 cal respectively. (the average heat capacities 8 cal mol^{-1} deg^{-1} and 7 cal mol^{-1} deg^{-1} can be used for this calculation). Hence the silicon and manganese contents of the metal must provide at least this quantity of heat during the rise in temperature of the system. Taking the calculated values for the heats of oxidation of these metals in solution in liquid iron, it can be seen that their thermal values as fuel elements are about the same for each mole of oxygen consumed but obviously much more *weight* of manganese is consumed than silicon to achieve this effect. The atomic weights of these elements are 54·93 and 28·06 respectively. For each mole of gas which is heated up to the final temperature, about $1/20$ g atom of silicon is consumed, and this represents the oxidation of the silicon content of about one mole of iron if the initial silicon content of the blast furnace metal has the typical value of 2 wt per cent.

THE BASIC OPEN HEARTH AND THE LINZ-DONAWITZ PROCESS

The basic open-hearth process which replaced the Bessemer process as the major source of liquid steel is being largely superseded today by the basic oxygen LD process. In the former, the metal was heated in a shallow hearth by an oil flame and the composition of the hot gases was sufficiently oxidizing to sustain the removal of carbon and other impurities. The evolution of carbon as carbon monoxide from the liquid metal which was covered with liquid slag caused enough slag–metal agitation to ensure a reasonably rapid approach to equilibrium. A further source of agitation was the evolution of CO_2 due to the decomposition of $CaCO_3$ which was added to make the basic slag. Furnaces containing hundreds of tons of liquid could be refined in 10–20 hours.

In the LD process, oxygen is injected from the top of the liquid iron as a high velocity jet, and this causes much more agitation than is found in the BOH furnace. Although the refining speeds are markedly greater in the top-blowing process than in the older process, one hundred tons of metal may be refined in 20 mins. the chemical aspects of the two processes are largely the same (Figs. 97a and 97b).

The change in the mean oxygen potential from that in the blast furnace to that which exists in the open hearth alters the nature of the equilibrium slag phase. The iron oxide content is now increased, and the slag–metal contact is enhanced during refining by the effect of the bursting gas bubbles of carbon monoxide which eject liquid metallic droplets into the gas phase. These are refined during their transit through the gas and slag on the way back to the metal phase. The production of CO bubbles in the metallic phase is now generally agreed to begin at crevices in the hearth of the furnace. The homogenous nucleation of bubbles in a liquid medium is very rare, and it is believed that air which is trapped in the uneven base of the furnace

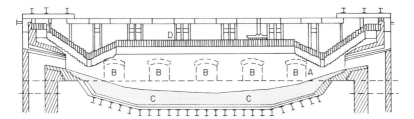

Fig. 97a. Schematic drawing of the Basic open-hearth Furnace.

A = Flame ports
B = Charging doors
C = Furnace hearth
D = Suspended roof

during charging acts as the origin of small bubbles during the refining period which then act during their passage through the metal as the heterogeneous nuclei of CO bubbles. Model studies of bubbles rising through liquid mercury which is covered with a layer of glycerine, to simulate the slag phase, suggests that bubbles over 1 cm in diameter are far from spherical in shape, and that the best description of the general shape is that of the cap of a sphere. This is a volume which is cut off by a plane intersection of the sphere somewhere between the centre and the tangent plane (Fig. 98).

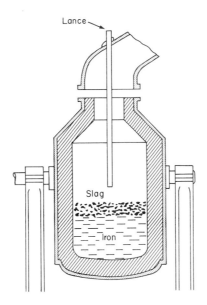

Fig. 97b. Schematic diagram of a top-blown Basic oxygen furnace.

Smaller bubbles than 1 cm diameter, are approximately spherical and the motion of these bubbles through the liquid metal may be calculated by the application of Stokes Law. For the spherical cap bubbles, a terminal velocity of about 20–30 cm s^{-1} is reached by bubbles which typically have a volume equivalent to that of a sphere of 1–5 cm diameter.

Experimental studies of the transfer of oxygen from solution in liquid silver to argon bubbles which were spherical cap in shape and of which the mean volume was 10 cm^3 gave a value of 0·042 cm s^{-1} for the mass transfer coefficient. This is about the magnitude which is to be expected for transfer from a liquid, $D \cong 5 \times 10^{-5}$ cm s^{-1}, across a boundary layer, $\delta \cong 10^{-3}$ cm to a well-stirred gas bubble.

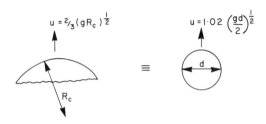

Fig. 98. Comparison of the spherical cap bubble of radius, R_c, and the equivalent volume spherical bubble having the same rising velocity, u.

The theoretical calculation of mass transfer coefficients between spherical cap bubbles and liquids has reached the state of advancement which can give a fundamental account in place of the simplified mass transfer approximation used above. Calderbank (1968) has derived a relationship for this situation between the dimensionless quantities used in mass transfer theory

$$N_{Sh} = 1 \cdot 28 \, (N_{Re} N_{Sc})^{1/2}$$

where

N_{Sh} = Sherwood number = kL/D

N_{Re} = Reynolds number = $Lu\rho/v$

N_{Sc} = Schmidt number = $v/\rho D$

L is a characteristic length of the bubble

The velocity, u, in the Reynold's number is the velocity derived by Davies and Taylor (1950).

$$u = 1 \cdot 02 \left(\frac{gd}{2} \right)^{\frac{1}{2}}$$

where g is the gravitational constant and d is the diameter of the spherical bubble of equivalent volume. Combining this equation with Calderbank's equality, and using d as the characteristic length L in the dimensionless numbers, we obtain for the mass transfer coefficient k, the relationship

$$k = 1 \cdot 08 \, g^{\frac{1}{4}} D^{\frac{1}{2}} d^{-\frac{1}{4}} \, \text{cm s}^{-1}$$

For bubbles of spherical cap shape, the value of d which must be used is the diameter of the spherical bubble of equal volume. For the experiment quoted above, where the mass transfer of oxygen to bubbles in liquid silver was studied, the equivalent sphere would be of diameter $(6/\pi \times 10)^{\frac{1}{3}}$, and the equation of Calderbank predicts a value of the mass transfer coefficient

$$k = 1 \cdot 08 \times (980)^{\frac{1}{4}} \times (10^{-4})^{\frac{1}{2}} \times \left(\frac{1}{2 \cdot 67}\right)^{\frac{1}{4}} \text{cm s}^{-1}$$

$$= 4 \cdot 73 \times 10^{-2} \, \text{cm s}^{-1} \text{ at } 1000°C.$$

The diffusion coefficient of oxygen in liquid silver which is needed for this study has been measured by Masson and Whiteway (1967) who used an electrochemical technique employing a solid CaO–ZrO_2 electrolyte, and by Mizikar, Grace and Parlee (1963) who used a conventional gas–liquid diffusion couple. The results of these two investigations are given in the equations

$$D = 5 \cdot 15 \times 10^{-3} \exp{(-9900 \pm 2600)}/RT \, \text{cm}^2 \, \text{s}^{-1}$$

and

$$D = (14 \cdot 7 \pm 1) \times 10^{-4} \exp{(-7100 \pm 3200)}/RT \, \text{cm}^2 \, \text{s}^{-1}$$

respectively. The equations, despite the differences between the preexponentials and the activation energies, yield closely similar values for the diffusion coefficient at 1000°C of $1 \cdot 05 \times 10^{-4}$ and $9 \cdot 04 \times 10^{-5} \, \text{cm}^2 \, \text{s}^{-1}$ using the mean values for the activation energies.

The use of a jet, which is placed above the liquid metal as a means of oxidation in the LD process, provides a very much more effective means for oxidation of carbon in steelmaking than that attained in the open hearth furnace. Furthermore, it is possible to use pure oxygen rather than air, and thus the significant thermal load of nitrogen which is carried in the Bessemer converter is eliminated. It will be remembered that heating nitrogen to the steelmaking temperature was by far the largest thermal requirement for the Bessemer process. It has been shown experimentally that the rate of decarburization of an industrial furnace charge of 140 tons by the LD process is slower during the first 10 mins of the refining period than during the second 10 minutes. This would be expected by comparison with the Bessemer converter where it was found that silicon and manganese oxidation precede the elimination of carbon from the melt.

The stirring rate which is achieved during the LD refining of iron is such as to lead to the formation of an emulsion between the metal and slag phases. The exchange of elements between metal and slag is thus very rapid, and the rate of elimination of all of the impurities is so rapid as to suggest complete utilization of the oxygen which can be supplied via the jet. The refining speed is therefore only determined by the rate of supply of oxygen.

THE CARBON-OXYGEN REACTION

The thermodynamics of the reaction between carbon and oxygen has been extensively studied at steelmaking temperatures, the experiments being carried out under atmospheres containing CO at around one atmosphere partial pressure. Carbon and oxygen have a mutual effect on the respective activity coefficients when present together and the value

$$\varepsilon = \frac{\partial \ln \gamma_C}{\partial X_O} \quad \frac{\partial \ln \gamma_O}{\partial X_C} = -20$$

represents an average value of the experimental results for the effect at 1600°C.

The activity coefficient of carbon alone in solution in liquid iron is also a function of carbon content, and this effect is given by the value 10 for ε_C^C. It follows that the simple mass action constant for the formation of CO from carbon and oxygen dissolved in iron

$$K = \frac{p_{CO}}{X_C X_O}$$

cannot be given a single value even for one temperature. The effect of carbon concentration is shown in the accompanying figure which was obtained by applying the effect of ε_C^C to the very dilute solution equilibrium constant where the activity coefficients of carbon and oxygen are reasonably constant. It should be remarked that even the oxygen saturation concentration in pure iron (0·79 atom per cent) is considerably lower than the carbon contents (the saturation limit is 21 atom per cent at 1600°C) which are normally encountered in steelmaking during the refining period, and so the major effect to be considered here is ε_C^C and hence the variation of the carbon activity coefficient. Figure 99 shows the variation of the activity coefficient of carbon in solution in iron at 1600°C as a function of composition with the infinitely dilute solution as reference state. Figure 100 shows the variation of the mass action constant which is obtained at one atmosphere pressure of $CO + CO_2$.

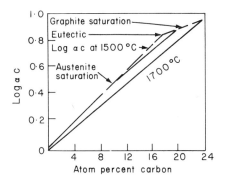

Fig. 99. The variation of the activity coefficient of carbon with carbon content in solution in liquid iron.

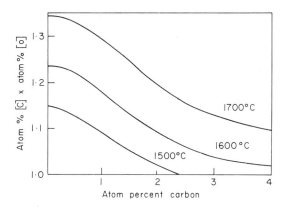

Fig. 100. The effect of increasing carbon content on the [C] [O] product in liquid iron ($p_{CO} + p_{CO_2} = 1$ atmos.)

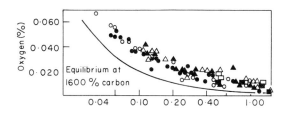

Fig. 101. Results of works trials showing closeness of approach to equilibrium in open hearth refining.

L

It is of considerable interest to discover how closely the refining process operates to conditions of thermodynamic equilibrium. It is found that the carbon and oxygen contents which are measured simultaneously during refining in the basic open hearth furnace are surprisingly close to the values which are at equilibrium with one atmosphere pressure of CO (Fig. 101). It is reasonable to assume, therefore, that calculations which involve the assumption of local thermodynamic equilibrium during steel refining will give a good approximation to practice.

Other refining reactions apart from carbon removal which are carried out in these processes may be written:

(a) $[Si] + 2[O] \rightarrow \{SiO_2\}$ [ ] = metal phase

(b) $2[P] + 5[O] \rightarrow \{P_2O_5\}$ { } = slag phase

(c) $[S] + \{O\} \rightarrow \{S\} + [O]$

It will be observed that in the first two reactions the transfer of an oxide to the slag phase requires both the impurity and oxygen atoms to be transported. This is because of the requirement of a sink for electrons when the impurity atom is ionized. The oxygen atoms are converted to oxide ions by this electron transfer, and the impurity atoms provide these. The removal of sulphur is a charge-exchange process as has been explained earlier, and the oxygen content of the metal tends to control the ease of transfer of sulphur from metal to slag.

THE MODELS FOR PHOSPHORUS REMOVAL

In all of the processes for refining in steelmaking, it can be seen that the oxygen content of the metal has an important effect. The determination of the thermodynamics of the dilute solution of oxygen in liquid iron has understandably played an important part in the physico-chemical study of steelmaking. The second aspect of the elimination of impurities, other than carbon, which is clearly very important is the thermodynamics of the oxides which are formed during refining and their interactions with the slag phase. In the light of the previous discussion, it is obvious that the silica activity can be lowered by working with a lime-rich slag, and the elimination of phosphorus is also aided by a high lime activity. The formal stoichiometric equation for the elimination of phosphorus may be written

$$4CaO + 2[P] + 5[O] \rightarrow \{Ca_4 P_2O_9\}$$

It is clear that the molecule of calcium phosphate does not exist as a distinct entity in the slag phase, but the real ionic description probably involves an anion or even a number of anionic phosphorus–containing species in the molten salt phase. It is interesting to discuss some of the attempts which have been made to "explain" the thermodynamics of phosphorus removal, since these typify the development of the physico-chemical description of slag systems.

The original way of rationalizing equilibria in chemical systems was by means of the mass action law. In order to understand the mechanism of a reaction, it was necessary to find the molecular species whose concentrations could be used in formulating the mass action constant. For this particular reaction, the most successful expression obtained by Winkler and Chipman (1946) involved the calcium tetra-phosphate $Ca_4P_2O_9$, and the mass action constant was obtained from the reaction

$$4\{CaO\} + 2[P] + 5[O] \rightarrow \{Ca_4P_2O_9\}$$

$$K = \frac{c\{Ca_4P_2O_9\}}{c^4\{CaO\}.c^2[P].c^5[O]}$$

Clearly in order to use this expression, it was necessary to obtain the concentration of the tetra-phosphate and that of the "free" lime which was available in the slag. This was done by means of a quasi "acid-base" calculation, in which oxides such as FeO, CaO, Na_2O, MgO would be "neutralized" by the requisite quantity of "acidic" oxide SiO_2, Al_2O_3, P_2O_5, etc. It thus became fashionable to speak of "acid slags" which had an excess of these latter oxides, or "basic slags" having an excess of metal oxide. The language was transplanted into the field of slag chemistry from the then highly-developed field of aqueous chemistry. By choosing suitable species, and points of equivalence between the various oxides (compositions of inter-oxide compounds), it was frequently possible to obtain a satisfactory mass action constant, such as that for the tetra-phosphate.

The alternative formulation which follows the ionic theory developed by Flood and Grjotheim (1952) takes no account of such "compound" formation, but takes the number of ionic structural units as a basis for calculation. These are the metal cations, oxide anions and the tetrahedral silicate and phosphate ions. Following the cyclic procedure which was outlined earlier for the evaluation of complex equilibria, we should write for the reaction

$$2[P] + 5[O] + 3\{O^{2-}\} \rightarrow 2\{PO_4^{3-}\}$$

$$\log K = \sum_i x_i \log K_i \qquad i = Ca^{++}, Fe^{++}, \text{etc.}$$

(see p. 211).

Data are therefore required for the reactions involving the cations of the slag such as

$$2[P] + 5[O] + 3\,CaO \rightarrow Ca_3\,(PO_4)_2$$

$$2[P] + 5[O] + 3\,FeO \rightarrow Fe_3\,(PO_4)_2 \text{ etc.}$$

Fellner and Krohn (1969) in a recent analysis of results for binary iron + calcium slags conclude that the terms in the Flood equation have the values

$$\log K_{Ca}^{2+} = 21 \pm 1 \text{ and } \log K_{Fe}^{2+} = 11 \pm 1$$

Because the relationship involves log K for the complex system, it follows that when the lime concentration in the slag is high, the term $x_{Ca^{2+}} \log K_{Ca^{2+}}$ is the predominant term and, in fact, the only one which need be considered.

The comparison between the two approaches to this particular slag equilibrium, the molecular and the ionic theories therefore rests on the data for the reactions

$$4\{CaO\} + 2[P] + 5[O] \rightarrow Ca_4P_2O_9$$

and

$$3\{CaO\} + 2[P] + 5[O] \rightarrow Ca_3P_2O_8$$

In the first equation only the 'free" lime content is to be used in the calculation, and in the second equation the equivalent fraction of the calcium ions is to be used with the cations of the solution only being taken into consideration.

Bookey (1952) has shown that the difference in standard free energy change between these two reactions is about 10 kcal per mol of compound. This represents a difference probably in log K of a little over one unit at 1600°C, and this difference is within the accessible experimental precision.

$$3CaO + P_2(g) + \tfrac{5}{2}O_2 \rightarrow Ca_3P_2O_8$$
$$\Delta G° = -537\,000 + 129.4T\,\text{cal mol}^{-1}$$

$$4CaO + P_2(g) + \tfrac{5}{2}O_2 \rightarrow Ca_4P_2O_9$$
$$\Delta G° = -541\,400 + 130.7T\,\text{cal mol}^{-1}$$

The principal difference between the two approaches lies, finally, in the functional use which is made of the lime content of the slag. For a typical slag containing 21 wt % CaO, 21 % SiO_2, 39 % FeO, 10 % MgO, 1.7 % P_2O_5 and 2.5 % MnO the mole fraction of "free" lime was found to be 0.01 by the use of the molecular approach, and the mole fraction of $Ca_4P_2O_9$ was calculated to be 0.014. The equilibrium constant at 1600°C for the molecular "reaction" is 4×10^9, and the temperature dependence is given by the equation

$$\log K = \frac{71\,700}{T} - 28.7$$

Turning now to the ionic model, it is necessary to obtain a value for $n_{O^{2-}}$ from a consideration of the distribution of the oxygen atoms amongst the various anions, SiO_4^{4-}, PO_4^{3-}, etc., before the value of the equilibrium constant can be obtained. It is then found that log K is a linear function of the calcium content of the slag, as required by the theory (Fig. 102) but the

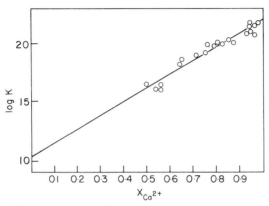

Fig. 102. The equilibrium constant for the equilibrium $2\underline{P} + 5\underline{O} + 3$ (Ca, Fe)O = (Ca, Fe)$_3$(PO$_4$)$_2$ (on a logarithmic scale) versus composition of the cation mixture. From the results of H. Knüppel and F. Oeters. *Stahl und Eisen* **81**, 1437 (1961).

———— = Calculated
O   O = Measured.

numerical value of the equilibrium constant must be checked against the ancillary data for the pure calcium phosphate-forming reaction before the theory can be said to be satisfactory. The value which is estimated from the slag data is about 10^{21} as previously stated, and the calculated value can be obtained in the following way. At 1600°C, using Bookey's data for the orthophosphate and data from Elliott, Gleiser and Ramakrishna (1963) for the solution of phosphorus and oxygen in liquid iron we can calculate the required constant.

$$3CaO + P_{2(g)} + \tfrac{5}{2}O_2 \rightarrow Ca_3(PO_4)_2$$

$$\Delta G^{\circ}_{1873} = -294600 \text{ cal}$$

$$P_{2(g)} \rightarrow 2[P]; \quad \Delta G^{\circ} = -58400 - 9.2\, T\text{cal}$$

$$\Delta G^{\circ}_{1873} = -75600 \text{ cal}$$

$$\tfrac{5}{2}O_2 \rightarrow 5[O]; \quad \Delta G^{\circ} = -140000 - 3.5\, T\text{cal}$$

$$\Delta G^{\circ}_{1873} = -146500 \text{ cal}$$

(The standard states for phosphorus and oxygen are here 1 wt%.)

For the reaction

$$3\,CaO + 2[P] + 5[O] \rightarrow Ca_3\,(PO_4)_2$$

$$\Delta G^\circ = -339\,600 + 142{\cdot}1\,T\,cal$$

$$\Delta G^\circ_{1873} = -72\,850\,cal$$

hence $\log K_{1873} = 8{\cdot}46$

Bookey, Richardson and Welch (1952) studied the equilibrium

$$4CaO + 2[P] + 5[O] \rightarrow Ca_4P_2O_9$$

for which they obtained the free energy change

$$\Delta G^\circ = -349\,100 + 145{\cdot}4\,T\,cal$$

$$\Delta G^\circ_{1873} = -76\,760\,cal \quad \log K_{1873} = 8{\cdot}96$$

This, in the light of the statement concerning the approximate equality of the stability of the tri- and tetraphosphate above, supports the calculation from the separate solution thermodynamics which is made above. Fellner and Krohn used the atom fractions for the phosphorus and oxygen contents of the metal phase in their calculation, and when the conversion of the standard state is made for the present calculations from 1 wt%, the agreement is satisfactory, although not exact.

Thus

$$1\,wt\%\,[P] = atom\,fraction\,0{\cdot}0178$$

$$1\,wt\%\,[O] = atom\,fraction\,0{\cdot}034$$

Hence K above should be multiplied by

$$\frac{1}{(0{\cdot}0178)^2(0{\cdot}034)^5} = \frac{1}{1{\cdot}42 \times 10^{-11}} = 7{\cdot}00 \times 10^{10}$$

Thus $\log K_{1873}$ should be increased to the value $19{\cdot}30$ for the formation of $Ca_3(PO_4)_2$.

DE-OXIDATION PRACTICE

The metal which is produced in any of the refining processes for steelmaking contains an important amount of oxygen after the refining step is completed. The reaction between carbon and oxygen can proceed in the subsequent casting stages unless steps are taken to reduce the oxygen content of the steel. A certain amount of "rimming action", the evolution of CO when the metal is cooling down in the mould, can be allowed when very low-carbon steel is required, but for most purposes, the steel is "killed" by the addition of a deoxidizing agent, usually aluminium. The deoxidation reaction corresponds to the formation of alumina by the homogeneous reaction between aluminium and oxygen atoms in solution in liquid iron

$$2[Al] + 3[O] = Al_2O_3$$

The alumina which is produced must float up through the liquid because of the large density difference, but because the addition is made in the ladle when the metal is cooling, it must frequently occur that fine particles of alumina are left dispersed in the billet.

The equilibrium constant for the deoxidation reaction may be written

$$K = \frac{a_{Al_2O_3}}{X_{[Al]}^2 \cdot X_{[O]}^3} \frac{1}{\gamma_{[Al]}^2 \gamma_{[O]}^3}$$

This shows that the final equilibrium oxygen content of the metal $X_{[O]}$ is determined not only by the stability of the oxide of the deoxidizer but also by the effect of the deoxidizing element on the activity coefficient of oxygen in solution in iron. The general trend is that elements which form oxides which are more stable than ferrous oxide, *lower* the activity coefficient of oxygen, whilst those such as cobalt and nickel raise the activity coefficient of oxygen, Silicon and manganese either singly or in combination find considerable use as commercial deoxidizing agents. The oxides of these elements have higher melting points than the temperatures which are normally achieved in steel-making (SiO_2, m.p. 1723°C; MnO, m.p. 1875°C) and thus the products of deoxidation by the elements are solid. When an alloy of silicon and manganese is used, a liquid manganese silicate can be formed, and this is found to yield more satisfactory products than is the case when the solid oxides are formed. This is thought to arise from the ease of agglomeration of a liquid product which is in favour of silicomanganese deoxidation practice, and this alloy system finds considerable practical application. The respective values for the interaction coefficients are:

$$\varepsilon_O^{Al} = -1300$$
$$\varepsilon_O^{Si} = -101$$
$$\varepsilon_O^{Mn} = 1$$

according to the experimental studies, and this sequence follows the decreasing stabilities of the oxides $Al_2O_3 > SiO_2 > MnO$. The deoxidizing elements interact with iron to differing extents, manganese forming Raoultian alloys, whereas Al and Si both interact strongly with iron. The partial heat of solution at infinite dilution of aluminium in iron is $-12,500$ cal and that for silicon is $-29\,000$ cal. The oxygen potentials of the sub-systems $[Al]-O_2-Al_2O_3$ and $[Si]-O_2-SiO_2$ are given by the equations:

$$\tfrac{2}{3}\Delta G^{\circ}_{Al_2O_3} = \tfrac{4}{3}\Delta\mu_{[Al]} - \tfrac{2}{3}\Delta\mu_{Al_2O_3} + \Delta\mu_{O_2}$$

and

$$\Delta G^{\circ}_{SiO_2} = \Delta\mu_{[Si]} - \Delta\mu_{SiO_2} + \Delta\mu_{O_2}$$

$$\tfrac{2}{3}\Delta G^{\circ}_{Al_2O_3}(1600°C) = -170\,\text{kcal} \quad \Delta G^{\circ}_{SiO_2}(1600°C) = -126\,\text{kcal}.$$

We may equate

$$\Delta\mu_{[Al]} = -12\,500 + 4\cdot575T \log X_{Al}$$

and

$$\Delta\mu_{[Si]} = -29\,000 + 4\cdot575T \log X_{Si}$$

and

$$\Delta\mu_{Al_2O_3} = \Delta\mu_{SiO_2} = 0$$

for pure oxide formation to obtain the following equations:
for Al deoxidation

$$-170\,000 = -16\,700 + 6\cdot10T \log X_{Al} + \Delta\mu_{O_2}$$

$$\log pO_2 = -\frac{33\,500}{T} - 1\cdot33 \log X_{[Al]}$$

and for Si deoxidation

$$-126\,000 = -29\,000 + 4\cdot57T \log X_{Si} + \Delta\mu_{O_2}$$

$$\log p_{O_2} = -\frac{21\,200}{T} - \log X_{[Si]}$$

Thus, when there is a mole fraction of aluminium or of silicon of 10^{-3} (0·1 atom per cent) in the deoxidized steel, the oxygen pressure finally reaches the value $1\cdot3 \times 10^{-14}$ atm and 5×10^{-8} atm respectively. The corresponding pressure for pure aluminium $-Al_2O_3$ is 10^{-20} atm and for pure silicon–silica it is 10^{-15} atm. Thus, the effects of the iron–aluminium and iron–silicon interactions are quite large in the overall effectiveness of the deoxidation reaction.

18

VACUUM REFINING OF STEEL

A method for the removal of small amounts of volatile impurities is to place the metal in a vacuum chamber where the evolution of volatile elements and gaseous species such as nitrogen and hydrogen and of carbon monoxide can occur. This process has advantages as a deoxidizing procedure, via the evolution of CO, over the chemical methods described earlier in that no solid product of deoxidation is formed which can remain in suspension in the melt after deoxidation. The process therefore promises to produce "cleaner" steel than that produced by the chemical method. Industrial vacuum systems which can reduce the total pressure above the contents of a ladle to pressures less than 10^{-3} atm and even down to 10^{-6} atm are now available and vacuum degassing is having considerable use at present. Under such a relatively high vacuum, the rate of evaporation of an element A is given by the Langmuir equation

$$m_A = 44.32 S_A p_A^0 \sqrt{\frac{M_A}{T}} \text{ g cm}^{-2} \text{ s}^{-1}$$

where S_A is the vaporization coefficient, p_A^0 is the vapour pressure of the pure element in atmospheres, M_A and T are the molecular weight in grams of the evaporating species and the absolute temperature respectively. As a good approximation the vaporization coefficient can always be assumed to be unity, and the evaporating species for metals are in most cases the atoms. For the common gases dissolved in a liquid metal, the molecules H_2, N_2, O_2 are the predominant vapour species but sulphur evaporates principally either as S_2 or monatomic S depending on the temperature and the partial pressure of sulphur which is exerted by the solution.

When an element is in solution, then the vapour pressure p_A which it exerts is related to the vapour pressure of the pure component, p_A^0, by the equation

$$p_A = \gamma_A X_A p_A^0$$

γ and X being the activity coefficient and mole fraction respectively. The calculation of the rate of evaporation from a solution is more complicated than that for the rate of evaporation of a pure substance because the vaporizing species must be replenished by diffusion from the bulk of the solution. There are obviously two possible rate-determining steps in evaporation of a component from a solution. If the transfer across the boundary layer is rapid, then local equilibrium exists between the surface layer and the bulk of the solution. The rate-determining step in the evaporation from a solution is then the surface process and the rate is given by the free evaporation equation

$$m_A = 44 \cdot 32 S_A \gamma_A X_A p_A^0 \sqrt{\frac{M_A}{T}} \ \text{g cm}^{-2} \ \text{s}^{-1}.$$

If the transfer across the boundary layer is rate-determining, then we have used the general mass transfer equation

$$J = k\{C_A^{\text{bulk}} - C_A^{\text{surface}}\} \ \text{g cm}^{-2} \ \text{s}^{-1}$$

to calculate the transfer rate.

The surface concentration at which these two processes are of equal magnitude is given by simultaneous solution with $X_A = X_A'$ on the surface. Then

$$J = k\{C_A^{\text{bulk}} - C_A^{\text{surface}}\} = 44 \cdot 32 S_A \gamma_A X_A' \, p_A^0 \sqrt{\frac{M_A}{T}}$$

The concentration term on the left-hand side is in the usual units, g cm^{-3}, and k is in units of cm s^{-1}. The atom fraction on the right-hand side X_A' corresponds to the concentration C_A^{surface} on the left-hand side, and hence for a dilute solution

$$C_A^{\text{surface}} = \frac{\rho_B}{M_B} \cdot M_A X_A' \ \text{g cm}^{-3}$$

where element B is the solvent of density ρ_B. Taking 5×10^{-5} cm^2 s^{-1} as a typical liquid diffusion coefficient and the boundary layer thickness for a well-stirred liquid of 10^{-3} cm yields a value for the mass transfer coefficient equal to 5×10^{-2} cm s^{-1} for an inductively-heated sample. When $X_A' < X_A$, then in the steady state

$$k \frac{\rho_B}{M_B} \cdot M_A(X_A - X_A') = 44 \cdot 32 S_A \gamma_A p_A^0 \sqrt{\frac{M_A}{T}} \cdot X_A'$$

and therefore

$$\left(\frac{X_A}{X'_A} - 1\right) = \frac{44 \cdot 32 M_B}{k\rho_B} \frac{S_A \gamma_A p^0_A}{(M_A T)^{\frac{1}{2}}}$$

If we now consider the vaporization of manganese from a dilute solution in iron, the activity coefficient is close to unity since Fe–Mn alloys are practically Raoultian. The vapour pressure of pure manganese at 1600°C is $4 \cdot 57 \times 10^{-2}$ atm, and ρ_{Fe}/M_{Fe} equals $0 \cdot 124$ g atom cm^{-3}.

$$\left(\frac{X_A}{X'_A} - 1\right) = \frac{44 \cdot 32}{5 \times 10^{-2}} \times 8 \cdot 064 \times \frac{4 \cdot 57 \times 10^{-2}}{(54 \cdot 94 \times 1873)^{\frac{1}{2}}} = 1 \cdot 02$$

Therefore, we may conclude that manganese depletion at the surface will occur to some extent during evaporation of the element from solution in iron at 1600°C. This has been verified experimentally by Ward (1963) who found an evaporation rate approximately one half of the calculated value, assuming that the surface concentration was the same as the bulk concentration.

This depletion equation can be used for all dilute solutes in liquid iron at 1600°C in the form

$$\left(\frac{X_A}{X'_A} - 1\right) = 1 \cdot 65 \times 10^2 \frac{S_A p^0_A \gamma_A}{(M_A)^{\frac{1}{2}}}$$

where p^0 is in atmospheres and M_A is in grams.

We can now solve for X'_A in a specific situation, and the condition for preferential removal of the solute to iron in the vacuum system, which constitutes refining, is given by the expression

$$44 \cdot 32 p^0_{Fe} \left(\frac{M_{Fe}}{T}\right)^{\frac{1}{2}} \leqslant k \frac{\rho_{Fe}}{M_{Fe}} \cdot M_A (X_A - X'_A).$$

since the activity of the iron solvent is virtually unity for a dilute solution of an impurity. This applies only when the loss of the solute is determined by transport across the boundary layer. When this is not so, and the surface constraint applies, the condition for preferential loss of the solute simplifies to the expression

$$p^0_{Fe} M^{\frac{1}{2}}_{Fe} \leqslant S_A p^0_A \gamma^0_A X_A M^{\frac{1}{2}}_A$$

To cite further examples of simple vaporization from dilute solution in liquid iron, we will consider dilute solutions of Cu, Si and Al. The vapour pressures of pure liquid copper, liquid silicon and liquid aluminium are $1 \cdot 14 \times 10^{-2}$, $8 \cdot 63 \times 10^{-6}$ and $1 \cdot 51 \times 10^{-3}$ atm respectively, and the activity coefficients for these dilute solutes are

$$\gamma_{[Cu]} = 8 \cdot 0 \qquad \gamma_{[Si]} = 7 \times 10^{-3} \qquad \gamma_{[Al]} = 3 \times 10^{-2}$$

We therefore have two elements of relatively high and similar vapour pressures but widely differing activity coefficients, Cu and Al, and two elements of approximately equal activity coefficients but widely differing vapour pressures, Si and Al. The right-hand side of the depletion equation for these three elements has the values 1.89, 1.88×10^{-8} and 1.44×10^{-3}. Hence, we may conclude from the equation that there will be some depletion of copper, but that the depletion is insignificant for silicon and aluminium evaporation.

THE ELIMINATION OF GASEOUS ELEMENTS

When a simple gas is removed from liquid iron by vacuum degassing, the solution equilibrium constant must be used in calculating the rate of loss. The appropriate equation can be obtained by combination of Sievert's law for the solution reaction

$$A_2 \rightarrow 2[A] \qquad K = \frac{X_{[A]}^2}{p_{A_2}}$$

with the mass transfer equation. Furthermore, unless all of the gas which escapes from the metal is pumped away efficiently, a back pressure of the gas will build up in time in the vacuum system. This is different from the situation which arises when metal atoms evaporate because these can condense on the cool walls of the vacuum chamber. It is therefore to be expected that the rate of removal of gaseous elements in practical systems will be less than that which would be calculated by assuming free evaporation.

A further complication of this process is that the desorption of a diatomic gas from the surface of a liquid can only occur after recombination of the atoms on the surface has taken place. Most gaseous species such as O_2, H_2, N_2, etc., obey Sievert's law, and are therefore in the atomic state in solution. The kinetics of the recombination step is usually not important for binary systems, but the rate is affected by the presence of foreign atoms which are strongly adsorbed to the surface. Oxygen and sulphur both lower the surface tension of liquid iron markedly (Fig. 103) and thus must be strongly sorbed. It is found that these elements slow down the degassing of other elements such as nitrogen from solution in iron, presumably because of the reduced availability of surface sites to the nitrogen atoms which must transverse the surface before recombining.

The maximum rate of evolution in the absence of recombination control can be calculated by means of the usual simple mass transfer expression if we assume that the surface of the melt is in equilibrium with the gas phase, and that diffusion in the gaseous state is not rate-controlling Then

$$J_X = k\{C_A^{bulk} - C_A^{surface}\}$$
$$= k\{Kp_{A_2}^{\frac{1}{2}} - Kp_{A_2}'^{\frac{1}{2}}\}$$

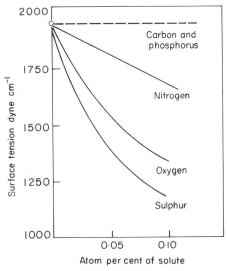

Fig. 103. The effect of dilute solutes on the surface tension of liquid iron at 1600°C. (P. Kozakevitch and G. Urbain *Mém. Sci. Rev. Mét.* **60**, 143 (1963).)

Here K is the Sievert's law constant, p_{A_2} is the pressure of A_2 under which the
material had been equilibrated before the application of the vacuum and
p'_{A_2} is the pressure of A_2 in the vacuum chamber. If this back-pressure is
small, then

$$J = kKp_{X_2}^{\frac{1}{2}}$$

The mass transfer coefficient, k, will usually be higher for dissolved gases
than for dissolved metals because of the higher diffusion coefficients which are
involved (Table XXXVI).

Experimental studies by Pehlke and Elliott (1963) of the adsorption and desorption of nitrogen by induction-stirred samples of liquid iron are consistent with the mass transfer control expression at any given temperature, but

TABLE XXXVI. Diffusion coefficients of solutes in liquid iron at 1600°C

Element	Diffusion coefficient ($cm^2 \ s^{-1}$)
Hydrogen	$13{\cdot}0 \times 10^{-4}$
Carbon	$0{\cdot}7 \times 10^{-4}$
Nitrogen	$1{\cdot}1 \times 10^{-4}$
Oxygen	$7{\cdot}0 \times 10^{-5}$
Manganese	$5{\cdot}5 \times 10^{-5}$
Cobalt	$6{\cdot}0 \times 10^{-5}$
Silicon	$3{\cdot}0 \times 10^{-5}$

the temperature coefficients appear to be too high for control by diffusion through a boundary layer. Whereas an apparent activation energy of only a few kilocalories would be expected for diffusion control, much higher values between 25 and 63 kcal were found, and the apparent activation energy varied between these limits as the oxygen contents of the metal was varied between 0·01 and 0·07 wt %. This suggests a significant role for oxygen at these levels in affecting the recombination kinetics of nitrogen atoms.

In a similar study of the dissolution of hydrogen by inductively-stirred samples of liquid iron, Boorstein and Pehlke (1969) found an activation energy for the process of about 9 kcal which is much closer to the anticipated answer for boundary-layer diffusion control. Corresponding studies of the desorption of hydrogen also yielded a simple diffusion control mechanism with a low temperature coefficient. It thus appears that nitrogen adsorption and evolution from solution in liquid iron are characterized by specific surface effects which are not accounted for by a simple mass transfer model.

The desorption of hydrogen into a vacuum system represents the maximum rate of evolution of a vapour species from liquid iron under vacuum. This maximum rate can be enumerated approximately as follows:

If the atom fraction of hydrogen in solution in liquid iron is X_H then the concentration in the bulk C_H^{bulk} is given by

$$C_H^{bulk} = \frac{p_{Fe}}{M_{Fe}} \cdot X_H = 0.124 X_H \text{ g atom cm}^{-3}$$

Mass transfer of hydrogen through liquid iron would proceed with a diffusion coefficient approximately 10^{-3} cm^2 s^{-1} and a boundary layer thickness 10^{-3} cm. The maximum rate of transfer which would therefore be achieved when the surface layer is totally depleted ($C_H^{surface} = 0$) is then

$$J_H^{max} = \frac{10^{-3}}{10^{-3}} (0.124 X_H) = 0.124 \times 10^{-2} X_H \cdot \text{g atom cm}^{-2} \text{s}^{-1}$$

For hydrogen at an atom fraction of 0·1 %, the maximum rate could therefore be about 10^{-4} g atom cm^{-2} s^{-1} and since the diffusion coefficients for other species are about 1/10 to 1/100 that of hydrogen, a value between 10^{-5} and 10^{-6} g atom cm^{-2} s^{-1} is a representative range for vacuum degassing at the maximum rate, for a number of typical solutes in dilute solution in iron.

CO EVOLUTION UNDER VACUUM CONDITIONS

Samarin (1961) has described some studies on a small scale industrial unit of the elimination of carbon from liquid iron by CO formation at 1590°C under vacuum conditions in an alumina crucible. The results which are shown in

Fig. 104 indicate that there is no advantage to be gained in lowering the pressure from about 10^{-2} mmHg to 5×10^{-6} mmHg in enhanced carbon elimination. This is the difference between the vacuum which can be achieved by mechanical pumps, at the upper limit, when compared with diffusion pumps which achieve a much higher vacuum.

It also appears true that the levels of carbon and oxygen which can be obtained approach the calculated level, assuming that thermodynamic equilibrium is achieved under a total carbon monoxide pressure of 0·1 atm. The carbon–oxygen relationship which emerges from the results show a higher carbon content than the equilibrium value at high carbon–low oxygen final contents, but the value is about the same as the equilibrium value at low carbon-high oxygen contents. This indicates that the transport of oxygen across the boundary layer between the molten metal and the bubbles of CO is the significant rate-determining factor under vacuum conditions as it is under normal atmospheric conditions.

A typical industrial-scale apparatus for vacuum degassing is shown in Fig. 105.

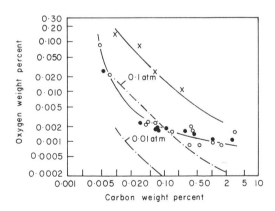

Fig. 104. The relationship between the oxygen and carbon contents of liquid iron in equilibrium with one atmosphere of $CO + CO_2$

————— and 0·1 and 0·01 atmos $CO + CO_2$ ——— · ———
● Results with a pressure of $1-5 \times 10^{-2}$ mm Hg in the system.
○ Results with a pressure of $5-7 \times 10^{-6}$ mm Hg obtained by vacuum refining. (After Samarin, loc. cit.)

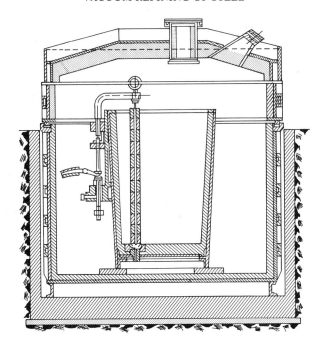

Fig. 105. Industrial apparatus for liquid-steel vacuum treatment (After Samarin, loc. cit.)

19

INERT GAS PURGING OF IMPURITIES IN STEEL

An alternative technique to the vacuum refining of liquid steel is the removal of the impurities in bubbles of an inert carrier gas. In this procedure, the simplest assumption to make is that each bubble becomes saturated with the equilibrium partial pressure of the impurity. The rate of loss of impurity to an inert carrier gas at one atmosphere partial pressure is therefore given by the transpiration equation

$$m_A = \frac{V_I}{22 \cdot 4} \cdot p_A \; \text{g atom s}^{-1}$$

where V_I is the volume in litres of the inert gas at NTP which is passed through the melt per second and p_A is the partial pressure exerted by the impurity A, in atmospheres, under equilibrium conditions. We are therefore making the simplifying assumptions that mass transfer to the purging gas is rapidly achieved by diffusion in the liquid boundary layer, that the bubble maintains a constant total pressure due mainly to the presence of the entraining inert gas, and that the temperature is constant throughout. In the practical situation, the process of saturation of the argon bubbles with hydrogen is one in which the transfer of hydrogen from the metal surface to the gas bubble is to be made against an increasing tendency to transfer in the opposite direction. This results from the partial pressure of hydrogen which will increase within the argon bubble with time. The net transfer across the liquid boundary layer will therefore be given by

$$J_{net} = J_H^{max} \left[1 - \left(\frac{p_{H_2}}{p_{H_2 eq}} \right)^{\frac{1}{2}} \right]$$

where p_{H_2} is the hydrogen partial pressure in the gas, and $p_{H_{2eq}}$. is that pressure which is in equilibrium with the melt. The value of J_{max} in this equation is that which was obtained above for free evaporation into a vacuum with boundary layer control in the liquid metal.

A further complication is that the bubble pressure will change, and hence expansion will occur, as the bubble rises through the liquid metal and the hydrostatic pressure of the liquid on the bubble is decreased.

Bradshaw and Richardson (1965) have made some computer-assisted calculations for the transfer of hydrogen from steel according to the boundary-layer control model. A bubble which is immersed a distance h below the surface of the liquid metal will rise with the velocity

$$u = -\frac{dh}{dt}$$

and have an internal pressure given by

$$P_t = \rho_s gh$$

where ρ_s is the density of liquid steel.

The change in the number of moles of hydrogen n_{H_2} in the bubble with time is given by

$$\frac{dn_{H_2}}{dt} = -u\frac{dn_{H_2}}{dh}$$

Now since

$$\frac{n_{H_2}}{n_{Ar}} = \frac{p_{H_2}}{P_t - p_{H_2}}$$

where n_{Ar} is the number of moles of argon in the bubble and by differentiation with respect to h as variable

$$\frac{dn_{H_2}}{dh} = \frac{n_{Ar}}{(P_t - p_{H_2})^2}\cdot\left(P_t\frac{dp_{H_2}}{dh} - p_{H_2}\cdot\frac{dP_t}{dh}\right)$$

and

$$\frac{dn_{H_2}}{dt} = \frac{dn_{H_2}}{dh}\cdot\frac{dh}{dt} = \frac{n_{Ar}\rho_s gu}{(P_t - p_{H_2})^2}\left(\frac{hdp_{H_2}}{dh} - p_{H_2}\right)$$

$$= k\frac{\rho_s}{M_s}[X_H^{bulk} - X_H^{surface}].ZV^{\frac{2}{3}}$$

where k is the mass transfer coefficient for hydrogen across the steel boundary layer at the bubble surface, and $ZV^{\frac{2}{3}}$ is the shape factor for the equivalent spherical bubble to the real bubble.

Since

$$V = \frac{n_{Ar}}{p_{Ar}} \cdot RT = \frac{n_{Ar} \cdot RT}{P_t - p_{H_2}}$$

and the terminal velocity can be written in the general form

$$u = A g^{\frac{1}{2}} V^{\frac{1}{6}}$$

where A is a constant (see p. 286), the equation above can be written in the form

$$\frac{dp_{H_2}}{dh} - \frac{p_{H_2}}{h} = \left(\frac{p_{H_2}}{h} - \rho_s g\right)^{\frac{3}{2}} \cdot h^{\frac{1}{2}} B (X_H^{bulk} - K_{H_2} \cdot \sqrt{p_{H_2}})$$

In this equation B is given by

$$B = \frac{Z}{A\sqrt{g}} \cdot \frac{\rho_s}{\rho_s g} \cdot RT \cdot \frac{k}{M_s} \cdot \frac{1}{(P_t^0 V^0)^{\frac{1}{2}}}$$

where P_t^0 and V^0 are the initial pressure and volume of the bubble. Since the number of moles of argon within the bubble remain constant

$$\left(\frac{RT}{n_{Ar}}\right)^{\frac{1}{2}} = \frac{1}{(P_t^0 V^0)^{\frac{1}{2}}}$$

yielding the final form of the differential equation

$$\frac{dp_{H_2}}{dh} - \frac{p_{H_2}}{h} + B \left(\rho_s g - \frac{p_{H_2}}{h}\right)^{\frac{3}{2}} \cdot (X_H^{bulk} - K_{H_2} \cdot \sqrt{p_{H_2}}) \cdot h^{\frac{1}{2}} = 0$$

At 1600°C we can evaluate B for hydrogen in steel with some values suggested by Bradshaw and Richardson, $Z = 4\cdot84$, $A = 0\cdot795$ and with $k = 1$ cm s^{-1} which was obtained here earlier for the mass transfer coefficient of hydrogen. The result for a bubble of 1 cm^3 volume which is introduced at an initial pressure of 1·5 atm is

$$B = 11\ 710$$

It should be noted in this calculation that the atmospheric pressure *above* the metal is computed in terms of the equivalent length of the metal column which would impose the same pressure. Therefore 1·5 atm is rendered as 220 cm height of liquid iron with density 7 g cm^{-3}. If the gas constant is used in cm^3 atm deg^{-1} ($R = 82$ cm^3 atm deg^{-1}) then $\rho_s g$ should be expressed in atm cm^{-1}. Again, for liquid steel, this value is $6\cdot956 \times 10^{-3}$ atm cm^{-1}.

Since the diffusion coefficients of other common gases such as nitrogen, are an order of magnitude smaller than that for hydrogen in liquid metals, it follows that a more typical value of B would be about 10^3. The computer

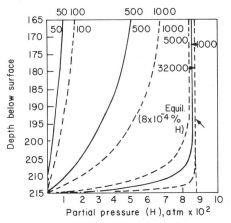

Fig. 106. Results for the calculated hydrogen pressure in a 1 cm³ argon bubble rising in molten steel (Bradshaw and Richardson.) The curves are for the given values of the constant, B, and the equilibrium H_2 content of the metal is 8×10^{-4} wt %.

calculations show that a bubble of argon which is released at a depth which is equivalent to a total pressure of 220 cm below the surface of a liquid steel bath containing hydrogen at an atom fraction of $4·5 \times 10^{-4}$ (8×10^{-4} wt %) will rapidly reach saturation with hydrogen (Fig. 106). This will be achieved within the first 10 cm of rise of the bubble. The equilibrium partial pressure of this solution of hydrogen is 9×10^{-2} atm at 1600°C.

Davenport (1968) has also made calculations of the mass transfer of hydrogen to argon bubbles in liquid steel. His results, which are shown in Fig. 107, indicate that the efficiency of transfer, or the approach to equilibrium, is

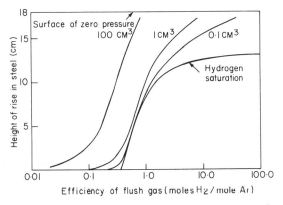

Fig. 107. Davenport's results for the efficiency of removal of hydrogen in argon bubbles from liquid steel at 1600°C.

practically unity for a small path of the bubble through the melt for small bubbles. When the bubble size is initially 100 cm^3, the efficiency is a good deal lower.

Further calculations for hydrogen transfer indicate that saturation of a bubble is also rapidly achieved with a mass transfer coefficient as low as one sixth of the boundary-layer control value which was used above. It may therefore be concluded that it is a good approximation to assume that equilibrium is achieved between a single bubble of initial volume of 1 cm^3 in passing through a layer of liquid metal which is deeper than 10 cm when boundary layer control in the liquid is rate determining. One further restriction is that the partial pressure of the material to be purged in the inert gas bubbles should not be too high, and the value used for hydrogen by Bradshaw and Richardson was 0·09 atm. Thus an impurity which would only exert a partial pressure at equilibrium in the inert gas bubbles of less than about 10^{-2} atm and which had a mass transfer coefficient which is typically about one tenth of that for hydrogen should transfer as readily as was demonstrated for hydrogen in these calculations.

This analysis of transfer to a single bubble may not be appropriate to a stream of bubbles rising through a liquid metal because now interference between bubbles may be significant. Thus if a bubble rises through a volume of liquid which has been to some extent denuded by the immediately preceding bubble, transfer to the latter bubble would not be equivalent to that for equilibration with the main bulk of the liquid. Such interference will obviously vary from one case another, and no general theoretical analysis could be expected.

THE KINETICS OF CO EVOLUTION

The equations which were developed for hydrogen evolution into an entraining argon bubble can also be used to describe the growth of a CO bubble during the steel refining reaction. The difference between operation under one atmosphere pressure and under vacuum conditions is now reflected through the addition of an extra path length in the liquid metal which is equivalent to one atmosphere in the former case. In both cases, homogeneous nucleation of bubbles is extremely unlikely and bubbles which are initially present in crevices in the container vessel are the bubble nuclei. Davenport showed that, under conditions of liquid boundary layer control with oxygen transfer as the rate-determining step, all very small bubbles $(10^{-1}-10^{-4} \text{ cm}^3$ initial volume) would grow to a limiting size of about 5 cm^3 after traversing a path length of 36 cm, and of about 13 cm^3 after a path length of 72 cm in liquid iron at 1600°C. These calculations are in good accord with the observations of industrial furnaces where a typical size is about 5 cm³ after a path length of about 40 cm. There is thus further evidence from kinetic calculations to suggest that rapid gas–liquid equilibrium is achieved under industrial conditions.

INERT GAS PURGING FOR DECARBURIZATION

The limitations on the removal of carbon which apply in the steelmaking processes largely come about because the carbon monoxide which is to be formed must achieve one atmosphere pressure in order to sustain bubble formation. Much lower levels of carbon content can be reached when CO is removed at a lower partial pressure in a bubble of some inert gas such as argon. This technique is particularly valuable in the decarburization of ferro-alloys such as ferrochromium when the initially carbon-saturated alloy contains an element with about the same affinity for oxygen as that of carbon.

In order to decarburize ferrochromium, it is important to avoid the costly simultaneous oxidation of chromium to form Cr_2O_3 as a separate phase. The lowest level of carbon content which could be achieved with this restriction is clearly that level at which the reaction

$$2[Cr] + 3CO \rightarrow Cr_2O_3 + 3[C]$$

$$K = \frac{a_{Cr_2O_3}a_{[C]}^3}{a_{[Cr]}^2 \cdot p_{CO}^3}$$

can occur. This can be controlled via the pressure of CO but a practical limit is achieved. This is when p_{CO} has reached too low a value to be economically attractive. The oxygen potential of the $Cr-Cr_2O_3$ sub-system of this process fixes the upper limit of the oxygen pressure for the avoidance of chromium oxidation in the alloy, and hence we can use the value obtained from the Ellingham diagram for this system, after applying the appropriate activity for chromium in the steel, to calculate the limiting CO pressure as a function of the carbon activity. At 1600°C, and with a chromium activity of 0·1 which corresponds to a typical stainless steel composition, the oxygen potential for Cr_2O_3 formation is -89 kcal mol^{-1}. Since the standard free energy of formation of 2 mols CO is -132 kcal at this temperature (see Fig. 4a, p. 15), it follows that for the reaction

$$2C + O_2 \rightarrow 2CO$$

$$\Delta G_{1873}^\circ = 2\left(RT\ln\frac{a_C}{p_{CO}}\right) + RT\ln p_{O_2} = -132\,000\,\text{kcal}$$

therefore

$$\log \frac{a_C}{p_{CO}} = \frac{-132\,000 + 89\,000}{2 \times 4\cdot575 \times 1873} = -2\cdot51$$

$$a_C = 0\cdot0031 p_{CO}.$$

If decarburization is to be achieved by bubbling an argon–oxygen atmosphere through the melt which then reacts to form a CO + Ar mixture, it would seem reasonable to assume as a basis for calculation that p_{CO} would be limited at a practical lower level of about 0·05 atm. This is mainly because of the difficulties of premixing a dilute O_2 + Ar gas mixture on the industrial scale.

This lower limit for p_{CO} yields a lower limit of the activity of carbon of $1\cdot5 \times 10^{-4}$. It will be seen immediately from the equation above that the lower limit of carbon activity which can be achieved by diluting the CO is directly proportional to this dilution factor, and hence it is only practical considerations of efficiency which determine the lower limit which is achieved.

This principle can obviously be applied for any alloying element in place of chromium, providing the appropriate alloying metal/metal oxide oxygen potential is applied to the calculation of the log a_C/p_{CO} ratio in the manner shown above. The way in which the final carbon contents will vary from one system to another depends on the oxygen potential at which the metal is oxidized and on the effect of the alloying element on the activity coefficient of carbon.

As an example of this effect, the oxygen potential for MnO formation at 1600°C of an alloy with aMn equal to 0·1 can be seen from the Ellingham diagram to be equal to $-99{,}000$ cal mol^{-1} O_2. This yields a relationship between a_C and p_{CO} for argon–oxygen decarburization

$$a_C = 0\cdot011 p_{CO}$$

Now the activity coefficients of carbon in dilute solution in ferrochromium and ferromanganese alloys at 1600°C for atom fractions X_{Cr} and X_{Mn} equal to 0·1 are 0·65 and almost one respectively with respect to carbon in solution in liquid iron. The interaction coefficients ε_C^{Cr} and ε_C^{Mn} have the values $-5\cdot0$ and $-0\cdot5$ respectively.

It follows that ferrochromium can be reduced in carbon content to a value 0·45 that of the corresponding ferromanganese alloy. This is because the two equations relating the activity of carbon with p_{CO} must be solved simultaneously and then, substituting for a_C, we have

$$\frac{a_C^{Fe-Cr}}{a_C^{Fe-Mn}} = \frac{0\cdot0031}{0\cdot011} = \frac{[\gamma_C X_C]^{Fe-Cr}}{[\gamma_C X_C]^{Fe-Mn}}$$

hence

$$\frac{X_C^{\text{Fe-Cr}}}{X_C^{\text{Fe-Mn}}} = 0.454.$$

The process for the decarburization of ferrochromium is in industrial production and is described in the article by Krivsky (1973). (See reference section).

20

THE REFINING OF LEAD

The metal which is produced in the blast furnace or the hearth process contains a number of deleterious impurities such as sulphur, arsenic and bismuth which must be eliminated in the refining process. The metal also contains metals of high value, silver and gold which must be recovered for their economic value. The refining of lead is a multi-stage process because of the presence of these impurities, and each stage is designed to eliminate one or two major impurities only.

The first stage of refining results mainly in the elimination of copper and tin from the metal. The product from the blast furnace is cooled down to a temperature which is a little above the melting point of lead (327°C) and held for a few hours. A separate phase, the dross, appears which contains a substantial amount of the copper impurity. Further, and virtually complete, elimination of copper may be accomplished by stirring elementary sulphur into the liquid metal. This process, the so-called "decopperizing process" is not properly understood from the physico-chemical standpoint at the present moment. This is because the free energy of formation of cuprous sulphide Cu_2S is approximately the same as that of lead sulphide PbS and, hence, no separation would be anticipated on purely thermodynamic grounds. The solution of the problem must rest in kinetic factors, and a study by Blanks and Willis (1961) showed that the presence of one of the minor impurities, silver, has a significant effect on the rate of elimination of copper from lead in this process.

It was found in these measurements of the rate of attack of sulphur vapour, which was transported over the sample in a stream of nitrogen, on pure lead and on lead containing 0·094 wt % Ag at 330°C that the gain in weight of the pure lead sample was approximately 100 times that of the Pb–Ag alloy in the same time of reaction.

It is therefore assumed that whilst the presence of a surface-adsorbed layer of silver on lead can reduce the rate of PbS formation, sulphur-rich CuS may be readily formed. This reduces the copper content of the metal down to such low values as 0·001 wt % with ease at temperatures slightly above the melting point of lead.

Although there have been claims that other impurities, e.g. antimony and tin, can also promote decopperizing, the effects of these elements were not observed by Willis and Blanks. It should not be thought that, under practical conditions, only copper is sulphidized in this process. There is always some formation of PbS accompanying the decopperizing of lead. Furthermore, if the dross is allowed to stay in contact with the lead bullion, then the formation of PbS goes on at the expense of the reversion of some of the copper into solution in liquid lead. This is the situation which would be anticipated on purely thermodynamic grounds, and taking account of the approximately equal free energies of formation of PbS and Cu_2S per mole of sulphur.

Clearly, metal from the blast furnace has been produced under a very low oxygen potential, and the second stage, which is similar to the refining of iron in steelmaking, is one in which a high oxygen potential is applied. The carbon solubility is very low in liquid lead, and so there is insignificant production of CO bubbles to agitate the liquid metal. At the low temperatures of lead refining, it is feasible to pass air through the metal from a steel lance which can be immersed so as to ensure the optimum contact between gas and metal, but the more general procedure is to allow stagnant gas-liquid contact. This oxidation stage brings about the elimination of arsenic and antimony, but not the other element from Group V which is an important impurity, bismuth. The ease of elimination of As and Sb from lead can be readily judged by comparison of the free energies of formation of the respective oxides, and the difficulty of removing Bi as Bi_2O_3 also becomes readily apparent. Any residual tin and copper are removed in this stage.

The solubility of oxygen in liquid lead at typical refining temperatures, around 600–700°C has been found to be between 2×10^{-2} and 10^{-1} atomic per cent, which is about one-tenth of the solubility of oxygen in liquid iron at steelmaking temperatures (0·7 atomic per cent at 1600°C).

The oxides of the elements which are eliminated in this stage of oxidation have approximately the same stability as does lead oxide (Fig. 108). The elements, As and Sb, are present in dilute solution as impurities of the major element, lead, and it follows that the oxides can only be formed at the oxygen potential at which PbO is formed when they, too, are formed in dilute solution. The slag phase consists therefore principally of PbO, which contains the oxides of the impurities as dilute constituents.

An estimation of the distribution of antimony and aresenic between metal and slag phases can be obtained by combining data for the free energies of formation of the oxides of these elements with that for PbO, and assuming that the liquid metal phases are Raoultian. The equations for oxide formation from the solid Group V elements, and liquid lead, are as follows:

$$\tfrac{4}{3}\text{As}(s) + O_2 \rightarrow \tfrac{2}{3}\text{As}_2O_3 \qquad \Delta G^\circ = -98\,300 + 30\cdot0T\,\text{cal}$$

$$\tfrac{4}{3}\text{Sb}(s) + O_2 \rightarrow \tfrac{2}{3}\text{Sb}_2O_3 \qquad \Delta G^\circ = -111\,300 + 41\cdot3T\,\text{cal}$$

$$2\text{Pb}(l) + O_2 \rightarrow 2\text{PbO} \qquad \Delta G^\circ = -107\,500 + 52\cdot60T\,\text{cal}$$

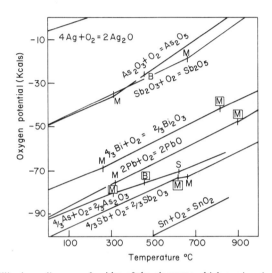

Fig. 108. The Ellingham diagram of oxides of the elements which are involved in lead refining.

These equations yield the results for the exchange reactions

$$\tfrac{4}{3}\text{As} + 2\text{PbO} \rightarrow \tfrac{2}{3}\text{As}_2O_3 + 2\text{Pb} \qquad \Delta G^\circ = 9\,200 - 22\cdot60T\,\text{cal}$$

and

$$\tfrac{4}{3}\text{Sb} + 2\text{PbO} \rightarrow \tfrac{2}{3}\text{Sb}_2O_3 + 2\text{Pb} \qquad \Delta G^\circ = -3800 - 11\cdot30T\,\text{cal}$$

At 700°C, assuming that $a_{\text{Pb}} = a_{\text{PbO}} = 1$ for a dilute solution of Group V metal and oxide in the metal and slag phases, the relationship between the activities of oxide and metal is given by the equations

$$\tfrac{2}{3}\log a_{\{\text{As}_2O_3\}} = 2\cdot87 + \tfrac{4}{3}\log X_{[\text{As}]}$$

and

$$\tfrac{2}{3}\log a_{\{\text{Sb}_2O_3\}} = 3\cdot32 + \tfrac{4}{3}\log X_{[\text{Sb}]}$$

If it is desired to remove arsenic and antimony by oxidation from lead, these equations suggest that at 700°C for a final content of As and Sb in the bullion of 0·1 atom per cent ($X_{As} = X_{Sb} = 10^{-3}$) the slag activities of these oxides are $2·0 \times 10^{-2}$ and $9·7 \times 10^{-2}$ respectively, at equilibrium. These results have been obtained by extrapolating the data for solid Group V elements and oxides up to temperatures where liquid phases exist. However, the length of the extrapolation is reasonably small, and the errors involved as a result of neglecting free energies of fusion are probably within the limits of the experimental errors for the solid systems.

One way in which the losses of lead as litharge during the process could be reduced would be by the addition of an oxide which would react very much more strongly with the oxides of the impurities than does PbO. The oxidation of these impurities could then occur at a lower oxygen potential than that at which PbO is formed, substantially at unit activity. Such a process has been used in which NaOH is added to form the sodium arsenates, antimonates, etc. This addition is difficult to make under practical circumstances, and does not seem to have found general acceptance, although the principle is obviously sound. In the Harris process, sodium chloride is added to the NaOH to reduce the temperature of fusion, and sodium nitrate has been added as a source of oxygen. The reaction involved in this process may be written formally as

$$2[Sb] + 4\{NaNO_3\} + 2[NaOH]\rightarrow2\{NaSbO_3\} + 4\{NaNO_2\} + H_2$$

for the elimination of antimony.

The removal of silver and gold from lead bullion cannot be made through the effects of non-metallic elements such as oxygen, sulphur and chlorine, because the affinity of lead for these elements is always greater than those of these impurities. An alternative method is to make use of the affinities of Ag and Au for liquid zinc for which Pb has a very small affinity.

The Parkes process applies the addition of zinc to liquid lead which is slowly cooled down to a temperature just above the melting point. A solid phase, which is rich in zinc but contains nearly all of the gold and silver impurities, and a small amount of entrained lead, separates from the liquid phase, and can be skimmed off the surface. The zinc is finally separated from a (Ag + Au)-rich lead phase by distillation at about 1200°C in a simple retort containing some charcoal to keep the oxygen pressure very low in the system.

The thermodynamics of this process can be simply summarized in terms of the data for the dilute solutions of silver in pure lead and pure zinc, and hence for the free energy of transfer of silver from lead to zinc.

$$Ag \rightarrow [Ag]_{Pb}; \qquad \Delta\bar{G}_{Ag} = 7500 - 7·0T + RT\ln X_{Ag}$$

$$Ag \rightarrow [Ag]_{Zn}; \qquad \Delta\bar{G}_{Ag} = -3400 + RT\ln X_{Ag}$$

$$[Ag]_{Pb} \rightarrow [Ag]_{Zn} \qquad \Delta\bar{G}_{transrer} = -10900 + 7·0T$$

From these data together with the integral free energy of mixing of the Ag + Zn solid system at 340°C, it is possible to show which solid phase of Ag + Zn would be in equilibrium with lead containing silver at a number of concentrations. (Fig. 109).

The free energy data for the Ag + Zn system were obtained from the experiments of Trzekiatowski and Terpilowski (1955) who used an EMF method, and Birchenall and Chang (1949) who measured the dew point of zinc vapour above the alloys. The results are used together with calorimetric ΔH values by Orr, Rovel and Hultgren (loc. cit.) who used the liquid tin calorimeter.

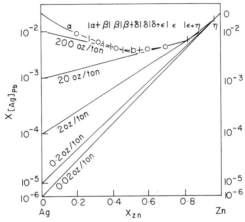

Fig. 109. Partial molar free energies of solution of silver in liquid lead, and the integral free energies of formation of silver-zinc alloys at 340°C.

There are not such complete results for the Au + Zn + Pb system which enable a complete calculation to be made as in the case of silver, but the thermodynamic data for the Au + Zn solid system show that the affinity of gold for zinc is greater than that of silver, and the partition of Au would be expected to be more heavily weighted in favour of the zinc phase. Hence, relatively complete removal of Au is to be expected.

The lead-zinc system shows a limited solubility of zinc in liquid lead, between the melting point of lead and that of zinc, 419°C. Thus, a not-insignificant amount of impurity is introduced into refined lead bullion during the Parkes process. This quantity of zinc may be removed either by chemical reaction such as oxidation or chlorination, or by distillation. In the former case, the limit of separation of zinc from lead is a matter of choice. This is because the attack of chlorine for instance on a lead-zinc alloy will lead to the formation of a $PbCl_2$–$ZnCl_2$ slag phase, and under conditions of local

equilibrium, the composition of the slag which will be formed at any given
Zn content of the metal phase is obtained from the equilibrium constant for
the exchange reaction

$$Pb + \{ZnCl_2\} \rightarrow [Zn] + \{PbCl_2\}$$

$$K = a_{[Zn]} \cdot \frac{a_{\{PbCl_2\}}}{a_{\{ZnCl_2\}}} = 1 \cdot 25 \times 10^{-5} \text{ at } 400°C.$$

Lead, being the major element, may be considered as having unit activity.
The activity of the dilute solution of Zn in Pb is given at 400°C by the equation

$$\log a_{[Zn]_{pb}} = 1 \cdot 19 + \log_{10} X_{[Zn]_{pb}}$$

which indicates an activity coefficient of nearly 15 for the dilute constituent.
At saturation, this reaches nearly 20, since the saturation solubility, as has
been stated earlier, is about 5 atomic per cent. The slag phase is probably
quite close to ideal in the thermodynamic sense, and thus by substitution in
the equilibrium constant

$$\log K = \log a_{[Zn]} + \log \frac{X_{\{PbCl_2\}}}{X_{\{ZnCl_2\}}}$$

$$-4 \cdot 90 - 1 \cdot 19 - \log X_{[Zn]} = \log \frac{X_{\{PbCl_2\}}}{X_{\{ZnCl_2\}}}$$

$$\log \frac{X_{\{PbCl_2\}}}{X_{\{ZnCl_2\}}} = -6 \cdot 09 - \log X_{[Zn]}$$

Since $X_{[Zn]}$ is much smaller than unity, and therefore $\log X_{[Zn]}$ is a negative
quantity, it follows that the lead content of the slag must *increase* as the zinc
content of the metal decreases. However, it can be seen that the zinc content
of the lead bullion can be reduced to negligible proportions, under equilibrium conditions before the lead loss becomes too serious. When the slag
phase contains equal mole fractions of $PbCl_2$ and $ZnCl_2$ the zinc atom
fraction in the metal phase is 8×10^{-7} at 400°C.

The removal of zinc from lead by distillation depends on the large difference
between the vapour pressures of the two metals. The heats of vaporization of
pure liquids are 30 and 46 kcal g atom^{-1} respectively, and hence the greater
difference between the vapour pressures is to be found at the lower temperatures. Although this would obviously lead to a better separation at lower
temperatures, a compromise must be struck since the rate of evaporation
is lower at low temperatures because of the smaller absolute value of the
vapour pressures. At the boiling point of zinc, which is 907°C for 760 mmHg
pressure, the vapour pressure of pure liquid lead is about 0·2 mmHg. Taking
consideration of the fact that the activity coefficient of zinc in dilute solution

in liquid lead is about 10, then the constant boiling composition at temperatures near the boiling point of pure zinc can be calculated as follows:
at the constant boiling composition

$$p_{Zn} = a_{Zn}p_{Zn}^0 = p_{Pb}^0[a_{Pb} \cong 1]$$

$$X_{Zn} = \frac{p_{Pb}^0}{p_{Zn}^0} \cdot \frac{1}{\gamma_{Zn}} = \frac{0 \cdot 2}{760} \times \frac{1}{10} = 2 \cdot 62 \times 10^{-5}$$

This shows that the zinc content of lead can be reduced to 26 parts/10^6 by distillation at 900°C.

There is obviously little to choose between these two processes for the dezincing of lead, and the choice of a preferred process depends on other than physico-chemical factors.

21

THE REFINING OF ZIRCONIUM

The metal which is obtained from the reduction of zirconium tetrachloride with magnesium in the Kroll process contains too high a concentration of some neutron absorbing elements, such as iron and silicon, to be useful in the construction of naval nuclear reactors. A further stage of refining must be added in order to eliminate these impurities, and the method which is most commonly employed is the Van Arkel iodide process. In this process, impure metal is converted to gaseous zirconium tetraiodide at low temperatures, around 250–300°C, and the iodide is decomposed on a hot zirconium wire at much higher temperatures 1200–1400°C. Due to the high chemical stability of ZrI_4, this compound is formed preferentially to those of the impurity elements, and thus a separation of zirconium from the impurities occurs via the gas phase.

The conditions for the optimum transfer of zirconium across the temperature gradient, and for separation of the element from its impurities have been studied empirically, but no physico-chemical analysis has so far been made. This is because the thermodynamic information for the $Zr-I_2$ system is far from complete.

Döring and Molière (1952) studied the rate of transfer of zirconium from a low temperature source to a metal filament at high temperature. These two temperatures could be independently controlled, so the effects of the source and filament temperatures on the rate of transfer could be studied separately. Figure 110 summarizes the results in which it can be seen that there is an optimum temperature for the source around 250°C corresponding to a wide range of filament temperatures from 1100–1500°C.

The effect of increasing total pressure in the gas phase has also been studied, by the addition of argon up to 1·5 mmHg pressure when the source was held at 280°C and the filament at 1450°C. As would be expected, it was found

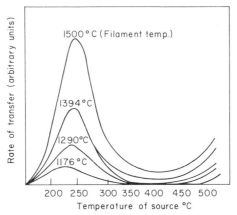

Fig. 110. The effects of filament and source temperatures on the transfer rate of zirconium in the Van Arkel reaction.

that the rate of transfer dropped, approximately linearly, to about 10% of the vacuum when 1·5 mmHg or argon was established in the apparatus (Fig. 111). The vapour pressure of solid ZrI_4 at the source temperature in these experiments, 280°C, was 2·5 mm, so the diffusion length was only halved if solid ZrI_4 was formed. The drastic effect of this addition of inert gas makes it appear probable, however, that the ZrI_4 pressure in the gas phase was somewhat less than this figure, and that a separate solid phase of ZrI_4 was not formed over the zirconium metal in the source.

Three solid iodides of zirconium, at least, have been prepared by classical chemical procedures, ZrI_2, ZrI_3 and ZrI_4. Of these three compounds only ZrI_4 is significantly volatile, and Sale and Shelton (1965) concluded from a

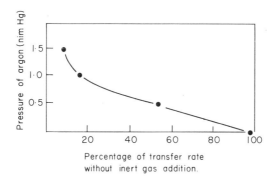

Fig. 111. The effect of argon additions on the rate of transfer of zirconium.

[Source at 280%C; Filament at 1450%C.]

M

Knudsen vapour pressure study, that specimens of the tri-iodide dispropor-tionate by volatilization of ZrI_4 thus

$$ZrI_3(s) \rightarrow ZrI_{2 \cdot 8 \pm 0 \cdot 3}(s) + ZgI_4(g)$$
$$\downarrow$$
$$ZrI_{2 \cdot 00}(s) + ZrI_4(g)$$

The interpretation of these vapour pressure results is not altogether certain, but it might be tentatively suggested that vapour pressure equations for the condensed phases would be as follows:

$ZrI_4(s)$ $\qquad\qquad\qquad \log p(\text{atm}) = \dfrac{-6450}{T} + 9 \cdot 17$

$ZrI_{2 \cdot 8} - ZrI_3$ $\qquad\qquad \log p(\text{atm}) = \dfrac{-8700}{T} + 9 \cdot 59$
$(T = 275–325°C)$

$ZrI_{2 \cdot 8} - ZrI_2$ $\qquad\qquad \log p(\text{atm}) = \dfrac{-26400}{T} + 34 \cdot 82$
$(T = 350–400°C)$

It is obvious that, as the solid phase becomes richer in zirconium, the ZrI_4 pressure decreases markedly. If it were possible to suppress the evapora-tion of the tetraiodide as, for example, in the iodination of zirconium metal in a closed system, then the sequence of phases which would be observed would be

$$Zr \mid ZrI_2 \mid ZrI_{2 \cdot 8} \mid ZrI_3 \mid ZrI_4 \mid ZrI_4 \text{ gas}$$

The formation of the solid phases would probably occur by diffusion either of zirconium outwards or of iodine inwards according to the parabolic rate law and the rate of formation of the products would increase rapidly with increasing temperature, as the diffusion coefficients correspondingly increased. If the rate of supply of iodine to the surface of the solid is such that $ZrI_4(s)$ is not formed, then the rate of vaporization of zirconium as $ZrI_4(g)$ would more nearly correspond to that appropriate for the lower pressures of the $ZrI_2–ZrI_{2 \cdot 80}$ or $ZrI_{2 \cdot 80}–ZrI_3$ two-phase equilibria. The more rapid consumption of iodine by the solid phases to form the lower iodides is thus a possible cause for the *decrease* in the rate of zirconium transfer as the temperature of the impure metal is increased above about 350°C.

The purification of the element which is affected in the Van Arkel process must be related to the high stability of zirconium iodides. The empirical studies of Bulkowski *et al.* (1950–51), which were made with binary zirconium alloys containing 0·1 wt% of selected impurities, demonstrate the chemistry of the purification process. Table XXXVII shows their results which indicate that elements such as nickel, chromium and nitrogen remain in the impure source whereas aluminium, titanium and hafnium are transferred with the zirconium.

The Ellingham diagram for the pertinent iodides gives a self-evident explanation of the effects which were observed, except in the case of carbon, which was probably transferred via the formation of CO either by reaction with oxygen which was dissolved in the metal, or by reaction with the residual oxygen content of the vacuum system.

TABLE XXXVII. The transfer of impurities during refining

Element	parts/10^6 in source material	parts/10^6 in refined material
Ni	1600	10
Cr	1000	20
Si	1100	130
Fe	1200	340
Al	1100	700
Ti	1000	1000

Since the source of zirconium is always kept at a low temperature, the diffusion of impurities through the feed material is unlikely to be significant. It follows that an impurity will become more concentrated near the surface of the feed material, as zirconium is preferentially iodinated, and therefore the thermodynamic activity of each impurity probably rises up to a value close to unity at the solid-gas interface during the refining process. It is pertinent then to consider the standard free energies of formation of the iodides as indicating the probability of formation of the impurity iodides, even although these data refer to the pure metals, whereas Bulkowski et al. studied the iodination of dilute binary solutions of impurities in zirconium.

In order to draw the Ellingham diagram (Fig. 112) for iodides, use was made of a number of approximations. In all of the cases shown, adequate data exist for the vaporization of either the solid or liquid iodide, but free energy of formation data for the solid iodides were entirely missing. The heats of formation of a number of the compounds have been obtained by means of aqueous solution calorimetry, e.g. ΔH°_{298} for AlI_3. The entropy contents of the compounds at 298°C were all estimated using Latimer's simple addition formula for the individual ions. The entropy contribution to be used for the iodide ion changes with the valency of the cation, and the values suggested by Kubaschewski have been employed.

Valency of cation	+ 2	+ 3	+ 4
S°_{298} I^-	13·6	12·5	13·0

The free energy change for the formation of the solids was then obtained by use of the approximation

$$\Delta G^\circ \cong \Delta H^\circ_{298} - T\Delta S^\circ_{298}$$

This is quite a good approximation in this particular example, since the results are only needed to treat the separation of zirconium from the impurities in the source of the Van Arkel process, and this is always operated below 500°C.

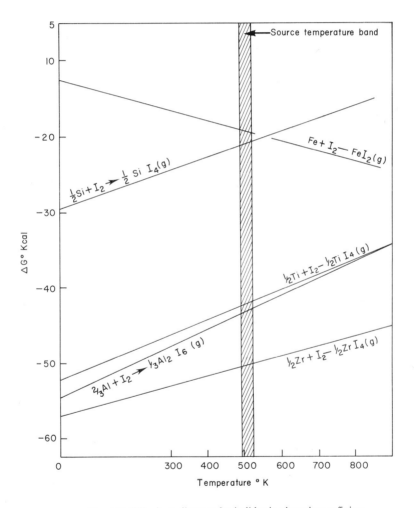

Fig. 112. Ellingham diagram for iodides in zirconium refining.

As an example of the procedure, the following data for AlI_3 were used

$$Al(s) + 3I(s) \rightarrow AlI_3(s) \qquad \Delta H^\circ_{298} = -74100 \text{ cal}$$

$$\Delta S^\circ_{298} = -3 \cdot 15 \text{ cal mol}^{-1} \text{ deg}^{-1}$$

$$3I(s) \rightarrow \tfrac{3}{2}I_2(g) \qquad \Delta H^\circ_{298} = +22400 \text{ cal}$$

$$S^\circ_{298}; I_2 \qquad 62 \cdot 25 \text{ cal mol}^{-1} \text{ deg}^{-1}$$

$$AlI_3(s) \quad 45 \cdot 5 \quad \text{cal mol}^{-1} \text{ deg}^{-1}$$

$$Al(s) \qquad 6 \cdot 8 \quad \text{cal mol}^{-1} \text{ deg}^{-1}$$

$$\tfrac{2}{3}Al(s) + I_2(g) \rightarrow \tfrac{2}{3}AlI_3(s) \qquad \Delta G^\circ = -64400 + 36 \cdot 5 \, T \text{ cal}$$

The vapour phase consists mainly of Al_2I_6 molecules at low temperatures, and making use of the results of Fischer, Rahlfs and Benze (1932) we find

$$2AlI_3(s) \rightarrow Al_2I_6(g) \qquad \Delta G^\circ = 30700 - 55 \cdot 8 \, T \text{ cal}$$

Hence for the reaction

$$\tfrac{2}{3}Al(s) + I_2 \rightarrow \tfrac{1}{3}Al_2I_6(g) \qquad \Delta G^\circ = -54200 + 17 \cdot 9 \, T \text{ cal}$$

Since the results which were used were not of the highest accuracy, each heat change has been rounded off to the nearest 100 cal and the entropy changes to the nearest $0 \cdot 1$ cal mol^{-1} deg^{-1}

The application of vapour phase transfer reactions in the refining and deposition of a number of refractory metals is discussed in detail in the book edited by Powell, Oxley and Blocher (1966).

22

ELECTRICAL METHODS OF REFINING

Electrical power may be used in broadly one of two ways in the preparation of pure metals. The first use involves the application of small d.c. potentials for the removal of impurities via electrolytes which are placed in contact with the primary metal phase. These will be referred to as electrochemical refining methods in the following discussion. The second role of electricity in purification is to provide a.c. heating, and this technique, now known as "electroslag refining" does not make use of electrolysis of the slag phase but merely provides a reactive liquid phase through which droplets of crude metal are caused to fall and become refined by normal slag–metal equilibration.

ELECTROCHEMICAL REFINING

A number of possibilities for this method are still only at the stage of laboratory experimentation, but the production of super-purity aluminium is an example of an application which has been carried out on the commercial scale for some time. In this process, impure aluminium is alloyed with copper to form a liquid alloy. This Cu–Al alloy which contains about 70 wt % Al functions as the anode of a cell in which the cathode is high purity liquid aluminium (m.p. 659°C). A number of electrolyte systems have been used, but all function as a path for the transport of aluminium ions from the crude to the pure metal. The principal components of the electrolyte are cryolite, Na_3AlF_6 and AlF_3 to which additions of barium and calcium fluorides are made. The process usually operates between 750 and 1000°C. Since all three of the phases involved are liquid at the refining temperature, it is clear that density considerations are important. The cell operates with the

Cu–Al anode at the bottom, and the Al cathode on top. Grube and Hantelmann (1952) made a laboratory study of this process, but used $BaCl_2$ in their electrolyte rather than BaF_2. The molar proportions in this electrolyte were: $1 \, BaCl_2$; $1 \, AlF_3$; $1.5 \, NaF$. The results of this study cover many aspects of the refining process. In a study of the transport of possible impurities. it was shown that metal can be transferred from liquid anodes of Mn–Zn, Fe–Sb, Si–Cu and Cu–Ag to a pure Al cathode through the electrolyte, the

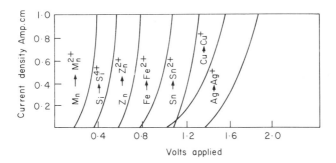

Fig. 113. Current–voltage curves for the transport of elements across the electrolyte $1 \, BaCl_2$; $1 \, AlF_3$: $1.5 \, NaF$ at 850°C. (After Grube and Hantelmann, loc. cit.)

first named metal being transported· The current–voltage curve for each element was obtained (Fig. 113) and it was found that manganese was transported at the lowest voltage, followed by Si, Zn, Fe and finally Sn, Cu, Ag, when the voltage applied to the cells was increased from 0 to 2 volts.

The EMF of the cell

Pure aluminium $\left| \begin{array}{c} BaCl_2 \\ NaF \; AlF_3 \end{array} \right.$ electrolyte $\left| \right.$ Cu–Al (70 wt %Al)

was found by these authors in an earlier study to be 6·6 mV at the operating temperatures of refining process which is considerably less than the lowest potential at which a significant transport of Mn occurs (0·2 V); hence, the refining reaction is readily explicable in simple terms.

Clearly when aluminium is being recovered from an anode, a substantial composition gradient of aluminium may build up across the boundary layer in the liquid metal at the anode/electrolyte interface. Then the EMF of the corresponding aluminium concentration cell would rise, since

$$E = \frac{RT}{3F} \log a_{Al}$$

where a_{Al} is the activity of aluminium in the anode surface. It is therefore possible that at high current densities of operation impurities could be transferred electrochemically in preference to aluminium unless steps are taken to remove the concentration gradient in the liquid anode. This is normally easily achieved in industry merely as a result of the convection currents which arise in the refining cell due to temperature and density gradients.

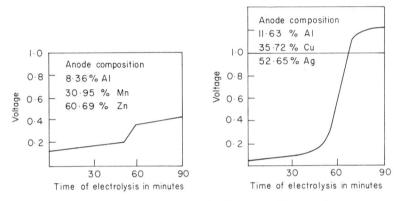

Fig. 114. Voltage–time curves for the electrolysis of anodes with a small atom percent of aluminium. (After Grube and Hantelmann, loc. cit.)

In their laboratory studies, Grube and Hantelmann were able to demonstrate that in a well thermally equilibrated system, the voltage needed to retain a constant current density ($0 \cdot 1$ A cm^{-2}) increased to a new level after a period of operation of about 50 minutes when alloys which contained less than 10 atom % Al alloyed with other metals were used as anodes. The change in potential corresponded to a transition from aluminium transfer to impurity metal. In line with their earlier findings of the applied voltage which was needed to transfer various impurities, it was found in these chronopotentiometric studies that the voltage change was smallest for alloys of aluminium with Mn and Zn, and largest for those with Cu and Ag (Fig. 114a and b).

REFINING BY MEANS OF MOLTEN SALT ELECTROLYTES

The alkali and alkaline earth metals have very high affinities for non-metallic elements such as oxygen, sulphur and selenium, and are thus potentially useful for the removal of these elements from liquid metals. The handling of Group IA and IIA metals in the industrial scale is hazardous, and their production as required by molten salt electrolysis has an attractive possibility for refining processes.

Ward and Hoar (1961) obtained almost complete removal of sulphur, selenium and tellurium from liquid copper by electrolysing molten $BaCl_2$ between the metal, which functioned as the cathode, and a graphite anode at 1100–1150°C. The barium salt was chosen because of its low vapour pressure, but the same reaction could be produced with any of the Group IA, IIA halides. The basic chemical reaction can be written for desulphurization of copper to yield BaS dissolved in $BaCl_2$ and the evolution of chlorine

$$[S] + 2\{Cl^-\} \rightarrow \{S^{2-}\} + Cl_2$$

The balance is made by the transfer of electrons from two chloride ions to form one doubly-charged sulphide ion. The decomposition voltage E required to bring this reaction about can be calculated from the following data and approximations.

$$E = E_o + \frac{RT}{2F} \ln \frac{\{X_{S^{2-}}\}}{a_{[S]}\{X_{Cl^-}\}^2}$$

where E_o is for the reaction

$$[S] + 2\{Cl^-\} \rightarrow \{S^{2-}\} + Cl_2$$

Temkin's rule is used for the molten salt and hence

$$a_{\{S^{2-}\}} = \frac{n_{S^{2-}}}{n_{S^{2-}} + n_{Cl^-}} = \{X_{S^{2-}}\}$$

For the copper solution of sulphur we shall use the 1 wt% standard state. Hence

$$RT \ln a_{[S]} = \Delta G^0 + RT \ln [\text{wt} \% S]$$

where ΔG° is for the reaction

$$\tfrac{1}{2}S_2 \rightarrow [S] \qquad K = \frac{[\text{wt} \% \, S]}{p_{S_2}}$$

At a temperature of 1100°C the standard free energy change for the reaction

$$\tfrac{1}{2}S_2 + BaCl_2 \rightarrow BaS + Cl_2 \tag{i}$$

has the value of approximately +75 kcal. Hence the standard EMF for the reaction is −1.62V. The free energy of solution of diatomic sulphur gas in liquid copper is given as a function of temperature by the equation

$$\tfrac{1}{2}S_2 \rightarrow [S] \tag{ii}$$

$$\Delta G^\circ = -29\,600 + 6\cdot30 T \, \text{cal}$$

And at $1100°C$:

$$\Delta G^{\circ}_{1373} = -21\,000 \text{ cal} \qquad E^{\circ} = 0.46 \text{ V}$$

It follows that the standard free energy change and the EMF for the refining reaction can be obtained by subtracting equation (ii) from equation (i) above

$$[S] + 2\{Cl^-\} \rightarrow \{S^{2-}\} + Cl_2$$

and using the values calculated above to yield the results.

$$\Delta G^{\circ}_{1373} = +96\,000 \text{ cal}; \qquad E^{\circ} = -2.08 \text{ V}$$

$$K = \frac{p_{Cl_2}}{\text{wt}\%[S]} \frac{X_{\{BaS\}}}{X_{\{BaCl_2\}}}$$

It can be seen from this discussion that the effect of the cation on the free energy change for this refining reaction appears mainly in the exchange reaction

$$\tfrac{1}{2}S_2 + MCl_2 \rightarrow MS + Cl_2$$

Furthermore, the position of equilibrium, that is to say the extent to which the refining reaction can be carried through, depends upon the chlorine pressure above the molten salt phase. In most experimental arrangements this is unimportant because the atmosphere will probably be an inert gas such as argon. However, if the refining reaction were to be carried out in the analogous manner using an oxide molten phase, then the important exchange reaction would be

$$\tfrac{1}{2}S_2 + MO \rightarrow MS + \tfrac{1}{2}O_2$$

which is identical with the reaction which occurs in the desulphurization of iron. Here the control of the partial pressure of oxygen in the atmosphere is very important.

ELECTROSLAG REFINING

A relatively new process for the refining of steel, after oxygen blowing, involves the re-melting of billets under a molten salt cover to form droplets of metal which are refined before reforming into a solid billet. The temperature for melting that part of the billet which is immersed in the salt phase is generated by using electrical resistance heating of the salt which acts as the conductor. A.c. heating can be used with the original billet, as one electrode and the re-fined re-consolidated billet, which is formed in a water-cooled copper mould, as the other electrode (Fig. 115). In order to reduce oxidation of the metal in

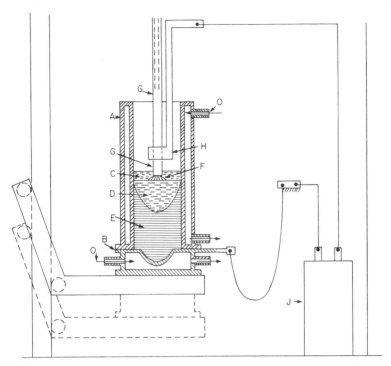

Fig. 115. The consumable electrode arrangement used in electroslag refining. (After Duckworth and Hoyle, loc. cit.)

the intermediate droplet phase, the electron transport number of the salt phase must be kept to a minimum. This can be achieved by minimizing the content of polyvalent ions in the salt. A redox process can occur by dissolution of iron as ferrous ion in the salt phase in the presence of polyvalent ions and the ferrous ion, on reaching the salt-air interface can facilitate the incorporation of oxide ions in the melt. The reaction steps are typically

$$Fe^\circ + 2M^{3+} \rightarrow Fe^{2+} + 2M^{2+}$$

The dissolution process for iron going into the salt phase is also facilitated by the presence of ferric ions through the charge-exchange reaction.

$$Fe^\circ + 2Fe^{3+} \rightarrow 3Fe^{2+}$$

These effects are minimized by the use of a stable alkaline earth fluoride, CaF_2, as the principal component of the salt phase. Because of the stability of CaF_2, the displacement reaction for the dissolution of iron also has a negligible effect at steel-melting temperatures

$$Fe + CaF \rightarrow FeF_2 + Ca \qquad \Delta H^\circ_{298} = +124\,kcal$$

Apart from the facts that calcium fluoride is readily obtainable and is a liquid electrolyte above the melting point (1418°C), this substance has the typically low vapour pressure for fused salt, which is found predominantly amongst the fluorides. The boiling point of the liquid is 2510°C and at 1600°C the vapour pressure only reaches 0·5 mmHg.

Calcium fluoride is the least volatile of the common alkaline earth fluorides. The salt has an electrical conductivity of 5 ohm^{-1} cm^{-1} at 1600°C, and thus is ideally suited as a resistance heating element since a higher resistance would necessitate the use of higher potentials in the heating supply. This is typically 20–50 volts at normal frequency and the energy consumption is of the order of 500 kWh per ton of metal refined.

Fluorides are subject to hydrolysis by water vapour in the air when melted in the normal atmosphere. The reaction with CaF_2 produces CaO which dissolves fortunately in liquid CaF_2 up to a mole fraction $X_{CaO} \cong 0.6$ at steel-making temperatures. The molten electrolyte can therefore accommodate a

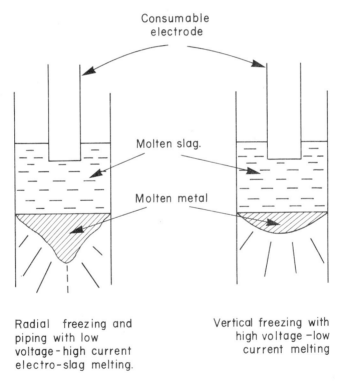

Fig. 116. The effects of the voltage–current characteristics of electroslag refining on the structure of the refined billet.

high CaO content and thus can produce a high CaO activity in the molten state. The eutectic in the CaF_2–CaO system lies at 1360°C and therefore the presence of CaO in the electrolyte reduces the fusion temperature by 58°C. The addition of Al_2O_3 to the electrolyte also reduces the fusion temperature considerably, to 1250°C, but raises the resistance of the electrolyte.

The physical problem which arises when the electrolyte resistance, and therefore the ohmic heating, is allowed to increase, is that the cooling of the metal after refining must be extremely fast to remove the possibilities of piping in the ingot. Clearly the optimum situation is when the product takes the form of a solid billet carrying a thin liquid lens of molten metal which is in contact with the electrolyte (Fig. 116). This requirement affects the choice of electrolyte in the industrial equipment, and a typical composition contains only 20% Al_2O_3 in CaF_2 + CaO. The addition of alumina has the chemical effect of reducing the CaO activity in the melt and thus a variation of slag function between the highly "basic" and 'acid" properties can be obtained.

Further physical effects of the presence of CaO and Al_2O_3 in the fluoride electrolyte are on the viscosity and density of the salt phase. Both additives lower the viscosity of CaF_2 in the range of 0–15 wt% addition but at higher contents the viscosity increases. The maximum reduction in viscosity is about 50% and the viscosity of the pure phase is around 5 centipoise at steelmaking temperatures. The density of the electrolyte increases monotonically on addition of CaO or Al_2O_3. In typical electroslag refining compositions, the density is 2·7 g cm^{-3} which should be compared with 2·55 g cm^{-3} for pure CaF_2 at 1450°C where accurate measurements have been made. The terminal velocity of liquid metal droplets through the salt phase will therefore be affected by the addition of reactive oxides to the CaF_2 salt phase. Considering the components of Stokes equation

$$V = \frac{2gr^2(\rho_{\text{metal}} - \rho_{\text{slag}})}{9v}$$

since the density of liquid iron is about 7·2 g cm^{-3} it is clear that the change in the viscosity, rather than density, is the major effect of the additives on the rate of fall, and hence the time of residence of the metal droplets in the molten salt.

THERMODYNAMICS OF ELECTROSLAG REFINING

In principle the refining reactions which can occur in this process do not differ from those under steelmaking slags. The major chemical difference in the electroslag process is that oxygen can now be virtually excluded, since most of the required carbon removal has been carried out in an earlier stage.

In typical studies of the effectiveness of refining, it has been found that better than 50% of the remaining sulphur and silicon are removed whilst the carbon, phosphorus and manganese contents remain substantially unchanged. It is because the process can be carried out with relatively minor effects of oxidation that electroslag refining is an attractive step for the purification of alloy

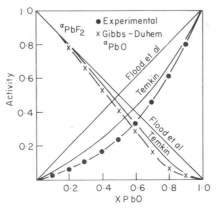

Fig. 117. A comparison of the observed activities with theoretical values for the liquid PbO–PbF$_2$ system. (After Sridhar and Jeffes, loc. cit.)

steels containing reactive elements with a high affinity of oxygen such as titanium and niobium.

There have been no direct measurements of CaO or CaF$_2$ activities in the binary liquid system at steelmaking temperatures but a study of the liquid system PbO + PbF$_2$ by Sridhar and Jeffes (1968) can be used to give results by analogy. These workers obtained measurements of the activity of PbO in this system by the use of an electrochemical cell with electrodes which are reversible to lead and oxygen. The PbO–PbF$_2$ mixture served as the cell electrolyte. It was shown that the results were in reasonably good agreement with Temkin's model (Fig. 117) and that the heat of mixing in the system is probably less than 1 kcal g atom^{-1} over the entire composition range.

According to the Temkin model, the activities of the two components are obtained from the following equations:

$$a_{PbO} = \frac{n_{O^{2-}}}{n_{O^{2-}} + n_{F^-}} = x_{O^{2-}}$$

the ionic fraction of oxygen ions and

$$a_{PbF_2} = \left(\frac{n_{F^-}}{n_{O^{2-}} + n_{F^-}} \right)^2 = x_F^{2-} \qquad x_{Pb^{2+}} = 1.$$

Since we may substitute

$$\frac{n_{O^{2-}}}{n_{O^{2-}} + n_{F^-}} = \frac{X_{PbO}}{X_{PbO} + 2X_{PbF_2}}$$

and

$$\frac{n_{F^-}}{n_{O^{2-}} + n_{F^-}} = \frac{X_{PbF_2}}{X_{PbO} + 2X_{PbF_2}}$$

we obtain

$$a_{PbO} = \frac{X_{PbO}}{1 + X_{PbF_2}} \text{ and } a_{PbF_2} = \left(\frac{2X_{PbF_2}}{1 + X_{PbF_2}}\right)^2$$

Sridhar and Jeffes further showed that their experimental results could be accommodated to a regular solution model with reasonable accuracy. The form which was chosen was therefore

$$RT \ln \gamma_{O^{2-}} = -1460 x_{F^-}^2$$

Kor and Richardson (1969) calculated the properties of $CaO + CaF_2$ melts by using the facts that the liquid was saturated with pure solid CaO at X_{CaO} equal to 0·26 at 1500°C and that the entropy of fusion of lime was 6·6 cal mol^{-1} deg^{-1} with a melting point of 2600°C. The free energy of fusion equation is therefore

$$CaO(s) \to CaO(l) \qquad \Delta G^\circ_{fusion} = 18960 - 6·6T \text{ cal}$$

$$\text{At 1773 K, } \Delta G^\circ_{fusion} = 7258 \text{ cal} = -RT \ln a_{CaO}$$

Here the activity of lime is for the solid with respect to the liquid as standard state. The activity is calculated to be 0·127 at lime saturation of the $CaO + CaF_2$ melt with the mole fraction of CaO equal to 0·26. The activity coefficient for CaO is therefore 0·49, and this result can be used to obtain a regular solution formulation following Sridhar and Jeffes

$$RT \ln \gamma_{O^{2-}} = -813 x_{F^-}^2$$

There is still some considerable doubt concerning the value of the saturation solubility of CaO in $CaO + CaF_2$, and the range of experimental values lies between 26 and 40 mol% CaO at 1500°C, but when this discrepancy is resolved, the broad picture will probably still show that Temkin's equation gives a good approximation to these results.

Kor and Richardson also established the sulphide capacities of the $CaO + CaF_2 + Al_2O_3$ slag system at 1500°C. It was found that the capacity remains approximately constant for a given $CaO:Al_2O_3$ ratio (Fig. 118) but increases when CaO is added to a mixture of CaF_2 and Al_2O_3 up to the liquid satura-

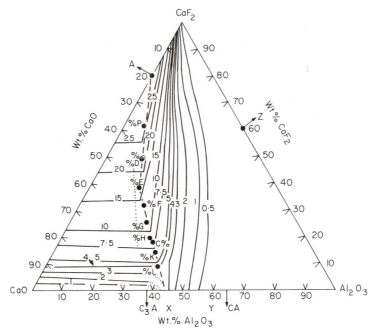

Fig. 118. Isosulphide capacity lines at 1500°C for mixtures of CaF_2 + CaO + Al_2O_3. The numbers on each curve indicate $C_s \times 10^3$. (After Kor and Richardson, loc. cit.)

tion limit. The sulphide capacities of the electro-slag refining systems are high when compared with most metallurgical slags, because of the high CaO activities which can be achieved. A typical sulphide capacity for electro-slag systems would be 10^{-2}, which is of the same order as the basic open hearth slags (see Fig. 75).

The high lime activities are also reflected in the high activity coefficients of other metal oxides in solution in this slag which were obtained by Davies and

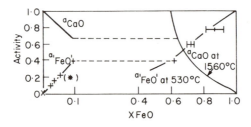

Fig. 119. Activities in the CaO–'FeO' system at 1450°C. (After Hawkins and Davies, loc. cit.)

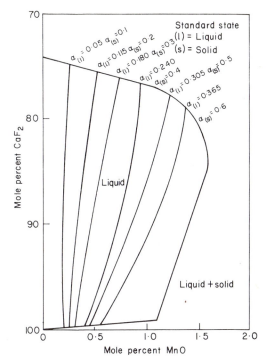

Fig. 120. Orthogonal plot of the ternary CaF_2–CaO–MnO showing liquidus and a_{MnO} values. (After Smith and Davies, loc. cit.)

co-workers. Typical ranges of values of the effect are to be found for FeO ($\gamma = 2$–20) (Fig. 119) and MnO ($\gamma = 50$–80) (Fig. 120). Generally speaking the highest values are found in the CaF_2-rich region of composition. Silica activity coefficients on the other hand are uniformly less than one and are as low as 0·1 in the CaO-rich region.

A good general account of this process and its application is to be found in the book by Duckworth and Hoyle.

REFERENCES

Alcock, C. B. and Richardson, F. D. (1958). *Acta Met.* **6**, 385; (1960). *ibid*, **8**, 882. Bond model for dilute solutes in binary alloy solvents.

Andrussov, L. (1950). *Z. Elektrochem.* **54**, 567. Interdiffusion in gaseous mixtures.

Birchenall, C. E. and Cheng, C. H. (1949). *J. Metals* **185**, 428. Thermodynamics of Ag + Zn System.

Bookey, J. B. (1952). *J. Iron Steel Inst.* **172**, 61, 66. Thermodynamics of solid calcium phosphates.

Bookey J. B., Richardson, F. D. and Welch A. J. E. (1952). *J. Iron Steel Inst.* **171,** 404. Equilibrium between phosphorus and oxygen in iron with $CaO-Ca_4P_2O_9$.

Boorstein, W. M. and Pehlke, R. D. (1969). *Trans AIME* **245,** 1843. Kinetics of the solution of hydrogen in liquid iron alloys.

Bradshaw, A. V. and Richardson (1965). *Brit. Iron Steel Inst.,* Special Report No. 92, **24.** Thermodynamics and kinetics of vacuum degassing.

Bulkowski, H. H., Beale, L. C., Sebenick, J. J., Campbell, I. E. and Gonser, B. W. (1950–51). *Battelle Mem. Inst. Reps.* **514, 522, 523.** Impurity transfer in zirconium refining.

Calderbank, P. H. (1968). *The Chemical Engineer* CE 220. Review of mass transfer from liquids to bubbles. See also Szekely and Themelis *loc. cit.,* p. 698.

Davenport, W. G. (1968). *Can. Met. Quart.* **7,** 127. Mass transfer to spherical cap bubbles in hydrogen removal and $C+O$ reaction in steel.

Davies, R. M. and Taylor, G. I. (1950). *Proc. Roy. Soc.* **A200,** 375. Rising velocity of spherical cap bubbles.

Döring, H. and Molière, K. (1952). *Z. Elektrochem.* **56,** 403. Laboratory studies of the zirconium refining process.

Duckworth, W. E. and Hoyle, G. (1969). "Electroslag Refining". Chapman and Hall, London. A general review of the science and engineering of the process.

Elliott, J. F., Gleiser, M. and Ramakrishna, V. (1963). "Thermochemistry for Steelmaking" Vol. II. Addison-Wesley Inc.

Ejima, T. (1966) & (1968). *Trans. Japan Inst. Met.* **7,** 133; **9,** 172.

Fellner, P. and Krohn, C. (1969). *Can. Met. Quart.* **8,** 275. Distribution of phosphorous between liquid iron and slags.

Fischer, W., Rahlfs, O. and Benze, B. (1932). *Z. Anorg. Chem.* **205,** 1. Vaporization of aluminium triiodide.

Flood, H. and Grjotheim, K. (1952). J. Iron Steel Inst. **171,** 64. Thermodynamic calculation of slag equilibria.

Grube, G. and Hantelmann, P. (1952). *Z. Elektrochem.* **56,** 1. Dissolution potentials of metals in $BaCl_2$-cryolite melts.

Hawkins, R. J. and Davies, M. W. (1971). *J. Iron Steel Inst.* **226.**

Hirschfelder, J. O., Curtiss, C. F. and Bird, R. B. (1954). "Molecular Theory of Gases and Liquids". Wiley and Sons, N.Y.

Jacob, K. T. and Alcock, C. B. (1972). *Acta Met.* **20,** 221. Bond model for dilute solutes in binary alloy solvents.

Jacob, K. T. and Alcock, C. B. (1973). *Acta Met.* **21,** 1011. Dilute solutions of Indium in Cu, Au and Cu + Au alloys.

Kor, G. J. W. and Richardson, F. D. (1969). *Trans. AIME* **244,** 319. Sulphide capacities of basic slags containing CaF_2.

Krivsky, W. A. (1973). *Met. Trans.* **4,** 1439. The Linde argon–oxygen process for stainless steel.

Masson, C. R. and Whiteway, S. G. (1967). *Can. Met. Quart,* **6,** 199. Diffusion of oxygen in liquid silver.

McCabe, W. L. and Thiele, E. W. (1935). *Ind. Eng. Chem.* **17,** 605. See also Szekely and Themelis, *loc. cit.,* p. 501. Distillation of binary mixtures.

Mizikar, E. A., Grace, R. E. and Parlee, N. A. D. (1963). *Trans. ASM* **56,** 101. Diffusion of oxygen in liquid silver.

Orr, R. L., Rovel, J. M. and Hultgren, R. (1963). *In* "Selected Values of Thermodynamic Properties of Metals and Alloys" (R. Hultgren *et al.,* Eds). Wiley and Sons, N.Y. Thermodynamic data for $Ag+Zn$ alloys.

Pehlke, R. D. and Elliott, J. R. (1963). *Trans. AMIE* **227,** 844. Solubility of nitrogen in liquid iron alloys.

Pfann, W. G. (1952). *Trans. AIME* **194,** 747. Principles of zone refining.

Powell, C. F., Oxley, J. H. and Blocher, J. H., Jr. (Eds) (1966). "Vapour Deposition". Wiley and Sons, N.Y.

Richardson, F. D. (1958). "Physical Chemistry of Metallic Solutions and Intermetallic Compounds". N.P.L. Symposium No. 9, London. H.M.S.O. Paper No. 6A. Thermodynamic principles in the refining of metals.

Sale, F. R. and Shelton, R. A. J. (1965). *J. Less Common Metals* **9,** 54. Vapour pressures of Zr iodides.

Samarin, A. (1961). *J. Iron Steel Inst.* **198,** 131. Vacuum refining of steel.

Schafer, K., Corte, H. and Moesta, H. (1957). *Z. Elektrochem.* **55,** 662. Experimental study of interdiffusion in gases.

Smith, P. N. and Davies, M. W. (1971). *Trans. Inst. Min. Met.* **80,** C87.

Sridhar, R. and Jeffes, J. H. E. (1968). *Trans. Inst. Min. Met.* **77,** C53. Thermodynamics of $PbO-PbF_2$ melts.

Thurmond, C. D. and Struthers, J. D. (1953). *J. Phys. Chem.* **57,** 831. Solid Solubility of Cu and Sb in Ge.

Trzekiatowski, W. and Terpilowski, J. (1955). *Bull. Acad. Polon. Sci.* **3,** 391. Thermodynamics of Ag + Zh system.

Turkdogan, E. T. (1965). "Steelmaking" (J. F. Elliott, Ed.), p 77, M.I.T. Press. Diffusivities of gases and metal vapours (review).

Ward, R. G. (1963). *J. Iron Steel Inst.* **201,** 11.

Ward, R. G. and Hoar, T. P. (1961). *J. Inst. Metals* **90,** 6. Electrolytic removal of O, S, Se and Te from molten copper.

Willis, G. M. and Blanks, R. F. (1961). "Physical Chemistry of Proc. Metallurgy", Part 2, p. 991. Interscience, N.Y.

Winkler, T. B. and Chipman, J. (1946). *Trans AIME* **167,** 111. Thermodynamics of phosphorus–oxygen reaction in steelmaking.

APPENDIX

Useful constants and conversion factors

1 atmosphere $= 1\cdot033 \times 10^4\,\text{kg m}^{-2} = 1\cdot013$ bar.

1 calorie $= 4\cdot184$ joule $= 4\cdot184 \times 10^7$ erg.

1 coulomb $= 2\cdot998 \times 10^9$ e.s.u.

1 electron volt $\equiv 23\cdot053\,\text{kcal mol}^{-1}$

Charge of the electron $= 4\cdot803 \times 10^{-10}\,\text{esu} \equiv 1\cdot602 \times 10^{-19}$ coulomb

Gravitational constant; $g = 980\cdot665\,\text{cm sec}^{-1}$

Gas constant $1\cdot987\,\text{cal deg}^{-1}\,1 \equiv 82\,\text{cm}^3\,\text{atm deg}^{-1}$

Avogadro's number $6\cdot0225 \times 10^{23}$

Boltzmann's constant $1\cdot3805 \times 10^{-16}\,\text{erg deg}^{-1}$

$\log_e 10 \equiv \ln 10 = 2\cdot3025$

$RT \ln a = 4\cdot575T \log_{10} a\,\text{cal} \equiv 19\cdot143\,T \log_{10} a\,\text{joule.}$

SUBJECT INDEX

Prandtl number, definition, 95
approximate value for gases, 96

Q

Quasichemical model for ternary alloy thermodynamics, 259–260

R

Ranz–Marshall, Nusselt number for particle-gas system, 98
Rectifying stage, in distillation, 275–276
Reduction reaction, general formulation, 111–112
Reflux ratio, 277–278
Refractory metal-oxygen systems, phase diagrams, 29
oxygen potentials in, 31
measurement of equilibria, 30
Regular solution, strictly, definition, 121
general definition, 122
Reverberatory furnace, 224
Reynolds number, definition, 95
values for gas flow, 96
particle, 98
Rotary kiln, disadvantage of in reduction, 105

S

Salts, molten halide exchange equilibria, 211–212
Schmidt number, definition, 142
average value for liquid metals, 142
Scotch hearth, process for lead, 153
Self-diffusion, measurement in liquids, 140–141
Semi-conduction, in oxides, 60–61
positive holes, 60
mobility of species and, 60
Shear planes, non-stoichiometry of oxides and, 58–9
Sherwood number, definition, 142
Silica, activity in slags, 117
lime activities in liquid, 117

activity measurement in slags, 117–118
activities in $CaO-SiO_2-Al_2O_3$ slags, 199
Silicon, distribution between metal and slag, 112–113
carbide, free energy measurement, 118
heat of solution in liquid iron, 196
reduction of dolomite, 236–237
effects in Bessemer converter, 283
deoxidation thermodynamics, 296
carbide, vapour phase transport, 39
Silicates, liquid activities and solid state thermodynamics, 120
Silico-manganese, in steel deoxidation, 295
Silver, removal from lead with zinc, Parke's process, 316–318
Sintering, reaction in leadmaking, 214–215
of cylinders, shrinkage equation, 70
of UO_2, 71–72
of ZnO, 72
Kuczynski's model equations, 66–70
by diffusion, equation for, 69
by vaporization-condensation, 69–70
Slags, ionic structure, 210
anion exchange equilibria, 212–214
phase in lead blast furnace, 216
Solid electrolyte, oxide, 11
Speiss, phase in lead blast furnace, 216
Stability diagrams for metal-sulphur-oxygen systems, 6–9
equilibria for construction, 7–8
differential sulphation, 8
Stagnant layer theory, 142
comparison with true profiles, 144
Stefan–Boltzmann law, 100
Stokes equation, 98
in fluo-solids roasting, 98–99
in slag–metal separation, 135
Stripping stage in distillation, 275–276
Structure, molten silicates, 137
Sub-regular solutions, definition, 123
pairwise bond model for, 124
Sulphates, Ellingham diagram for, 22
SO_3 decomposition pressures, 23
Sulphation, of cobalt oxide, kinetics, 87 9
of cuprous oxide, kinetics, 79
of nickel oxide, kinetics, 80